International Series in Operations Research & Management Science

Founding Editor

Frederick S. Hillier

Volume 360

The book series **International Series in Operations Research and Management Science** encompasses the various areas of operations research and management science. Both theoretical and applied books are included. It describes current advances anywhere in the world that are at the cutting edge of the field. The series is aimed especially at researchers, advanced graduate students, and sophisticated practitioners.

The series features three types of books:

- Advanced expository books that extend and unify our understanding of particular areas.
- Research monographs that make substantial contributions to knowledge.
- Handbooks that define the new state of the art in particular areas. Each handbook will be edited by a leading authority in the area who will organize a team of experts on various aspects of the topic to write individual chapters. A handbook may emphasize expository surveys or completely new advances (either research or applications) or a combination of both.

The series emphasizes the following four areas:

Mathematical Programming: Including linear programming, integer programming, nonlinear programming, interior point methods, game theory, network optimization models, combinatorics, equilibrium programming, complementarity theory, multiobjective optimization, dynamic programming, stochastic programming, complexity theory, etc.

Applied Probability: Including queuing theory, simulation, renewal theory, Brownian motion and diffusion processes, decision analysis, Markov decision processes, reliability theory, forecasting, other stochastic processes motivated by applications, etc.

Production and Operations Management: Including inventory theory, production scheduling, capacity planning, facility location, supply chain management, distribution systems, materials requirements planning, just-in-time systems, flexible manufacturing systems, design of production lines, logistical planning, strategic issues, etc.

Applications of Operations Research and Management Science: Including telecommunications, health care, capital budgeting and finance, economics, marketing, public policy, military operations research, humanitarian relief and disaster mitigation, service operations, transportation systems, etc.

This book series is indexed in Scopus.

Zhaohan Sheng

Outline of Complex Systems Management Theory—Based on Irreversibility of Reductionism Thinking

Springer

Zhaohan Sheng
School of Management and Engineering
Nanjing University
Nanjing, Jiangsu, China

ISSN 0884-8289 ISSN 2214-7934 (electronic)
International Series in Operations Research & Management Science
ISBN 978-981-97-3359-0 ISBN 978-981-97-3361-3 (eBook)
https://doi.org/10.1007/978-981-97-3361-3

This Springer imprint is published by the registered company Springer Nature Singapore Pte Ltd.
The registered company address is: 152 Beach Road, #21-01/04 Gateway East, Singapore 189721, Singapore

Preface

The "complex"[1] of the physical world has been long discovered by human beings and is well perceived. As a result, the concept of "complexity" has been used as a general term for various complex physical phenomena.

On this basis, people use the speculative concept of a "complex system" as the "matrix" for generating various complex phenomena. This approach aims to clarify the different meanings of a multitude of "complexity terms" and to explore the common origin of complex phenomena (complexity).

This has greatly promoted the development and progress of world science and technology over the past few decades, which is an incredible achievement.

In the 1970s and 1980s, Hsue-Shen Tsien, a renowned Chinese system scientist, integrated the achievements of modern natural science research with the Chinese philosophical concepts of the unity of heaven and humanity and the unity of knowledge and action. He deeply explored the self-causes and origins of complicated systems as phenomenon entities[2] in the socio-economic field, that is, the ontological attributes of the real existence of complicated systems. He abstracted these attributes in the concept of "complexity" with "irreversibility of reductionism" as its essence. That is, the "irreversibility of reductionism" (complexity) is the origin and essence of the phenomenal existence of complicated system in the socio-economic field.

If a system has the attribute of "irreversibility of reductionism," it signifies that such systems have the origin and self-rootedness of complex phenomena. The ontological attribute of such a so-called complex system is theoretically identical to complexity, containing both the entity and essence of complex existence. Therefore, complex systems can serve as academic symbols representing the ontological attributes of complicated systems.

Hence, it is evident that the "complexity" and "complex systems" in the socio-economic field are entirely different in their fundamental meanings from those in the physical world.

[1] "complex" refers to people's vague and indistinct perception of the complex nature of the physical world.

[2] Phenomenon entities are a category of entities that can be characterized as a certain phenomenon.

Hsue-Shen Tsien's academic thought has provided a "ballast stone" for further expanding complex systems thinking into the field of management science and guiding Chinese scholars in exploring the knowledge hierarchy of complex systems management in recent decades, featuring distinctive Chinese characteristics.

Hsue-Shen Tsien's systems thinking theoretically formed a cognition and methodology based on complex systems. This original and independent thinking addresses the applicability and logic of using complex system paradigms in the field of management studies. Moreover, through the paradigm shift of complex systems thinking, we can construct an academic management system based on the paradigm of complex systems. This is the complex systems management in the socio-economic field.

Complex systems management, rooted in the irreversibility of reductionism, encompasses the management of complex scenarios, behaviors, and issues in the socio-economic sphere. It embodies the essence of the complex socio-economic field within the "complex system" of systems science. Therefore, from a scientific connotation perspective, complex systems management theory is a theory about the essence of complex management in the socio-economic field.

Complex systems management has become an important part of the management science field. This new progress in the academic field of management studies has shattered the mechanistic worldview and reductionism methodology that have long dominated management philosophy thinking. It provides a new paradigm for management activities and behaviors to meet the complex challenges of today's new technological revolution and social transformation, endowing management with new dynamics of co-evolution in the complex system era.

Generally, complex systems management can be seen as an intersection of systems science and management science. Moreover, as traditional management thinking and academia face numerous complex challenges and demonstrate inadequacy, the integration of Chinese ontological philosophy with modern science and technology as a new management paradigm has provided management academia with fresh dynamics. This paradigm of complex systems management promotes the modernization of current management thought and capabilities.

The academic value of this original research work by Chinese scholars lies in, in a general sense, transforming the ontological thinking of complex systems in systems science into the practical application of complex systems management in management science. It transforms people's perceptions of "complex" into actions related to "complex," achieving management studies that encompass ideas, perspectives, and approaches for dealing with "complex." In a certain sense, complex systems management is not only about understanding the ability to solve complexity problems but also focuses on exploring how to develop the capabilities to acquire problem-solving skills.

Following Mr. Hsue-Shen Tsien's complex systems thinking, Chinese scholars have transitioned from studying "complex systems" in systems science to "complex systems management" in management science [1]. In doing so, they have begun constructing an independent knowledge system for complex systems management that embodies Chinese characteristics and reflects an overview of Chinese philosophical thought.

This system distinctively embodies the characteristics of the highly open cross-domain, interdisciplinary, and extensive knowledge and technology integration of the current "big science" era. It involves human contemplation of systems science and responses to significant scientific issues, integrating natural science, social science, and humanities. It also encompasses its historical origins, practical significance, and the logical structure of its theoretical framework.

Nanjing, China Zhaohan Sheng

From How We Understand ‘Complex’ to How We Tackle It

I

People have long discovered and perceived “complex.”

What is the meaning of “complex”? In the Chinese context, the term “complexity” combines the characters “复,” which means “many,” and “杂,” which signifies “chaos.” The Ming Dynasty poet Yinglin Hu once remarked in his Poetry Thicket: Ancient Forms Part I (诗薮•古体上), a classic Chinese work on poetic theory, the sentences and language of the “Sao” (a type of lyrical verse) and the “Fu” (a type of descriptive poem) are not vastly different. However, their styles and structures are quite distinct: the “Sao” is intricate and unordered, while the “Fu” is organized and systematic.

For ages, humans have identified and perceived various “complicated” phenomena and challenges in diverse management practices. Common colloquialisms like “difficult to explain, unpredictable, and susceptible to decision-making errors” are often employed to vividly and intuitively convey their understanding of these “complicated” management circumstances and issues.

With the continuous development of human society, management activities and problems have generally become more complex rather than simple. People not only need to deal with “complex” in management practice and summarize their experiences, but also begin to study the theory of complex management at a theoretical level.

This book primarily delves into the theories of complex management. In a general sense, since the subject of study is “complex management,” wouldn’t it be more straightforward and to the point to use the title of “Complex Management”? If that was the case, the fundamental approach to writing would largely involve collecting some typical complex management issues. It would then examine how people view, analyze, and resolve them, clarify related methods and techniques, summarize some experiences and insights, and conclude with some lessons as reflections. That would essentially shape the content of this book.

This approach would work well as a practical summary or a collection of typical cases about complex management. However, if it is considered as a theory of complex management, the following practical problem arises: First, what is "complex management"? According to common sense, "complex" is people's intuitive perception of something, referring to their vivid feelings of complex resulting from direct contact with that thing. Therefore, the perception of "complex" is bound to vary from one person to the other or over time. For the same thing, person A may feel it is "complex" while person B may consider it "not complex." Even for the same person, their perception of whether something is "complex" may vary with their level of expertise. For example, a mathematics problem that seemed "complex" to those in elementary school may appear "not so complex" to those in middle school. While in high school, this mathematics problem may be considered as "not complex."

In this way, if we write a book on the theory of complex management, it would be challenging because interpretations of whether a particular problem is "complex" may vary significantly from one reader to the other. This ambiguity further extends to whether the corresponding theoretical research can be considered a "complex management" theory. Such confusion primarily derives from conducting theoretical research directly based on the subjective perception of "complex." Theory, being a rational cognition, requires the rational condensation and abstraction of perceptions rather than conducting theoretical research solely at the perceptual level.

Furthermore, it is well recognized that any management theory belongs to the subjective realm, stemming from the rational construction of human management thinking in explaining the reasons behind management activities and phenomena. In a general sense, management theory is a systematic and logical body of knowledge established by people based on philosophical thinking in the practice and thinking activities of management [1]. Therefore, if we were to write a book on complex management theory, we need to adhere to the following fundamental principles:

1. It should be based on a wide range of concrete and complex management activities and problems that have emerged in various domains of human management practice, such as social, economic, technological, and engineering fields.
2. We should distill and abstract the essential and regular aspects of these activities and problems, recognize their universal nature, and make them the focus of our theoretical research.
3. Management theory should possess clear philosophical thinking, which forms the logical basis for theoretical research and becomes the essence and foundation of the theory.
4. The contribution of management theory to both academia and practice is manifested in providing management principles. These principles need to be transformed into knowledge, making management knowledge the tangible carrier of management principles. The process of transforming knowledge should follow a top-down approach based on philosophical thinking and ensure systematic and logical coherence from the bottom up. Together, they form an independent knowledge system. The management knowledge system should have a stable

and complete theoretical chain to reflect the systematic, logical, expansive, and self-consistent nature of theoretical research.

The history of scientific development has revealed that people have long discovered and intuitively perceived various complicated phenomena and problems in different fields. However, to study "complex" management theory, as aforementioned, it is necessary to refine and raise various perceptions of "complex" to the rational level and to form representations of the essential attributes of "complex."

Experience shows that this matter is not simple. This is primarily because the various real-world forms of "complex" are diverse and multifaceted, making them hard to distill. It is crucial to transcend specific fields and manifestations, think on a higher philosophical level, and leverage human wisdom. On this issue, people have spent decades before making progress on characterizing the essential attributes of "complex." Without this step, studying the theory of "complex management" would inevitably lack the core rational thoughts about "complex," making it impossible to write a work on "complex management theory."

This makes it clear that in modern human management practices, people often use the adjective "complex" to describe management phenomena and issues that are "multifaceted and miscellaneous" in everyone's mind. To study the theory of "complex management," we must delve beyond the superficial and external intuitive perceptions of "complex." It is essential to distill and abstract the inherent essence of "complex," extracting its fundamental attributes. These attributes consist of the foundational, consistent, and expansive nature of this theoretical research.

Of course, so far, we do not know what the essential attributes of "complex" are, nor do we know the linguistic symbol that represents these essential attributes. In the academic discourse system, the linguistic symbol representing this essential attribute should not merely be the creation of a new term without a theoretical foundation. Instead, it needs to possess both the logical reasoning that represents the essential attribute of "complex" and the value and significance of a conceptual identifier in the knowledge system.

Understanding the essential attributes of "complex" in the field of management and identifying the conceptual identifier for this essential attribute serves as the logical starting point and core of complex management theory

II

The development path of management theory often involves both necessity and contingency. For a specific theory, its development path may be relatively smooth, while for another specific theory, it may be more tortuous. The formation process of complex management theory, as a new theory, is relatively tortuous.

In the process of human scientific development, complex phenomena and theoretical research in natural science fields such as physics, chemistry, biology, mathematics, and even meteorology emerged much earlier and matured much more extensively than in management. Nearly a century ago, scientists discovered and began in-depth research on "complex" phenomena in different natural disciplines, resulting in many theoretical research achievements.

For contemporary management studies, as a "*latecomer*" contemplating complex management theories, there are many insights and references to draw upon. However, it is imperative to adhere to the principle of "*respecting precedence.*" This means acknowledging the historical facts, scientific consensus, and fundamental conclusions that have emerged in the history of human scientific development surrounding the study of "complex."

Therefore, in the first instance, we need to briefly introduce some important facts in the history of human scientific research on "complex" phenomena. This is of great significance for understanding the logical progression of the content of this book.

Since the early twentieth century, scientists from various countries have discovered, through experimental research in fields such as physics and chemistry, new **complex phenomena** that contradict traditional scientific knowledge. For example, non-living material molecules, under certain conditions, can arrange themselves in an orderly manner as if following the commands of a conductor. In the history of human scientific research, numerous unexpected complex phenomena have been discovered in the fields of physics, chemistry, meteorology, and other natural sciences. Scientists, on the one hand, intuitively perceive that these phenomena are so complex and incredible, but on the other hand, are temporarily at a loss and can only use the term "**complexity**" to collectively describe such complex phenomena in various natural science disciplines (please note that here "complexity" is a linguistic symbol collectively referring to complex phenomena in multiple natural science fields!). They have also created certain new concepts (i.e., terms related to complexity) based on their specialized knowledge in their respective fields to vividly describe the unique forms of their own complex phenomena. Over the decades, dozens of such "complexity terms" have emerged like a profusion of flowers. However, overall, a unified and essential concept about "complex phenomena" has not yet been suggested. This simply proves that it is intricate to distill the underlying mechanisms of various complex phenomena in natural science disciplines. In other words, it is genuinely challenging to identify the shared essential attributes of complex phenomena across multiple natural disciplines. Consequently, all scientific research that utilizes different complexity terms to describe the unique forms of their own complex phenomena has been collectively referred to as "complexity science."

Scientists persistently explore complexity in the realm of natural sciences, occasionally altering their approaches. For instance, they no longer transcend beyond simply expanding the scope of complexity in a multidisciplinary, horizontal manner. Instead, they delve into a vertical exploration, probing the origins and root causes of these complex phenomena. At a philosophical level of reflection, scientists hypothesize the existence of a kind of "substance" that can give rise to complexity (complex phenomena) and refer to it as a complex system."

In April 1999, the American "Science" magazine published a special issue on complex systems. The editor explicitly pointed out that since they were studying complexity, why use the term "complex system"? The primary reason was to avoid the term "complexity," which might be ambiguous or muddled. They hoped to add "complex system" alongside the term "complexity," so when researching complex phenomena in specific areas, they could use a term that would implicitly embody the notion of complexity, yet not be rigidly defined as such, allowing for more nuanced and flexible applications in research. However, for more abstract studies spanning multiple specific fields of complex phenomena, they would use "complex system." It's evident that the "complex system" in such context is a speculative concept based on the origins of complexity rather than. For this reason, when dealing with a particularly complex phenomenon, it is still necessary to understand the "complex system" in terms of its "complexity," meaning that specific features and attributes must be ascribed to the complex system. It is similar to "putting on" unique hats for the complex system. Consequently, for a period of time, within systems science, there were definitions of complex systems such as stochastic systems, self-organizing systems, nonlinear systems, and adaptive systems allowing complex systems to elucidate certain forms of complexity more effectively.

It must be acknowledged that the development of viewpoints on complex systems and related in-depth research have played a significant role in promoting the development of modern human science. From a macro perspective, the nonlinear systems theory, dissipative structures theory (Prigogine), synergetics (Haken), catastrophe theory (Rene Thom), fractal theory (Mandelbrot), chaos theory (Lorenz), adaptive systems theory (Holland), complex networks, and others, which have made significant academic advancements during this period, have provided valuable sources for the paradigm shift and the birth and development of new disciplines since the second half of the last century. From a micro perspective, the introduction of genetic algorithms, evolutionary algorithms, the development of the swarm software platform, agent-based system modeling, artificial life and artificial societies described using agents, and others have greatly enhanced and deepened human understanding and analytical capabilities of complex phenomena in natural systems. For example, in 2021, the Nobel Prize in Physics was awarded to three scientists from the United States, Germany, and Italy for their contributions to "physically modeling the Earth's climate, quantifying its variability, and reliably predicting global warming" and "the discovery of interactions of disorder and fluctuations in physical systems from atomic to planetary scales," which have helped to gain deeper insights into the **nature and evolution of complex physical systems**.

It is evident that since the concept of complexity here is primarily derived from complex phenomena discovered in scientific experiments in the fields of physics, chemistry, and other natural sciences, the understanding of complex systems concepts also needs to be rooted in complexity. This indicates a strong dependence on this complexity and the understanding of complex systems on the perception of materialistic ontology. Such a strong dependence means that philosophical thinking based on phenomenology, emphasized empiricism, and verifiable facts has a significant influence on this understanding of complexity and the perspective of complex systems.

Overall, in the past few decades, complexity science and complex systems theory, gradually developed in various natural science fields, have achieved significant scientific accomplishments and played a crucial role in advancing modern scientific development. In addition, the scientific community has actively explored and applied "complexity terms" and complexity (complex systems) science to study the complex management issues of human socio-economic systems. This has deepened our understanding and analytical capabilities of complex socio-economic phenomena. This has provided the field of management with many new inspirations and possibilities.

III

The path of human scientific exploration has never been singular. While Western scientists have made important achievements in complexity science and complex systems research, Chinese systems scientist Hsue-Shen Tsien, with his rich logical reasoning from systems engineering practice, Chinese philosophical thinking, and autonomous systems science framework construction theory, has opened up another path to understand the essential attributes of "complex" and complex systems. Importantly, this methodology has represented a new "tool" for people to cognize the essential attributes of "complex" in the field of management using a new paradigm of thinking.

Hsue-Shen Tsien initially advocated not treating complexity and complex systems as a matter of speculative definitions. Instead, he suggested starting from the perceptual aspect of human understanding of natural laws, capturing the inherent essence of "complex," and establishing a closer connection between complexity and people's social and practical activities. Starting from the basic cognitive points of "complexity" as the opposite of "simplicity" and "wholeness" as the opposite of "part," he believed that the essence of a system lies in the whole. In comparison to its parts, the "wholeness" is relatively complex, while the "parts" are relatively simple. Therefore, the whole of a system must be more complex than any of its parts. If the parts of a system are complex, the system as a whole must also be complex. Conversely, if the whole system is complex, its parts may be simple, and even if they are complex, the overall complexity will be reduced. This is the fundamental law that the complexity of a system decreases gradually from top to bottom.

Although the logic of "*complexity of the whole is greater than complexity of the part*" is simple and intuitive, it holds significant theoretical significance. If a system is perceived as complex as a whole, we can try to decompose the system into several parts, which will naturally be relatively simple (not complex), making the research much more convenient. If the study of the parts is still considered complex, further decomposition can be done until each part is no longer considered complex. At this point, if the knowledge of each part (including sub-parts) can be summarized and integrated from the bottom to the top, it becomes the cognition of the entire system. It can be concluded that the originally considered complex system is not "truly" complex because its "complexity" can be eliminated through top-down decomposition and **reversible** integration.

This insight into the "*complexity of the whole is greater than the complexity of the part*" has inspired us in epistemology and methodology. We can perceive that certain so-called complicated systems can be degraded and eliminated in complexity using a methodology called **reductionism**, which involves top-down decomposition and bottom-up integration. Therefore, it is believed that such seemingly complex systems are not inherently complex. In contrast, if the complexity of a system cannot be eliminated through top-down decomposition and bottom-up reversible integration, i.e., reductionism is "inapplicable" or "inappropriate" to it, it indicates that the system indeed possesses an essential (profound, genuine, inescapable) complexity. In this instance, the system exhibits an inherent complexity that fundamentally sets it apart from those systems that lack inherent complexity.

In this way, Hsue-Shen Tsien has provided us with a touchstone for extracting and abstracting the essence of "complex" through a novel approach to thinking. As long as things (phenomena) possess the attribute of "irreversibility of reductionism (i.e., reductionism is not applicable)," they can be fundamentally differentiated from seemingly complex but ultimately reducible things (phenomena).

Based on this, Hsue-Shen Tsien further pointed out that if a system exhibits phenomena or issues of "irreversibility of reductionism," we can say that this system possesses the attribute of "irreversibility of reductionism" and label such systems as complex systems. In this way, the original concept of systemic wholeness can be categorized into a type of simple additive wholeness (simple wholeness) and another type of non-simple additive complex wholeness (irreversibility of reductionism). The latter serves as a distillation and abstraction of the "complex" essential attributes within the system, commonly referred to as complexity.

Based on Hsue-Shen Tsien's scientific thinking about complexity and complex systems, and following the coherence of the aforementioned theoretical logic, it can be clearly stated that **complexity, complex wholeness, and irreversibility of reductionism are conceptually identical**. They represent **the essential attributes of "complex" in the sense of systems. Specifically, it can be regarded as the identifying concept and linguistic symbol of the essential attributes of "complex."**

Returning to the scientific term or linguistic symbol of **complexity**, previously, in the field of natural sciences, scientists broadly defined various complex phenomena as complexity and speculatively regarded complex systems as the source of complexity (complex phenomena). However, we define "complexity" as the essential attribute of complex phenomena and a representation of complex system attributes. This reveals the significant difference in epistemology and cognitive thinking paradigms between the two. Therefore, in many rigorous theoretical research contexts, it is necessary to clearly explain the different theoretical connotations of these two concepts under the same term and avoid a simplistic interpretation that would blur the semantic clarity of basic concepts.

In summary, in addressing philosophical and cognitive issues related to complexity and complex systems, Hsue-Shen Tsien combines the latest empirical achievements of modern science with the rational spirit of humanity's ontology in Chinese philosophy. This exploration forms a new approach to rational thinking about complexity and complex systems, integrating modern scientific civilization with traditional Chinese cultural civilization. This new approach reveals, through a new thinking paradigm, the essential universal properties of "complex" phenomena and problems that differ from complexity terms.

IV

In the above sections, we have introduced and interpreted two different main epistemologies that have been formed in the scientific community over the past few decades regarding complexity and the view of complex systems. Today, our task is to explore a complex management theory that focuses on the essential attributes of "complex" as the major research focus. In doing so, we need to determine which of the two epistemologies of complexity and complex systems mentioned above is more appropriate to construct complex management theory.

Here, the term "appropriate" mainly refers to which epistemology of complexity and complex systems is most compatible and relevant to the authenticity of management studies, the uniqueness of management scenarios, the orientation of management issues, and principles of management research. Of course, as science advances, if we discover new connotations and views of the essential attributes of complex and complex systems beyond the two mentioned epistemologies, we should propose more appropriate innovative thinking that advances with the times. Therefore, our choice of the epistemology of complexity and complex systems is based on the relatively limited theoretical literature of complexity and complex systems in current people's understanding, and we need to constantly make new academic innovations and knowledge changes that advance along the times.

Overall, regarding the two different cognitive paths of complexity and complex systems mentioned above, the first path is mainly based on the complexity derived from natural science experiments and empirical evidence, as well as complexity terms and views of complex systems. They are more closely related to the natural principles and laws in their respective disciplines. Therefore, when it comes to revealing complex phenomena and laws in the field of natural sciences, this epistemology is more applicable, relevant, and capable of demonstrating a high level of proficiency. In fact, in recent decades, researchers have achieved a series of breakthrough scientific research results based on this epistemology. This has greatly promoted scientific development in various fields of natural sciences over the past half-century and continuously demonstrated powerful scientific strength. At the same time, this view of complexity and complex systems has also been applied to the study of complex management phenomena and problems in the field of socio-economics, providing multiple viewpoints and metaphors for understanding complex phenomena and problems in socio-economic systems and yielding novel results. It enriches our cognition, associations, and inspirations regarding socio-economic complex systems. However, real-world socio-economic management systems are much more complex than physical, chemical, and other natural systems. Complexity terms originate from the specialized knowledge of respective natural disciplines, making them difficult to encompass and cover the complex phenomena and connotations of socio-economic management systems in the world. It is also challenging to accurately reveal the underlying management phenomena and behavioral rules guided by the realities of socio-economic management systems. Not only do we encounter difficulties in encapsulating the myriad complex and unique scenarios in socio-economic management using terms of complexity, but these terms often impose "restrictions" on the "flexibility" inherent in real-world management scenarios and behaviors. This can result in the oversimplification of intrinsically dynamic and multifaceted management problems, stripping away the richness and vibrancy typically associated with real-life management complexities.

On the other hand, the latter approach, based on Hsue-Shen Tsien's epistemology of complexity and complex systems derived from the philosophical perspective of irreversibility of reductionism, establishes a cognitive understanding of the essential attributes of real-world problems. This approach not only prevents the fragmentation of thinking caused by the limitations of complexity terms but also avoids forcing the complex systems view into a fragmented dilemma. In particular, "irreversibility of reductionism" or "complex wholeness" belongs to the cognitive level of philosophical thinking, which provides great expandability for this understanding of complex systems in theoretical research, allowing for various complex management issues in today's management field. This is exactly what we need in the study of complex management because we need to fully integrate diverse social and humanistic elements and real scenarios into our management activities. Without these connotations and scenarios, management will be "lifeless" and "cold." "Irreversibility of reductionism" demonstrates high-level management philosophical thinking and the potential to promote the development of management theories. It also allows us to further understand the philosophical and theoretical significance of Hsue-Shen Tsien's epistemology of complexity and complex systems in addressing various complex management challenges in management studies.

The above comparison is only based on which epistemology of complexity and complex systems aligns more suitably for the characteristics of management practice and the research paradigm of management studies. It is not a pure intellectual opposition to compare the scientific significance of the two epistemologies of complexity and complex systems themselves, nor is it meant to negate one of them with the other. On the contrary, since both have the original intention to explore the connotations of complexity and complex systems, from an academic perspective, they should complement each other and play their respective roles under the premise of distinguishing different contexts.

Why do we need to discuss the different understandings and cognitive paths of complexity and complex systems in the scientific community, as well as the characteristics and scope of different epistemologies of complex systems, in such a detailed and meticulous manner? This is mainly because there have been significant differences in the scientific community for decades in terms of thinking and cognitive paths regarding complexity and complex systems. Since this book focuses on the theoretical research of "complex systems management," even if we have not yet deeply understood the scientific connotations of the concept of "complex systems management," the term "complex systems" is clearly present in this new scientific term. Therefore, we must respect the previous explorations of complex systems and the academic logic and basic consensus formed by previous studies on complex systems.

Since the two cognitive paths of complexity and complex systems in the past did not have direct relevance to management studies, in such a context, "complex systems management" is clearly a synthesis of the "complex systems" category in the field of systems science and the "management" category in the field of management science. Does this synthesis have any relationship with the two types of complex systems proposed by previous scholars? If so, which type of complex systems does it refer to? If not, what is the connotation of the "complex systems" in this case? Otherwise, how can we clarify the rationality and coherence of the concept of complex systems management from an academic perspective?

Of course, it is not always necessary to meticulously identify and choose a concept in certain situations and contexts. For example, if a college student understands "complex systems management" as "the management of complicated systems" or even confuses "complex systems management" with "complex management systems," we may not be too demanding. However, if such conceptual ambiguity occurs in academic research on complex management theory, or even in an academic theoretical book《*Complex Systems Management*》, it is definitely inappropriate. That is why we have spent a considerable amount of space in the earlier part of this book to carefully explain and analyze the process of the scientific community's understanding of complex systems and their respective scientific connotations and to clearly state the reason why **complex systems management** in the field of management refers to a specific view of complex systems.

V

Every vigorous management theory research in the world starts and ends with the aim of addressing and resolving significant management issues faced by human society, reflecting a clear orientation toward the times. In the current era, there is a widespread demand across various fields of human endeavor to study issues of complexity. That is, there is a need to study "complex issues" whose essential attributes are of a complex holistic nature [2]. Consequently, contemporary cutting-edge management theory research will inevitably be guided by complexity issues, encompassing the identification of problems, problem-solving approaches, scientific analysis of problems, and the development of problem-solving methodologies. If this were merely a matter of a specific topic, a particular context, or a specific level, we could rely on existing developed theories or borrow from other disciplines to complete it by integrating experiences and techniques at the methodological level. However, the depth, richness, and profound academic foundation of complex issues in the field of management practice necessitate the construction of a new, complete theoretical chain within a comprehensive disciplinary system, academic system, and discourse system [3].

Therefore, it is not to be assumed that once we understand the "complex systems" involved in studying "complex systems management," we have a complete understanding of "complex systems management." That is not the case. We merely provided an interpretation of the basic semantics of "complex systems management." In other words, we only defined the "complex systems" when studying complex systems management. However, "complex systems management" as an integrated scientific concept cannot be simply understood as "complex systems + management." This indicates that we have not yet revealed the essential connotations of "complex systems management."

Let us review the formation context of the two aforementioned types of epistemologies on complex systems. The first type follows the logical cognitive path: various complex phenomena-complexity-complexity terms-complexity science-complex systems. The second type follows the logical cognitive path: complex problems-complexity as a representation of the essential attributes of "complex"-complex systems. It is evident that in this context, the essential connotation of complex, as an adjective, becomes a noun, "complexity," and it predates the appearance of the concept of "complex systems" while guiding its formation. Hence, in the context of investigating "vivid and intuitive" complex problems that frequently emerge in management practice, and acknowledging that the essence of these problems is encapsulated by the term "complexity," it becomes imperative to study the fundamental attributes—namely, the complexity—of such issues. This investigation is necessary from the standpoint of rational thinking, as it seeks to deepen our understanding of the complexity inherent in these complex managerial challenges [3]. The concept of complex systems proposed by Hsue-Shen Tsien exemplifies the irreversibility of reductionism in complex problems existing in socio-economic systems. Thus, in management practice, what should be considered more fundamental and profound than the concept of complex systems is the issue of complexity with the two characteristics of "complex wholeness" and "irreversibility of reductionism." Therefore, **if we propose a new discipline in management studies that investigates complex holistic problems, the most natural, essential, and logical term would be "complexity management."**

Why is this essentially complexity management referred to as complex systems management? In the past few decades, the concept of complex systems has been favored over complexity because it gives people a sense of ontology, entity, and carrier. Moreover, the "complex system" is the substance that produces "complexity." Therefore, people have started to refer to the management activities addressing the issue of complexity with the two characteristics of "complex wholeness" and "irreversibility of reductionism as complex systems management." Moreover, when a new scientific concept does not originate from a strict academic starting point, and there are multiple possibilities that are widely understood, a less precise definition and a certain level of tacit acceptance can become the academic path for the formation of a new concept. The reason "complex systems management" is more commonly accepted and "traditioned" in the academic community compared to "complexity management" today is largely due to this.

VI

From general and simple management to complex management, and finally, to complexity management (complex systems management), it signifies important changes in the attributes of the managed objects, the cognition of managers toward the managed objects, the paradigms of research questions, and research methods. Therefore, studying complex systems management requires us to establish a thinking paradigm of complex systems. Specifically, after we establish new concepts such as complexity, complex wholeness, and complexity management, we must conduct research on how to achieve cognitive transformation in this new field of management from a philosophical perspective. This is referred to as a **paradigm shift** in scientific philosophy.

A thinking paradigm shift is the manifestation of a scientific paradigm shift at the level of thinking and stages of thinking. The fundamental purpose of the thinking paradigm shift in complex systems is to apply the principles of complex systems thinking, knowledge frameworks, and discourse systems to reveal and analyze the basic characteristics, laws, and principles of managing complex wholeness problems in complex systems management activities. It aims to provide corresponding methodological systems and enhance the ability to solve complex wholeness problems. For example, in the field of management, the traditional reductionism proposed by Descartes is no longer adequate for addressing the growing number of problems arising from the emergence of complex holistic attributes. Therefore, new methods must be created to address the problems of irreversibility of reductionism. As a result, transforming Descartes' reductionism thinking paradigm becomes one of the starting points for the thinking paradigm shift in complex systems [3].

The thinking paradigm shift in complex systems for complexity management involves a series of key points that are related to the cognitive principles and basic behavioral guidelines on how to perceive and deal with complex wholeness problems. After establishing the basic points of the thinking paradigm shift in complex systems, further management details can be supplemented through unique scenarios or contexts of complex systems. This not only constitutes one of the core abilities to analyze complex wholeness in real management activities and to master such complexity but also enables us to further adapt traditional management academic concepts into new knowledge of "complexity management" through this ability.

Under the thinking paradigm shift in complex systems, "complexity" as a core concept possesses particularly strong vitality. When combined with other discourse in the specific context of managing complexity problems or management thinking, it can further provide rich connotations of complex wholeness management activities and behaviors. For example, managers initially experience a complex, intuitive sense of a type of practical problem that is challenging to express clearly, analyze thoroughly, and predict accurately, thus difficult to identify causes, make decisions, develop solutions, and propose plans. Based on this, the irreversibility of reductionism is extracted as the essential attribute of these problems, thus abstracting complex problems into **complexity problems** at the level of rational cognition. The new connotation of complexity problems requires managers to adopt a complex system thinking paradigm, which is referred to as **complexity thinking**, throughout the entire management process. Consequently, the various forms of complexity (complex wholeness) inherent in management activities due to the existence of complexity problems are referred to as **management complexity**. Correspondingly, the management model based on complexity thinking is referred to as **complexity management**. It is an integrated, holistic concept that encompasses complexity problems, complexity thinking, and managing complexity.

The complex systems management discussed in this book is a specific type of complexity management with its own unique specifications

From the above, it can be seen that as people's understanding of the essential attributes of management activities and problems continues to evolve, management thinking, management models, and management methodologies are constantly developing and enriching. This, in turn, enhances the analytical and control capabilities of the complexity of managed objects. It is a fundamental phenomenon and law of the development process of human management cognition.

VII

We have already pointed out the specific and unique characteristics of complex systems management in theoretical terms. However, any scientific concept must have theoretical consistency and logical coherence. In the field of management, the synthesis of the words "complex systems" and "management" into "complex systems management" should not be understood literally as "the management of complicated systems," "managing complicated systems," or "complex management systems." It is not simply adding management content to complex systems or applying some complex systems thinking in management. Instead, it represents a fundamental shift in the realm of management, leading to a new understanding of complex systems and management.

Firstly, based on the analysis of complex systems mentioned earlier, "complex systems management" in this book is a complexity management paradigm primarily based on Hsue-Shen Tsien's irreversibility of reductionism and complex systems thinking while incorporating other academic thoughts on complex systems. Philosophically, Hsue-Shen Tsien's complex systems thinking paradigm constitutes the core of complex systems management. In practice, it primarily deals with management activities and processes demonstrating "complex wholeness" problems in complex socio-economic engineering systems. Academically, it is a methodological-driven research paradigm within the field of complexity management.

Complex systems were originally a concept in the field of systems science. It appears within the new comprehensive concept of "complex systems management" primarily to provide thinking principles and logical starting points for understanding and managing complex holistic problems in terms of epistemology and methodology. In this sense, we can consider a complex systems epistemology as a philosophical view and guidance for exploring the paradigm of complexity management. Therefore, when faced with various challenges of complex problems and feeling insufficient in terms of cognitive understanding, analysis, and control, there is a pressing need for new, higher level philosophical thinking to provide precise and powerful guidance and support. In such situations, the complex system epistemology offers a convincing thinking paradigm and contributes to the innovation of management concepts, techniques, and methods, forming a complexity management paradigm with new capabilities.

Therefore, from a philosophical perspective, it is more accurate to understand complex systems management as a comprehensive reconstruction and advancement of the management paradigm when traditional management thinking faces significant challenges caused by complex holistic problems. It is an assistance provided by the complex systems philosophical view and a demonstration of the value of management philosophy. This relationship embodies a profound philosophical logic and an inherent quality: "complex systems" plus "management" equates to "complex systems management."

VIII

Management science is a theoretical, methodological, and applied system formed at the scientific level based on human management practices. Within the system of management science, various relatively stable and distinctive fields have emerged due to different management ideologies, paradigms, themes, content, methodologies, and methods [3].

In this sense, complex systems management has already developed its own basic logical framework, thinking principles, practical thinking, basic paradigms, core concepts, methodology, and method system. It has established a logical starting point, ecological environment, and endogenous driving force for the self-growth of a new discipline within management science. As a result, complex systems management has laid the foundation for a new discipline within management studies.

However, despite this progress, there is still a need to deepen and complete the construction of the disciplinary system, academic system, and discourse system within this field. It is necessary to integrate the "three systems" into an interconnected and mutually reinforcing whole. The integration of complex systems science and management science needs to be achieved to ensure the applicability of the complex system paradigm in the context of management science. This requires the necessary logical prerequisites for interdisciplinary research methodologies in complex systems management. Particularly, it is essential to anchor complex systems management within the academic ecosystem of management science based on the current reality. These aspects require gradual improvement and advancement. Therefore, until the aforementioned aspects are adequately developed and comprehensive, the discipline of complex systems management cannot be deemed developed within the realm of management science. On the contrary, it strongly demonstrates the transformative connotation and development essence as a new discipline within management science, requiring substantial original and systematic knowledge innovation and transformation.

Undoubtedly, among the aforementioned tasks in the construction of academic systems, the most important and comprehensive one is to follow the basic rules and paradigms of theory formation and construct the theoretical framework of complex systems management. The theory of complex systems management is a systematic and logical framework established by incorporating knowledge as its fundamental element through practical and cognitive activities in complex systems management. This framework enables people to systematically describe and analyze various phenomena in the practice of complex systems management, facilitating a profound understanding of the essential characteristics and general laws of management problems and behaviors. This is because the system has been endowed with the qualities of systematic and logical research object attributes.

As a new discipline within management science, the process of forming the theory of complex systems management must follow the complete theoretical and logical chain of "thinking principles-core concepts-basic principles-scientific topics-method system." The thinking principles have been determined through the transfer of complex systems thinking paradigms, while the overall design and implementation of each link, such as "core concepts-basic principles-scientific topics-method system," need to be gradually completed.

In the core concepts section, the core concepts listed in this book are distilled and abstracted from the complex holistic attributes of important aspects and elements in complex systems management activities. Even if some scientific terms of these concepts, such as multi-scale and scenario, are also used in other disciplines, they have been expanded and redefined here in terms of conceptual connotations. These concepts not only embed the complex holistic attributes of complex systems management activities but also incorporate behavioral guidelines for complex systems management.

Furthermore, the theory of complex systems management is a rational cognition of the essence and laws of management activities and problems, including the basic elements of management environment, subjects, objects, goals, and behaviors. In other words, as the core concepts of the theory, they should not be "isolated" or "fragmented." They should not only provide comprehensive coverage of complex systems management activities and issues but also have a close logical connection with each other. In theory, the core concepts should not only be derived from the practice of complex systems management activities but also reflect the logical relationships between concepts. Otherwise, people cannot build basic principles and scientific topics in theory based on concepts and their combinations.

In the section on fundamental principles, the principles of complex systems management must revolve around the fundamental and universal elements of management subjects and complex wholeness, fully revealing the basic laws of subject behaviors and objects in complex systems management activities. Whether this can be achieved is the main criterion for measuring the academic quality of the basic principles in the theory of complex systems management. The basic principles proposed in this book are derived from the practice of complex systems management and closely revolve around the fundamental management elements of subjects and complex wholeness. They fully reveal the logical relationships, causal relationships, and correlations between complex systems management activities and subject behaviors, as well as the basic guidelines for subject behaviors.

Academic research in management is generally guided by scientific questions. So why doesn't this book list several typical scientific questions in complex systems management but instead focuses on foundational topics? This is mainly because a topic may represent a category of problems that have similar or approximate attributes or features. The purpose of proposing scientific topics is not necessarily to obtain specific solutions to a particular topic but rather to attract academic attention and research interest in a category of common or approximate problems. This is particularly true for complex systems management, which is a new discipline formed by the transfer of complex systems thinking paradigms. It reflects the exploratory mode of "feeling the stones while crossing the river" in the early stages of complex systems management research. Just like when we don't have enough abilities to step precisely on "a particular" stone at every step, it is important and necessary to explore "some" of the stones that can be stepped on.

Proposing scientific topics often requires thinking at a higher level than specific scientific questions. For example, the construction of the discourse system of complex systems management knowledge, the essence of management, or robust management, as proposed in this book, are all topics that have certain principles of thinking and methodological significance. In any case, when constructing the theory of complex systems management, any proposed scientific topic has a strong exploratory nature. As the theory gradually improves, scientific topics will become more mature and diverse.

IX

Complex wholeness issues in complex systems management also referred to as the irreversibility of reductionism issues are poised to bring about substantial alterations in the associated management techniques. Therefore, adaptive responses are required in the methodology and system of complex systems management. Given the complex wholeness nature of the problems, the methods and techniques used in complex systems management are inevitably interdisciplinary and cross-sectoral, demonstrating an extremely rich diversity and adaptability. As a result, it is neither feasible nor necessary to enumerate all the specific methods and techniques valuable for managing complex systems due to their vast number. However, it is evident that among the myriad methods and techniques, there are new, crucial ones that adequately address the complex wholeness.

For instance, within the framework of complex systems management methods and techniques, there should be abstract representations and descriptive methods for complex wholeness. This is what we call modeling techniques for complex wholeness. Since complex wholeness is an abstract attribute, the most tangible entities in complex systems management that can best characterize this attribute are management problems or scenarios of management activities. Thus, modeling complex wholeness can be achieved by scenario modeling.

The importance of scenario modeling in complex systems management is not limited to the aforementioned logical foundations. Throughout the entire theory of complex systems management, we repeatedly "encounter" the presence of scenarios in different levels and topics. For instance, "scenario" serves as the realistic carrier of complex wholeness in complex systems management and is a core concept in the theory. In the basic principles, there are principles based on scenarios and principles related to the contextualization of unique scenarios. In foundational scientific topics, there are discussions on essential management, scenario-based robust management, and the acquisition of knowledge about unique scenarios. These instances repeatedly highlight the significance of scenarios and scenario modeling techniques. Therefore, it is appropriate to consider scenario modeling as a hallmark technique in complex systems management.

Specifically, the essence of complex system scenario modeling is to use situational language and modeling symbols such as when, where, who, what, why, the result, the reason, and how to provide a holistic description of a fully and clearly defined scenario. This becomes an abstract representation of the elements, structure, and dynamics of that scenario.

The academic implication of scenario modeling can be further summarized as follows: a modeling agent abstractly represents and describes scenarios and their evolution within a certain spatiotemporal scope of complex systems management activities. In simpler terms, the modeling agent acquires "segments" of the complex holistic aspects of scenarios at various crucial spatiotemporal junctures. It then applies fundamental dynamic principles rooted in scenario evolution to assemble an organized "sequence of segments" that approximates a particular comprehensive scenario.

So, what kind of technological path and operational process can be designed to model scenarios? It starts with people's initial perception of scenarios: complex system scenarios are the holistic form of human complexity management activities and processes in the real world, encompassing all the details of these activities and processes. The information (data) of these details is the trace of the state and evolution of complexity management activities and process scenarios. Under the current information technology conditions, scenario traces can be traced, observed, described, measured, and further quantified into mappings or projections of complexity management activity and process scenarios in data space (traces become data). The information (data) characteristics of scenarios can be summarized as complex wholeness, "infinite" information, heterogeneity, and containing certain values. This closely aligns with the characteristics of big data (volume, velocity, variety, and depth). Therefore, it can be considered that the methodology of big data-driven approaches is suitable for scenario modeling research, especially for studies involving three-dimensional (dimension, scale, granularity) scaling, cross-domain correlations, holographic scenarios, and scenario-oriented modeling.

This provides us with a perspective on scenario modeling driven by big data [3]. It involves perceiving, recognizing, experiencing, acquiring knowledge, theorizing, associating, hypothesizing, and speculating about the complex wholeness of scenarios and collecting data through multiple channels and digitizing it into big data. By applying the principle of scenario computability, various means, such as conducting independence tests on the data sequences of scenario-related elements, are used to form a set of essential scenario elements under certain dynamic mechanisms. Then, based on big data, techniques such as spatial network, event time series, synthesis of heterogeneous data in scenario space, as well as scenario spatiotemporal state description and analysis of scenario evolution dynamics, are employed. The machine's unlimited computational capability and other statistical analysis abilities are utilized to perform data modeling analysis and value discovery. Finally, in the scenario space, historical reenactment, real-time reproduction, or future exploration of complex system scenarios at different time frames can be achieved, used for reflection on the past, simulation of the present, and prediction of the future.

This allows to identify an actionable technological path, which is to achieve scenario modeling through big data-driven approaches. Below are the basic ideas and key techniques of big data-driven scenario modeling as designed in this book.

The theoretical logic of big data-driven scenario modeling primarily stems from the identity of big data and scenario attributes. That is, both big data and scenarios share common essential characteristics and logical starting points. This clarifies the theoretical logic for realizing the value of big data scenarios with the goal of scenario modeling through a certain paradigm.

In terms of reductionism, scenarios can be decomposed into a set of discrete scenario elements (information). Thus, the information and data sets derived from real scenario elements can be seen as representations of the fragmented information set of scenarios. The richer and more complete the scenario information contained in the data set, the more effectively it can represent the scenario. Although there is no one-to-one and complete correspondence between scenario data sets and scenarios, it is undeniable that big data sets obtained through data processing contain rich holistic scenario value.

Under this academic thought, the basic idea of big data-driven scenario modeling is that, using big data as the primary resource and based on generalized algorithmic thinking, computers "calculate" in the digital space to produce virtual digital mirrors of interactive and immersive real scenarios. These replicas might represent a new scenario image, recreate a fragmented scenario, serve as a "pillar" of knowledge, use big data to "draw the framework (dragon) (画龙)," and pay attention to minor details "(dotting the eyes) (点睛)." In any of these transformation processes, big data realizes its "externalization" of scenario value.

It should be noted that since there is no unique mode of "big data-driven" transformation, the final form and quality of modeling may vary. This is closely related to the modeling agent's goals, perspectives, model design, and modeling methods.

Furthermore, to truly drive big data and externalize its scenario value, specific technical routes, key technologies, and process designs are required. This is equivalent to proposing a "transformer" that converts the potential scenario value of big data into an "externalized" form, making the "driving" process a mechanistic paradigm.

From the perspective of the philosophy of science, the "driving" associated with "big data" is largely based on modern computer technology and achieved through the design of "algorithm," systematic steps, processes, and specific operational procedures.

X

The above is the main academic thinking and theoretical content of this book. Based on the definition and selection of the connotation of complex systems, this book discusses the essential attributes of complex systems management. Through the transfer of complex systems thinking paradigms, the author condenses and abstracts a wide range of new types of management activities in today's real world in terms of complex holistic attributes, forming a new discipline of complex systems management in the field of management.

Although the connotations of these management activities exist in different fields, manifesting in different forms and having different scientific issues, they share common essential attributes in their research objects. This gives rise to a common academic logical starting point, common thinking principles, and a basic academic paradigm, which is a statement of great academic significance. Just as decades ago, when people were temporarily unable to cope with the intricate "complexity," they could only create numerous "complexity terms" to deal with it. This time, we are attempting to use the thinking of complex wholeness and the paradigm of complex systems management to deal with the challenges of complexity issues in the social-economic field. In essence, it is expected to provide a theoretical thinking principle and knowledge system framework for management practitioners and scholars from different fields and disciplines, enhancing their ability to understand and manage complexity issues.

Especially in the era of "big science" and "big management" of complex systems, there is often a lack of management resources, experience, and capabilities. From a theoretical perspective, the academic and philosophical thinking of complex systems management aims to enhance individuals' ability and adaptability to acquire resources for solving complexity management problems at a higher level.

Such "acquisitive ability" and "adaptive capability" are precisely the scarcest and most precious in addressing and managing the widespread complexity challenges in various fields today.

An independent knowledge system of complex systems management is constructed (as shown in Fig. 1).

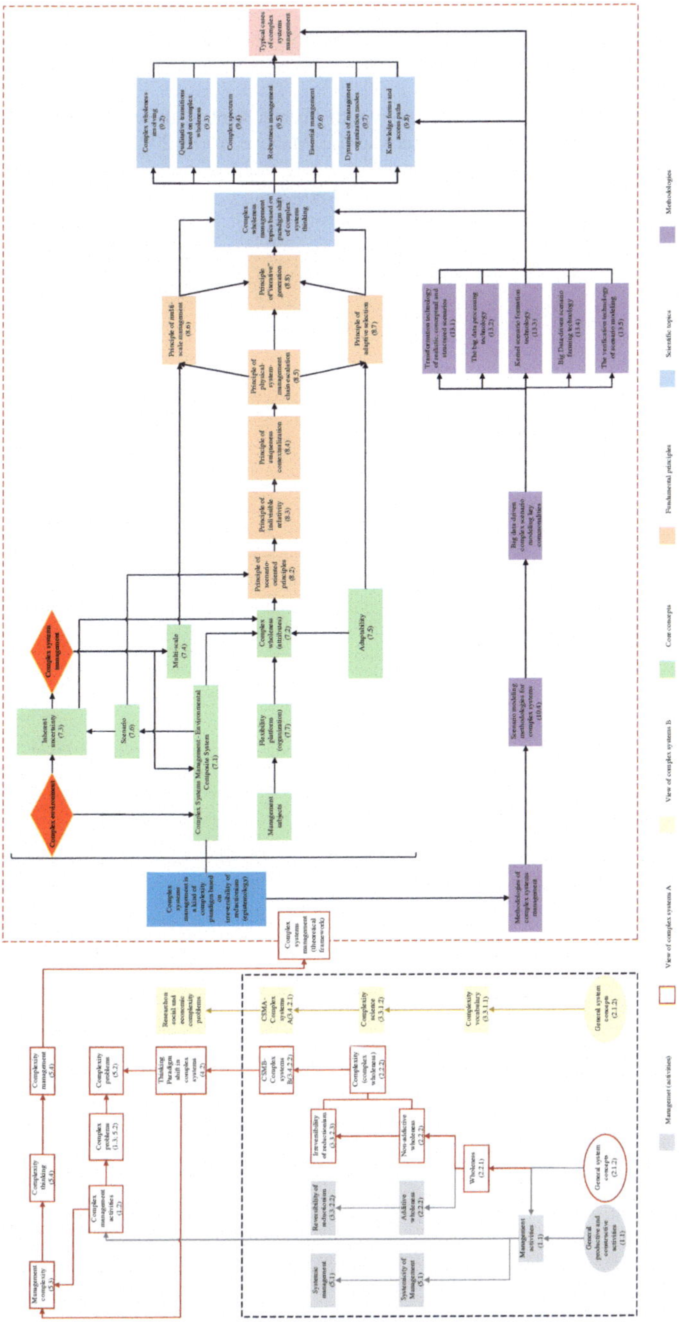

Fig. 1 Logic diagram of constructing an independent knowledge system of complex systems management

References

1. Sheng Z. From systems management to complex systems management – written on the occasion of the 30th anniversary of the founding of the Journal of systems m anagement [J]. J Syst Manag. 2022;31(06):1031–4.
2. Yang X, Gao Z, Sheng Z, et al. Complex system management is an important part of the management system with Chinese characteristics [J]. Manage World. 2022;38(10):1–24.
3. Sheng Z, Yu J. Complex systems management: a new field of management with chinese characteristics [J]. Manage World. 2021;37(06):36–50+2.

Contents

Chapter 1
Management and Complex Management

1.1 The Essence of Management

Since ancient times, humans have engaged in practical activities such as hunting, fishing, gathering fruits, farming, and constructing houses, roads, and bridges for survival and development in specific environments [1]. It is self-evident that any productive and constructive activity has its purpose. For instance, planting crops is for harvesting food to satisfy hunger. Hunting not only provides meat but also animal fur for making warm clothing. Additionally, building simple houses offers shelter from wind and rain and defending against wild animals. Therefore, productive and constructive activities of human beings are a category of activities with clear purposes. The initial purpose of these activities was to obtain or create something that fulfills certain needs for humans. Some things naturally exist in the world, such as the fruits of plants, while others, like houses and bridges, are creations of human beings. These creations naturally fall into the category of "artifacts." In this book, the understanding of the purpose of human beings' productive and constructive activities mainly refers to the creation of such "artifacts."

Any productive and constructive activity of human beings requires certain resources. For example, planting crops requires land, hunting requires spears, and constructing houses requires cutting down trees with axes. It also requires knowledge of the most suitable timing for planting and harvesting, as well as expertise in constructing sturdy houses. The essential elements, including land, materials, tools, and knowledge, involved in productive activities constitute the components of those activities. The "artifacts" that serve a certain purpose, formed by these elements through these activities, are the tangible manifestations of the purpose of productive and constructive activities. This indicates that human beings' productive activities are a complete process, from integrating relevant elements to forming artifacts, encompassing practicality, process, and entirety. Among these qualities, entirety and purpose become the most fundamental qualities of human productive and constructive activities.

Z. Sheng, *Outline of Complex Systems Management Theory—Based on Irreversibility of Reductionism Thinking*, International Series in Operations Research & Management Science 360, https://doi.org/10.1007/978-981-97-3361-3_1

Initially, people's productive and constructive activities were very simple and small in scale, often undertaken and completed by individuals or a few people. However, with the development of human society, the scale, scope, and complexity of these activities has continuously increased. For instance, hunting a large animal necessitates the collaboration of a group of individuals, while felling a large tree and transporting it to a construction site demands the combined efforts of many people. When an individual or a very small group cannot complete a task, it becomes a collective endeavor. At this point, people's productive and constructive activities begin to involve division of labor and cooperation. In this scenario, some individuals no longer directly participate in these activities. Rather they specialize in planning, organizing, dividing labor, and coordinating activities based on the environment and expected goals of the activities.

Over time, people gradually realized the positive impact of specialized planning, organizing, labor division, and coordination in leveraging collective strength and wisdom for better task completion. This is because they make the process of activities more orderly and effective, ensuring the achievement of productive goals. Thus, these activities evolved into a distinct category of practical approaches in productivity, forming the basis of what we now understand as management activities.

In general, **management activities**, or simply **management**, refer to a category of activities within human productive and constructive activities. These activities involve planning, organizing, and allocating resources, as well as arranging and coordinating various relationships in order to achieve productive and constructive goals, thereby making the process more orderly and efficient.

Based on the above definition, the following in-depth interpretations can be made for the connotation of management activities:

(1) Human beings' management activities stem from fundamental human productive and constructive activities and are closely linked to the creation of artifacts. The purpose of human productive and constructive activities is to create various artifacts, while management plays the role of integrating, regulating, coordinating, and constraining the relationships between people, as well as between people and objects, and among objects themselves. Therefore, human beings' productive and constructive activities depend on management, and vice versa.
(2) Human management activities originated from serving the most basic human productive and constructive activities. As human society has progressed, management activities have continuously evolved and taken on various forms in all practical, productive, and constructive human activities, playing a significant role.
(3) Each specific productive and constructive activity of humans is specific. No productive activities are identical in the world, neither are two completely identical management activities. In other words, any specific management activity in the world will vary depending on various factors such as specific activities, time, location, individuals, events, and circumstances.
(4) This indicates that even for two highly similar productive and constructive activities, management will have different content and characteristics. This variance

is due not only to the unique situational details of each specific activity but also the fact that management involves one group of people as the orchestrators and another as the recipients. Therefore, the varied psychological and behavioral tendencies, along with differing values, between managers and those being managed, shape the nature and features of management activities. In other words, management is a people-centric and adaptable activity, where the focus on objects should not overshadow the human element. People are always the dominant core element in managing various factors.

(5) Management activities encompass a wide range of practical content. They should be guided by actual needs. These include: what tasks need to be accomplished, what problems need to be solved, and what techniques need to be employed. Emphasis should be placed on feasibility and operability, as well as performance and cost. Therefore, the choice of different management models and methods should be based on their suitability and efficiency rather than being extravagant and redundant.

(6) Most importantly, since management activities support productive and constructive activities, each of which has a specific purpose and manifests as a complete process, every management activity inherently maintains its own integrity and process completeness of the process. For example, any management activity consists of various components, including a management environment, clear management objectives, management subjects, management objects, management organizations, various management resources, and management problems to be addressed. This results in an orderly and comprehensive process. The essence of management activities lies in comprehensive thinking and a process based on this integrity and completeness, which enables the realization of predetermined purposes (goals) of management activities. Therefore, integrity and purposefulness are the two fundamental qualities of any management activity [2].

1.2 Complex Management Activities

As the scope and scale of human productive and constructive activities expand, encompassing more elements and diverse internal connections, people perceive and form sensations and impressions of "simple" and "complex." This perception is through direct contact with these activities and the sensory organs (eyes, ears, nose, tongue, body), generating direct sensations in the brain. For example, building a bridge over a small river is a simple constructive activity, while constructing a bridge over the sea is a complex constructive activity.

From the perspective of human cognition, our understanding of "complex productive and constructive activities" generally goes through the following two stages. The first stage involves directly perceiving and understanding the external connections and surface phenomena of "complex productive and constructive activities." This stage, characterized by directness and vividness, belongs to "vivid intuition" in cognition terms. The second stage of understanding employs abstract thinking to understand the internal connections and essential laws of "complex productive and constructive activities." Characterized by indirectness and abstraction, this stage belongs to "abstract thinking" in cognition. Completing this stage requires the use of concepts, judgments, and reasoning.

We can understand how "complex productive and constructive activities" create a "vivid intuitive" impression of "complex management activities," by analyzing several basic elements of management activities.

(1) **Management Environment**. The real-world environment of "complex productive and constructive activities" often mirrors the respective management environment. The world is currently undergoing the most extensive and profound changes, and people's management concepts, behaviors, and values are undergoing unprecedented changes in response to environmental changes. These changes will have a series of profound impacts on the reorganization of the relationship between management models, management elements, and management functions. For example, environmental protection and green, low-carbon aspects are becoming increasingly important goals in management activities.
(2) **Management Subjects**. Management subjects refer to the group of stakeholders who have decision-making power, property rights, supervisory rights, and discourse rights in management activities. Although we generally refer to them as management subjects, they are a group of stakeholders called the subject group rather than individual entities. With the continuous expansion and enrichment of management scale and content, the composition of the management subject group is becoming more diverse. Generally, within the management subject group, a "dominant subject" emerges leadership and decision-making power, and greater discourse rights forming a hierarchical structure within the group. At the macro level, some subjects form decision-making bodies for major issues. At the meso level, some subjects self-organize into strategic partners. At the micro level, some subjects establish operational rules and processes based on operational and coordination requirements. The interconnection and communication across these different levels not only require the subject group to have stronger leadership and coordination capabilities but also call for mechanisms to prevent any deviation from the subject's own behavior. Additionally, in the face of complex management environments and tasks, the subject group needs to have self-organization and adaptive capabilities.
(3) **Management Organization**. It is difficult for management subjects to construct a structurally rigid management organization all at once when dealing with complex management tasks. Similarly, they also do not possess the ability to solve and control all management issues throughout the complex management process. This requires management organizations to demonstrate sufficient structural "flexibility" and "adaptive capabilities" during the management process. It calls for the development of a flexible management "platform" model by the management subjects. Such a model can have stronger adaptive functions through different approaches, such as changing the composition of the subject

and altering management mechanisms and processes to replace traditional rigid and static management organization models.

(4) **Management Objectives**. Under the guidance of complex production and construction objectives, it is necessary to fully consider how to design management objectives considering multiple domains, levels, dimensions, and scales. This involves integrating and refining objectives, requiring enhanced skills in goal decomposition, analysis, and synthesis throughout the management process. For example, some management objectives have different spatial and temporal scales, which requires management subjects to consider compressing large spatial and temporal scales into smaller ones or vice versa. Similarly, when conducting comprehensive evaluations of the entire objective group, it is necessary to coordinate and handle conflicts between objectives at different scales.

Through a brief analysis of several basic elements of management activities, such as environment, subjects, organizations, and objectives, it becomes clear that "complex productive and constructive activities" prompt subjects to generate a series of new "vivid intuitive" perceptions and responses to these management elements. These perceptions and responses profoundly influence the corresponding management activities in various ways, thus enhancing the complexity of management activities in service of "complex productive and constructive activities."

Therefore, with the development of management practices, people have gained intuitive experiences of "simple management" and "complex management." These experiences correspond to the complexities inherent in "complex productive and constructive activities."

1.3 Complex Management Problems

The practice has shown that the most important aspect of the "vivid intuitive" physical experience of "complex management" is the perception that "complex problems" arise in management. Since all management activities are guided by management problems. Therefore, we will now analyze "complex problems" within "complex management" in detail.

Logically speaking, in "complex management activities," the complexity of the management environment, subjects, organizations, and objectives will inevitably lead to the emergence of "complex management problems." Conversely, if a management activity does not have complex problems, it cannot be considered "complex."

Of course, not all issues in "complex management activities" are "complex," nor are they equally complex. This requires an analysis of all the issues in "complex management activities."

In this regard, we can classify the corresponding management issues based on the degree of complexity in productive and constructive activities and the degree of complexity in the environment (as shown in Fig. 1.1). The problems in the A region can be considered simple issues because both the degree of complexity in productive and constructive activities and the degree of complexity in the environment are low, and they can be solved using known experience and knowledge. The problems in the B region, as shown in the figure, arise from a higher degree of complexity in the environment, leading to noticeable uncertainty or dynamic correlations. As for the problems in the C region, the high degree of complexity in the internal structure of productive and constructive activities can lead to enhanced interdependencies among elements, resulting in imprecise cognition and unclear causal relationships. Therefore, for management issues in B and C regions, new management rules can be formulated, and known experience and knowledge can be utilized to solve them. For a subset of issues that exhibit strong correlations, techniques such as systems engineering can be employed for resolution. This indicates that there is a considerable portion of issues (A, B, C regions) in complex management activities that can be addressed through a combination of conventional management practices and systems engineering.

However, management issues in the D region, as shown in the figure, represent highly complex problems in terms of both productive and constructive activities and environmental complexity. These issues encompass challenges such as designing management organizational platforms for heterogeneous subjects, generating iterative and uncertain decisions, conducting complex risk analysis and prevention, and coordinating multi-agent and multi-objective control in production and creative sites. Generally, these complex issues cannot be resolved by simply using conventional methods applicable to the A, B, and C regions. Instead, they require new thinking and new approaches for effective resolution. We refer to them as **complex management problems** or simply **complex problems**.

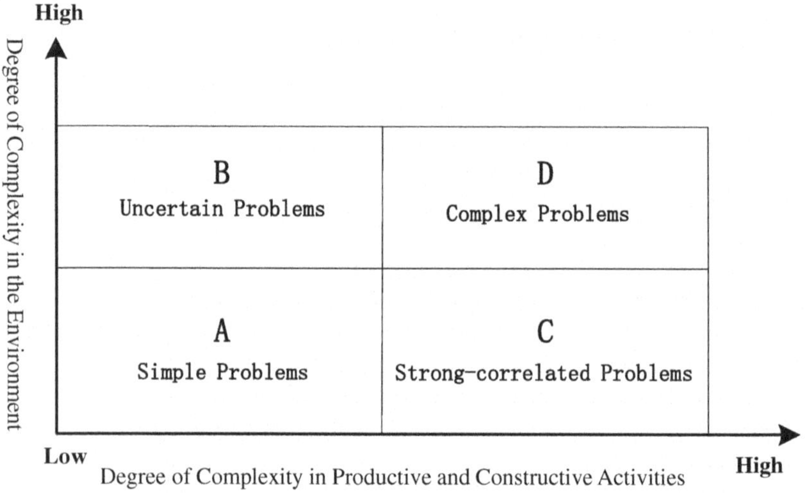

Fig. 1.1 Classification of problems in complex management activities

In general, the complex problems in complex management activities possess the following new characteristics:

Firstly, the boundaries of these problems are often ambiguous and not entirely clear. Within the problem, in addition to the definite input/output relationships between elements, there are also somewhat uncertain or even highly uncertain associations. Besides the explicit and known relationships, there are implicit and hard-to-determine relationships. Moreover, some relationships or associated elements that we identify may change during the management process due to changes in certain factors.

Secondly, these problems are generally difficult to be fully described using a clear structured approach or model. For example, complex problems often involve elements of engineering technology, socio-economics, human behavior, and cultural values. Engineering technology elements are primarily governed by natural sciences and technical principles, and they can be described in a structured manner. Elements in the socio-economic domain are mainly governed by social or economic laws and can be described in a semi-structured manner. On the other hand, human behavior and cultural value elements can often only be described in a non-structured manner. As a result, these management issues must be described using structured, semi-structured, and even non-structured approaches simultaneously. This not only greatly increases the difficulty of problem description but also adds challenges in integrating and synthesizing different types of descriptions.

Furthermore, solving these problems will involve knowledge from multiple disciplines and fields. According to human cognition patterns, the understanding of these problems will inevitably be a process from ignorance to knowledge, from limited knowledge to more comprehensive understanding, from an unrepresentative perspective to a holistic view, and from superficial to profound insights. Therefore, in generating solutions for these complex problems, we will exhibit a continuous "trial and error" exploration process. In this process, management solutions are usually not optimized in one go but determined through multiple comparisons, adjustments, and improvements of different alternative solutions based on the depth and precision of problem understanding. Generally, this process follows an ordered path from relatively vague to relatively clear, from relatively partial to relatively comprehensive, and from lower quality to higher quality intermediate solutions, iterating and approximating until converging to the final solution.

Consequently, during the solution formation process for complex problems, many new interfaces and connections will inevitably emerge and intensify. For instance, there will be a need for more coordination and communication among managers, more frequent revisions and comparisons of previous and subsequent plans during the iterative process of management solutions, and an emphasis on the effective integration of different types of information. This involves a holistic evaluation (considering overall costs, timeliness, quality, etc.) and optimization of the plan.

Based on the analysis above, management issues in complex management activities can be broadly divided into three levels. The lower two levels mainly correspond

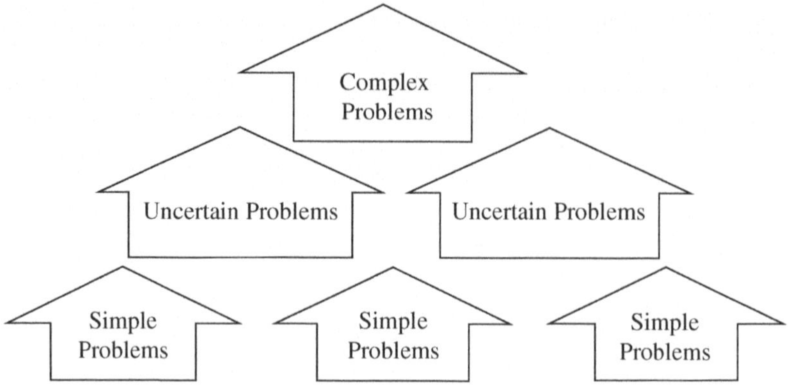

Fig. 1.2 The framework of problems in complex management activities

to the aforementioned A, B, and C issues, while the uppermost level primarily represents D issues. The integration of these three levels of issues forms a complete system of management issues within complex management activities Fig. 1.2.

The problem system in complex management activities categorizes all management issues into three types: complex problems, uncertain or strongly correlated issues, and simple issues. This classification primarily stems from the properties and characteristics of the problems themselves, and it is closely related to the cognitive abilities of managers. For example, two managers at different levels will have different perceptions. A high-level manager might see fewer complex problems, while a lower level manager might think the opposite. Even for the same individual, as his/her capabilities improve over time, he/she may perceive a decrease in both the number and complexity of complex problems. An extremely experienced and high-level manager might even believe that there are hardly any complex problems at all.

The above understanding enlightens us that based on the perceptual cognition of complex productive and constructive activities and the interpretation of their elements, we can form a certain intuitive perception of the external connections and superficial characteristics of "complex management problems" in "complex management." This perception is developed by correlating the "complex production artifacts" with "complex management." Of course, the "complex" in "complex management problems" refers to an intuitive perception of the external appearance and phenomena of the problem. It is a natural language expression in layman's terms of this perception, and it has not yet revealed the essential connotation and attributes of "complex problems."

References

1. Sheng Z. Management: from systemic to complex. J Manag Sci. 2019; 22(03):2–14.
2. Laszlo E, Min J. The origins of general systems theory from the writings of Berta Langfi (Introduction to the book on general systems theory). Chin J Sys Sci. 1993;02:60–4.

Chapter 2
Systems and Complicated Systems

2.1 Overview of the Concept of the System

2.1.1 Initial Thinking About the System

In the history of human understanding of the world, more than 2000 years ago, outstanding thinkers from both China and ancient Greece independently pondered and explored how the universe, as a whole, is constituted. For instance, during the Spring and Autumn period in China, Lao Tzu proposed that the "道 (Taoism, or Tao)" is the origin of all things in the world and is the "ancestor of all things." The pattern of the creation of all things is "*Tao begat one, one begat two, two begat three, and three begat all things*." In Lao Tzu's philosophy, the "Tao" is the force behind all laws of the world. This force first gives birth to "one," which is "existence." This "one" then divides into "two," which represents "阴(Yin) and 阳(Yang)." The combination of Yin and Yang forms "三(three)" (many), which represents all things in the world.

The ancient Greek philosopher Heraclitus believed that the entire universe is composed of the smallest, indivisible particles called atoms and the "void" between them. He further stated that atoms are infinite and diverse in quantity, and the collision of different atoms gives rise to various states and forms encompassing all things.

It can be said that over two thousand years ago, these remarkable thinkers, with their astonishing speculative abilities, constructed a subjective "vast world" that almost approached the limits of the macrocosm and the microcosm. They showcased the greatness of human wisdom. However, while praising the power of speculation, people also realized that speculation does not necessarily require objective facts or practical experiences as prerequisites. Therefore, when it comes to understanding natural phenomena and activities such as production, construction, and management that transform the objective world, relying solely on speculation can sometimes be inconsistent with reality or can only provide vague and general "ideas" that fall short of guiding thoughts into action. As mentioned above, in understanding the totality of

Z. Sheng, *Outline of Complex Systems Management Theory—Based on Irreversibility of Reductionism Thinking*, International Series in Operations Research & Management Science 360, https://doi.org/10.1007/978-981-97-3361-3_2

all things, humans can overcome the limitations of speculation only through scientific experiments and numerous new discoveries. This approach provides a solid logical starting point for comprehending the whole.

Since the early twentieth century, with the rapid development of various disciplines like physics and biology, scholars from different fields have continuously put forward new ideas when considering the wholeness of things. For example, in 1925, British philosopher Whitehead proposed the theory of organism over mechanism, suggesting that only by viewing living organisms as organic wholes can we explain complex life phenomena [1]. Subsequently, scientists from the United States and Germany proposed expanding beyond the field of biology from multiple disciplinary perspectives to study the wholeness of things. This involved not only examining how things are constituted as a whole but also exploring the ways in which the relationships between different elements of the whole affect the state, behavior, and function of the whole. This marked the beginning of what later became known as "general systems theory" in the scientific community.

During this period, Ludwig von Bertalanffy, an Austrian theoretical biologist of Canadian origin, made important contributions to the concept of "wholeness." As early as the 1920s, he advocated treating the biological organism as a whole to discover the organizing principles at different levels of the organism. He introduced the concepts of coordination, order, and purpose into the study of organisms, forming a comprehensive view, dynamic perspective, and hierarchical perspective of studying organic entities. In 1937, Bertalanffy first proposed the concept of general systems theory during a philosophical discussion, signifying the beginning of people's contemplation and exploration of the common issues of wholeness, interconnectedness, and functionality in various fields. It also marked the beginning of the era of systems science [2].

2.1.2 Interpretation of the Concept of System

The evolving practices in various fields of human endeavor have led people to increasingly recognize that things generally exist as interconnected wholes and different wholes are interrelated, forming larger-scale wholes. To understand the objective laws governing this universal phenomenon, a specialized scientific term is needed to define its attributes, and this is the birth of the abstract concept of "systems."

1. Ludwig von Bertalanffy's Definition of Systems

In a general sense, Bertalanffy defined a system as a "**complex of interacting elements**." The term "elements" refers to important (essential) components that constitute an objective entity and maintain its characteristics as the necessary minimum units. They are both the indispensable basic units that make up a system and the fundamental reasons for the system exhibiting certain characteristics. "Interacting" can be understood in a general philosophical sense, encompassing all forms of relationships, connections, and influences. "Complex" refers to the integration of elements into a whole, and its core meaning is the synthesis of a complete and unified "set." Clearly, this definition of a system has broad applicability. For example, all-natural systems such as planets, solar systems, and primitive forests, as well as all artificial systems, including material-based systems like cities and engineering and non-material-based systems like knowledge and culture, fall within the scope of Bertalanffy's definition of systems.

Bertalanffy's definition of systems implies that there are connections between the elements (parts) that make up the system. In other words, "association" is the fundamental "component" of the system. This is because:

(1) Association implies that the elements in a system are different, heterogeneous, and even diverse. If a system only contains elements of the same type, their relationships can only be trivial and lacking substantive meaning, and a new "whole" cannot be formed.
(2) The term "association" expresses a universal meaning of connection, linkage, relevance, and coherence, but its specific forms can vary greatly. It can be physical, such as a bolt and a nut; it can be correlational, like the temperature in Summer and the sales of ice cream; and it can be logical, such as the same words or characters having completely different meanings when used in different contexts to form sentences.
(3) Association is of significant importance to a system because the sum of various ties constitutes a structured system with ordered space and actions. The significance of system structure includes two main aspects. Firstly, the structure of a system is the physical basis and realization mechanism of the overall effectiveness and functions of the system. Secondly, even if there are problems with local relationships within a system, in many cases, the overall structure can still maintain and ensure the main functions of the system. For example, if a person loses a finger, their overall bodily structure can still support their major life functions.

2. Hsue-Shen Tsien's Definition of System

The systems most closely related to human survival and development are artificial systems. The key difference between artificial systems and natural systems lies in the fact that pure natural systems exist independently of humans or are entirely unrelated to human purposes, whereas artificial systems are created by humans and reflect the deliberate intentions of humans. These intended functions of artificial systems, which are essential to humans, are referred to as the system's "function."

Hsue-Shen Tsien, a Chinese systems scientist, recognized the importance of artificial systems to the socio-economic and engineering development of human society, particularly the significance of the intended functions of artificial systems in human productive and constructive activities. Consequently, he provided the following definition of systems: "An organism composed of interacting and interdependent components that combine to perform specific functions." The term "organism" refers to the interconnectedness and interactions among the components, similar to the human body, where all organs and tissues collaborate and support each other, forming a unified ensemble with life characteristics.

Notably, Hsue-Shen Tsien's system concept introduces the term "function." This represents the transformation of humans' intended functions for artificial systems into the objective effectiveness and capabilities of these systems. By emphasizing the functional attributes of systems, the definition highlights the practical value of artificial systems to humanity, reflecting the closer relationship and greater significance of artificial systems to humans. Fundamentally, the history of human development involves continuously empowering non-existent artificial systems and continuously creating, realizing, utilizing, and improving artificial systems, leading to the progress and advancement of human society. Therefore, Hsue-Shen Tsien's system definition, which incorporates the functional attributes of artificial systems, better aligns with the essence of human productive and constructive activities and management activities.

2.2 Interpretation of the Overall System

2.2.1 The Functional Connotation of the Overall System

Wholeness is the core implication of the system concept. Philosophically, "wholeness" is relative to "parts." The basic meaning of the term "system" is that "wholeness is composed of parts, and parts are subordinate to the wholeness." However, this is not enough. The concept of "system" also includes the important component of "interrelation." Therefore, a system also implies that "wholeness is formed by the mutual interactions and connections of the parts according to a certain order." In philosophy, this is referred to as "organic connection." Thus, a system is recognized as a new unity (wholeness) formed by the parts through organic connections. The formation of a system is necessarily a spontaneous or driven process, either by some force or a combination of both, in which some mechanisms or laws govern the formation of wholeness.

Systems can generally be classified into two categories. One category is pure natural systems, which exist independently of humans, such as primitive forests and Antarctic glaciers. The other category consists of systems that do not exist in nature and are created by humans, such as nations and companies. Why do people create artificial systems? The main reason is that people create artificial systems to utilize certain functions at the systemic level to satisfy their practical needs.

In ancient times, people initially used the natural functionalities of entities directly to meet their needs, such as a naturally lying tree trunk over a stream that could help people cross the river easily; natural caves could protect people from the severe cold. However, when natural entities no longer provided the functions that people needed, or when the quality of these functions couldn't meet their requirements, people would endeavor to create these functions themselves. However, any specific function in reality must be attached to a specific entity and cannot exist independently. Therefore, in order to obtain a function of utilitarian value, people must create an artificial system that embodies this function. For instance, if a river is too wide and tree trunks are not long enough, people would design bridges for large rivers based on the characteristics of small bridges over streams. In places without caves, they would build houses with cave-like functions to withstand wind and rain.

In this way, in order to survive and improve living conditions, people must create artificial systems with various functions. The functions of artificial systems are both the empowerment of people in those systems and the corresponding behavior or characteristics of the systems at the systemic level. Therefore, the functionality of an artificial system is the holistic value created by people through productive and constructive activities. The functions and values of artificial systems often surpass the original functions of natural systems. Human wisdom creates infinite functions and values which did not exist in nature. These functions and values then facilitate the progress and development of human society.

Generally speaking, the functionalities of a system refer to the representation of the collective behaviors, characteristics, features, attributes, and capabilities of the artificial system. They reflect the value pursued by the creators of the artificial system. In simple terms, the reason why people design and create an artificial system is to utilize and benefit from its advantageous functionalities. Therefore, in many contexts, the concepts of system properties, qualities, values, roles, and functionalities often have inherent similarities in their connotations.

With further research on systems, it has been discovered that if humans are included in the environment of an artificial system, the concept of system functionality can be considered as the system's impact on the environment. As long as the system is not completely closed, there must be some kind of relationship between it and the environment. This is how humans perceive the system's functionality by humans in the environment. Therefore, every system has functionality.

Furthermore, systems not only have beneficial and useful functions for the environment but can also have harmful or unnecessary functions. For example, artificial systems like companies have the function of providing various valuable products for society, but they also generate various pollutants that harm the environment. This is also a function of the company's impact on the environment. In particular, when a company both provides products and pollutes the environment, it is not always possible to completely pursue the benefits while avoiding the harm. Instead, we can focus on environmental protection, promote low-carbon and green production methods, and reduce the negative impact of companies on the environment while still harnessing their beneficial functions.

In summary, systems can be described as a complete entity with new traits and functions formed under a specific order and organization. This is the functional connotation of wholeness in the system concept.

2.2.2 The Emergent Connotation of System as a Whole

As aforementioned, the concept of a system implies that "the whole is formed by the interconnection and interaction of its parts according to a certain order." Therefore, the way in which the parts are connected to form a whole directly influences the form and characteristics of the overall system. The simplest form of order is mere spatial movement and aggregation, where the whole merely represents the movement of spatial positions or just a "sum" in quantity without any new qualitative changes. For example, when several classes of students gather on the playground, it is a form of connection where the parts (classes) come together through spatial movement, and the relationship between the "whole" on the playground and the "parts" of each class is simply expressed as the total number of students equaling the sum of the students in each class. We express this type of system characteristic, where the "whole equals the sum of its parts" with a simple equation, namely $1 + 1 = 2$. It is understandable that if the whole of a man-made system is merely formed by such a simple order of aggregation, it cannot create any new, significant functions for humans. Consequently, the practical value of such a man-made system would be limited.

Obviously, we hope that systems, especially man-made systems, can exhibit new characteristics and functions at the whole level that their individual elements or local parts do not possess. In this case, the whole of the system must not be simply formed by a mere "sum" of its parts, and we express it with another equation, namely $1 + 1 > 2$. This means that, in this case, the "order by which the parts are interconnected" must involve mechanisms or actions that enable the system to form new characteristics and functions as a whole, resulting in the emergence of new "qualities" at the system level. The reasons why this new "quality" appears and how it arises are often the starting point for studying, creating, and optimizing systems.

The emergence of "qualities" at the system level, which did not exist in the system's individual parts, is akin to a spring suddenly appearing from underground. Therefore, people refer to such phenomena where $1 + 1 > 2$ as "emergence." Literally, "emergence" implies the phenomenon of " coming into existence from nothingness" or transitioning "from zero to one." The phenomenon of "emergence" can be beneficial or harmful to humans, and we need to exercise control over it.

For example, robots, as man-made systems, can improve work efficiency and replace humans in dangerous and repetitive tasks. However, if robots autonomously develop an antagonistic consciousness toward humans, they can become extremely dangerous forces opposing humanity. In other fields, such as chemical pharmaceuticals and gene editing, the ambivalence of "emergence" in man-made systems is evident, requiring scientific precautions.

It can be seen that the mechanism behind the emergence of system characteristics is very complex. One intuitive reason is that through interconnections and permeation between the microelements or parts of the system, new "qualities" are formed during the process of combination into a new whole. The emphasis here is on the "new whole," which highlights that transcends its original components. It is this new entity that leads to the emergence of new mechanisms and functions. For example, in the field of life sciences, cell fusion, also known as cell hybridization, refers to the phenomenon where two or more cells as constituent elements (parts) permeate and combine to form a new cell. In the fusion of animal cells in vitro, certain inducing substances are often used to create a new hybrid cell line, and these hybrid cells generally exhibit strong vitality and vigorous proliferation. These represent the new "qualities" of the system.

From the above, it can be seen that the new "qualities" of the system as a whole often embody the core values of the system and the purpose of creating man-made systems. For us, it is not just about observing the phenomenon of new qualities emerging at the system level but also about understanding the mechanisms through which new qualities "emerge from nothing" from the microscopic level to the macroscopic holistic level.

In summary, we have outlined two basic types of the system as a whole. One type is the simple aggregative wholeness represented by $1 + 1 = 2$, which mainly exhibits aggregative functionality. The other type is the non-aggregative wholeness represented by $1 + 1 > 2$. "Non-aggregative wholeness" is a crucial concept in this book. Firstly, "wholeness" refers to the attributes and functionalities that the system possesses at the whole and global levels but are absent in its constituent parts or local components. This primarily represents the characteristic attributes of the system at different hierarchical levels. Secondly, "non-aggregative" represents the characteristic attributes of the system among its multiple parts, including their interactions and relationships. The fusion of these two concepts indicates that a certain type of system, based on the general systemic wholeness, exhibits complex attributes and functionalities vertically and horizontally. Additionally, it forms complex systemic characteristics with new connotations between these dimensions. This is different from the general systemic wholeness and can be referred to as "complex wholeness," which represents the emergent connotations of the system as a whole.

From the above elaboration on the connotation of systemic wholeness, it can be understood that a system's "non-additive wholeness" is equivalent to "complex wholeness." This type of wholeness of the system is very important to the text: it is precisely this "complex wholeness" possessed by the management object that becomes the logical starting point and new paradigm for complex systems management.

2.2.3 Systemicity

From the concept of a system to the interpretation of the system's entirety, including the functional connotation of the system's entirety and the emergent connotation of the system's entirety, our understanding of the system has gradually evolved from an external, superficial level to an internal, connotative level. Notably, natural and social sciences are categorized based on their specific research subjects. The concept of a system, regardless of the specific field or the unique characteristics of the specific issue it studies, transcends these categories. It views them as abstract "systems" for observation and research. This characteristic determines the cross-disciplinary nature of the system concept and forms the inherent attributes and laws of the system.

In philosophical terms, the most fundamental quality of a thing is its essential attribute, which is the abstract representation of the unique, basic, and stable properties of the thing. The essential attribute determines the fundamental quality of the thing that makes it what it is and differentiates it from other things. It embodies the inherent determinism that distinguishes the thing from others.

Based on the aforementioned explanations of systems, it is not difficult to see that, in a general sense, systems have the following inherent and unique main characteristics (attributes):

(1) Diversity (every system consists of multiple elements);
(2) Correlation (there are interrelationships among different elements);
(3) Wholeness (the existence, behavior, and functionality of a system are complete and consistent);
(4) Functionality (the overall state and behavioral characteristics of a system have certain roles and meanings).

These four characteristics (i.e., diversity, correlation, wholeness, and functionality) are generally referred to as the fundamental attributes of systems. Among them, the most important attribute is the wholeness of the system (or the wholeness and functionality), as it encompasses diversity, correlation, and functionality. Therefore, **wholeness is often regarded as the fundamental attribute of a system, a concept known as systemicity**.

Systematicity, especially the wholeness within it, has brought about profound changes in our way of thinking and approach to problem-solving. This is mainly manifested in the following:

(1) Often considering the object of study as a whole, analyzing the interrelationships, structures, functionalities, and changes and evolutions among its elements, correlations, and environment.
(2) In addition to employing traditional methods of analysis, decomposition, and dissection in problem-solving, paying attention to analyzing and maintaining the interrelationships, wholeness, and connections with the external environment.

The concept of systems, along with systemic thinking and the resulting changes in thinking, provide us with a profound and accurate understanding of the world, management activities, and managerial behavior. They offer significant guidance and insights for us.

2.3 Complicated Systems

In the previous sections, we have interpreted key "components" of the system concept, such as elements, correlations, structures, wholeness, and functionality, providing a relatively complete and clear understanding of this concept. With the development of artificial systems and the advancement of theoretical research on systems, people's comprehension of the concept of a system is gradually becoming more profound.

It is easy for people to intuitively perceive that systems can be categorized as "large" or "small." For example, a city is a system and a big city is a "large system," while a small city is a "small system." Systems can also be classified as "simple" or "complicated." Generally, a university system is considered "complicated," while in comparison, a secondary school system is seen as "simple." The distinction between "large" and "small" in systems is relatively easy to understand, but what constitutes a "complicated system" and how complicated systems are formed are not as straightforward and obvious.

Firstly, describing a system as "complicated" may seem straightforward since "complicated" is a descriptive adjective. This statement is expressed in a colloquial and vernacular manner. Its implication is that when people see, hear, and perceive various aspects of a system, they form intuitive impressions and perceptions in their minds regarding its external, explicit, and superficial phenomena. For example, the basic meaning of this statement includes the understanding that the system is composed of many different parts, the number of elements is relatively large, and the relationships between some parts of the system are complicated and disordered. Similarly, it is difficult to explain certain phenomena and characteristics of the system. Therefore, people naturally perceive the system as "complicated."

Next, we will introduce the meaning of the statement "this system is complicated" from the perspective of the connotation of the system concept, which is much more profound.

Firstly, when we say "多" (many) regarding elements, it doesn't simply mean a large quantity. Rather, it is more likely to refer to a wide variety of element types. Secondly, when we mention "无序" (disorder or lack of order), it doesn't necessarily mean true disorder. Rather, there is a wide variety of relationships between the elements. For example, within a system, the relationships between elements may involve linear relationships, nonlinear relationships, deterministic relationships, nondeterministic relationships, stationary relationships, non-stationary relationships, positive feedback relationships, negative feedback relationships, and so on. If a system possesses numerous types of relationships and is driven by multiple mechanisms simultaneously, it will appear "disordered and complicated" when observed from the outside. This means that perceiving multiple types of relationships leads us to "perceive the system as complicated."

Secondly, if we cannot adequately explain the overall characteristics or functionalities of a system based on the behaviors and characteristics of its micro-level elements, whether due to the complexity of the underlying mechanisms or a lack of clear understanding, people will "perceive the system as complicated."

Furthermore, the characteristics and features of a system generally exhibit dynamic changes. If the future state or changes of a system involve unpredictable situations such as instability, abrupt changes, or turning points, or it is difficult to explain the causes and laws of these situations, we tend to view the system as complex or complicated.

The above analysis shows that a system can be considered complicated, relatively complicated, or extremely complicated. This perception is influenced by both the inherent reasons of the system itself and the level of understanding and capability of the observer. In general, both of these factors are often present simultaneously.

The fundamental purpose of scientific research is to uncover the objective laws behind things or phenomena. If we attribute the complexity of a system mainly to the observer's level of understanding, then the study of "complicated systems" would vary from person to person and may lead to conflicting conclusions. Therefore, in order to reveal the inherent objective laws of the research object itself, we tend to believe that a system is perceived as "complicated" not primarily due to the observer's limitations but rather because the system itself possesses certain inherent and essential attributes. When we state "the system is complicated," it means that the inherent and essential attributes of this system are being reflected. If our understanding reaches this point, it means that our understanding of "the system is complicated" has transitioned from the vivid and intuitive "plain language" stage to the second stage of abstract rational thinking.

Now, let's consider a simple example: The likelihood of an electronic component completing a specified function or capability under prescribed time and conditions is referred to as reliability. Let's assume that the reliability of an electronic component is 0.9 (as shown in Fig. 2.1). If four identical components are arranged in a series structure, the reliability of that system is $0.9^4 = 0.6561$. On the other hand, if the four identical components are arranged in a parallel structure, the reliability of that system is $1-(1-0.9)^4 = 0.9999$. Additionally, the reliability of two other combinations of series-parallel configurations is 0.8019 and 0.9801, respectively.

This example demonstrates that the causal relationship between the specific structure and functionality of a system is an inherent and essential attribute of the system. It is precisely this attribute that allows us to perceive the system as complicated in terms of the relationship between structure and functionality. By leveraging this attribute, we no longer strive to manufacture the best quality products solely using

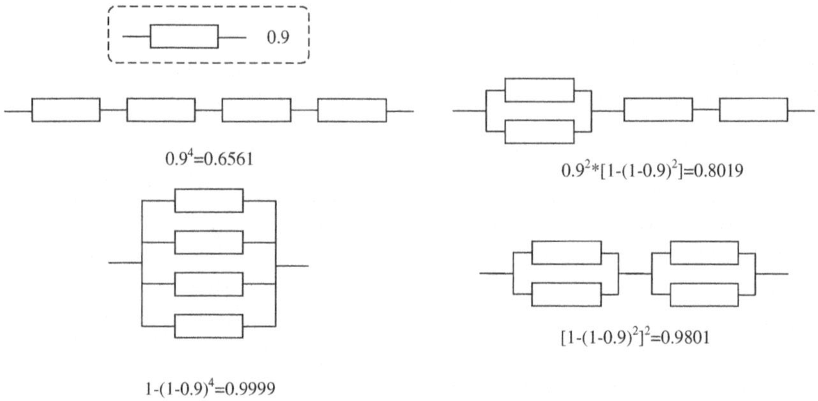

Fig. 2.1 An instance of system reliability

top-tier components, but rather through proper coordination and planning, even with standard components. At the same time, this example also helps us understand what it means to empower artificial systems and how to bring forth new "qualities" at the overall level of artificial systems.

Of course, this example only represents a relatively simple illustration at the physical level of systems, using a structured approach to intuitively perceive the phenomenon of "complicated systems." Our more important task involves using abstract and refined thinking, employing new concepts to understand and define the inherent and essential attributes of "complicated systems." We aim to grasp the true essence of what "complex" means within a system. This will be the main content of Chap. 3, "Complex Systems and Complexity," which follows.

References

1. Whitehead T. Science and the Modern World. Business. 1959.
2. Laszlo E, Min J. The origins of general systems theory from the writings of Berta Langfi (Introduction to the book on general systems theory). Chin J Syst Sci. 1993; (02):60–64.

Chapter 3
Complexity and Complex Systems

This book primarily focuses on the study of complex systems management, which is a holistic scientific concept formed by the deep integration of the concepts of "complex systems" and "management." It embodies an integrated scientific connotation. Therefore, one should not perceive "complex systems management" merely as the combination of "complex system" and "management." Otherwise, it would lack the new qualitative aspect of being "neither this nor that." Similarly, "complex systems management" should not be interpreted as "managing complicated systems" because "a complex system" represents a layman's intuitive perception and colloquial speech. Otherwise, "complex systems management" would merely be a colloquial term as well. Similarly, it would be a mistake to misinterpret "complex systems management" as "managing complex systems" or "a complex management system" without understanding the broader and deeper meaning intended by this term.

As Chinese modern thinker Hsue-Shen Tsien said, terminology in academia should not be used casually; it requires clear and precise meanings. Clearly, to understand the scientific connotation of this new concept of "complex systems management," it is necessary to clarify several preliminary questions related to complex systems and management:

1. Given that complex systems are distinct from complicated systems, what is the difference between the two?
2. Complex systems are a concept within the field of systems science. Does "complex systems management" entail the application of complex system paradigms to management? Is this application both rational and logical?
3. When "complex systems" and "management" are integrated into "complex systems management," what kind of new "quality", transcending conventional categories, emerges in theoretical terms?
4. Does the concept of "complex systems management" have internal consistency? What role and practical significance does it have for the development of contemporary management studies?

Z. Sheng, *Outline of Complex Systems Management Theory—Based on Irreversibility of Reductionism Thinking*, International Series in Operations Research & Management Science 360, https://doi.org/10.1007/978-981-97-3361-3_3

Clearly, these questions have direct and fundamental theoretical significance for us to have a precise and profound understanding of the new concept of complex systems management. Without addressing these questions, "complex systems management" risks becoming an ambiguous term without theoretical foundation logic, internal consistency, and extensibility.

The content of this chapter will revolve around complexity and complex systems.

3.1 Inherent Attributes and Concepts

According to Sect. 2.3, the term "complicated system" refers to our perception of the external connections and superficial characteristics of systems. This understanding is characterized by its directness and vividness, belonging to the first stage of knowledge acquisition, which is "vivid intuition." At this stage, people generally use plain, natural language. Following this is the understanding of the inherent connections and essential features of such systems. This understanding is characterized by its indirectness and abstractness, and it belongs to the second stage of cognition, which is "rational thinking."

Attributes refer to the characteristics that belong to a certain type of thing (object). If some attributes are present in the object as well as in others, they are termed as shared attributes. Conversely, attributes unique to an object and not found in others are referred to as its specific attributes. Attributes that define the object's distinct identity, distinguishing it from others, are referred to as its essential attributes. Attributes that define the object's distinct identity, distinguishing it from others, are referred to as its essential attributes. In philosophical terms, essential attributes are the most basic qualities of the thing, namely the abstract properties that are unique, fundamental, and stable to the thing. Essential attributes are also known as fundamental attributes.

Thus, in the second stage of "complex systems" cognition, we cannot use "colloquial language" as we did in the first stage to describe "complicated systems." Such language would be unable to express, convey, and communicate the essential features of such systems, nor could it be used to refine people's abstraction of their attributes.

A theory is the result of rational thinking, which, besides being based on natural language, also requires the use of specialized scientific language to accurately and profoundly refine the specific attribute that the theory intends to reflect. The foundation of this scientific language in a theoretical system is concepts. Each concept is a condensation and abstraction of the essence and internal relationships of things. Therefore, concepts can propel people from concrete thinking to abstract thinking and, on this basis, become a language tool for communication and transmission among members of the scientific community, constituting the "building blocks" of theoretical scientific discourse. The primary expression of concepts is scientific terminology [1].

As human understanding deepens, at the initial stage of understanding "complicated systems," the corresponding concepts generally only reflect the non-essential attributes of "complicated systems." However, as the understanding advances to a deeper stage, the corresponding concepts can gradually reflect the essential attributes of "complicated systems."

For example, "The car is getting faster and faster" is "colloquial language," while a physicist would say, "The car's acceleration is positive," and a mathematician would say, "The second derivative of the car's displacement with respect to time is greater than zero." These are all scientific language expressions. Acceleration and second derivative are scientific terms; they are concepts, not colloquial language. It is precisely these scientific terms and concepts that constitute the theories of physics and calculus.

Based on the above analysis, it is not difficult to understand that the task before us is how to abstract and refine the attributes of "complicated systems" and define them using a scientific term, namely a concept. This requires us to repeatedly comprehend the main phenomena, behaviors, characteristics, and forms of "complicated systems" and, based on this, refine their essential features and formation logic. It is not simply a matter of creating a "new term."

This is a delicate task because only by deeply understanding the essential attributes of "complicated systems" and forming scientific concepts can we truly understand the essence of "complicated systems." In addition, once the essential attributes of "complicated systems" are abstracted, they should first be self-consistent, meaning they should not cause conflicts or confusion in the existing theoretical system. Secondly, as a subset of "systems," the attributes of "complicated systems" should be in harmony with the attributes of systems. Furthermore, based on the aforementioned self-consistency and harmony, they should have their own academic vitality and contribute to new knowledge.

In the following content of this book, "attributes" generally refer to "essential attributes" or "basic attributes."

3.2 Various Reflections on the Attributes of Complicated Systems

3.2.1 Is Emergence an Attribute of Complicated Systems?

In reviewing Chap. 2, the following content is closely related to complicated systems and also provides inspiration for refining their attributes:

The essence of a system is to exhibit holistic phenomena and behaviors at the systemic level. There are two basic types of the wholeness of a system: one is additive wholeness where $1 + 1 = 2$, mainly manifested in the aggregative functionality of system components; the other is non-additive wholeness where $1 + 1 > 2$, mainly manifested in the emergent functionality of system components. The non-additivity refers to the characteristics and functionalities that emerge at the systemic level, which individual elements or local parts do not possess [2].

Clearly, from the perspective of human cognitive patterns, the former type of wholeness is "simple," while the latter is "not simple," but "complex." A system is perceived as a "complicated system" precisely because it exhibits non-additivity, which allows people to "see" complex phenomena such as "something from nothing," "from one to many," or "from zero to one."

So, can we consider emergence as an attribute of "complicated systems?" Theoretically speaking, although the phenomenon of emergence does make the system feel "complex," it causes intuitive confusion which stems from the instantaneous phenomenon of "something coming out of nothing" at the holistic level of the system when the mechanism of emergence is not yet understood. If the mechanism related to emergence is revealed, emergence is just the final result of the system evolving according to certain rules. This result has a definite principle and mechanism before its "instantaneous" appearance; it's just that people are not aware of it.

For example, for a long time, chemists conducting chemical experiments could only observe the final result of the chemical reaction at the macro level. In a sense, this result is equivalent to the overall emergence from the evolution process of the chemical reaction system. Although chemists can explain the basic principles of the reaction process, they do not precisely know the complete mechanism of the reaction process details that govern the interactions between chemical elements and how they lead to the formation of emergent phenomena during the reaction process. That is to say, they are unclear about the origins and courses of emergence. Three American chemists, focusing on chemical reactions, established the "multi-scale model of complex chemical systems," allowing information technology to assist in observing the behavior of atoms and electrons during chemical reactions. The basic idea is that, given the extremely fast speed of chemical reactions and electrons instantaneously jumping from one atomic nucleus to another, such high-speed behavior at the microscopic level is imperceptible to the naked eye.

With the assistance of the multi-scale model, when simulating how a drug couples with the target protein in the body, the chemists used computers to perform quantum calculations on the atoms involved in the interaction between the target protein and the drug, while using classical physics models to simulate the larger proteins which do not require such high precision. In this way, with the help of computers, the mysterious "emergent" results of the chemical reactions became "completely transparent," and chemists could "observe" the entire mechanism of a drug's effect within the human body at different scales. This indicates that a complex indirect causality exists between the lower level element behaviors of complex material systems and the holistic level. When people lack clear knowledge about these interactions, they can only see certain emergent scenarios with uncertain causality at the holistic level. These three chemists were awarded the Nobel Prize in Chemistry in 2013 for their outstanding contributions.

This case provides valuable insights into how we understand the emergence in systems: Emergence is considered as a complex phenomenon, and due to the lack of clear understanding or insufficient knowledge about the multi-scale interactions between the micro level and macro level of systems, a cognitive gap or ambiguity often occurs. However, with the development of scientific technology, people's knowledge has become increasingly abundant. What was once seen as the "moment" of the emergence of results can now be understood in terms of the underlying mechanisms and principles that lead to that moment, thus unraveling the complex puzzle of emergence.

Therefore, emergence is indeed one of the characteristic phenomena at the systemic level of complicated systems. However, it largely stems from intuitive perception when people's cognitive abilities are limited. Therefore, deeper analysis is required to consider emergence as an attribute of complicated systems, which is not necessarily a characteristic of all complicated systems.

3.2.2 Other Attempts at the Attributes of Complicated Systems

There have been attempts to distill the attributes of complicated systems using alternative methods: For example, starting from the fundamental components of a system to reveal its attributes. Here are a few examples:

1. The form of relationships

The overall structure of a system is formed by the interconnections between its parts according to a certain order. This indicates that how elements within a system are correlated plays a fundamental role in the overall behavior of the system. Can we extract the attributes of complicated systems from the form of relationships between elements?

For instance, let's take nonlinear systems as an example. Nonlinear systems are characterized by the presence of nonlinear relationships within the system. Generally, these systems are described using nonlinear mathematical models. This methodological approach determines that the research on such systems is based on the premise of not satisfying the superposition principle, where at least one of the characteristics of the system's output response, state response, or state transition does not satisfy the superposition principle. Following this logic, criteria such as system stability, attractors, and self-excited oscillations have been established. However, even if these criteria have led to the discovery of some complex phenomena, it is difficult to consider them as universal criteria for the attributes of "complicated systems" in a general sense.

Another example is stochastic systems. The essence of scientific research is to reveal the correlation between the realization of conditions (cause) and the occurrence of events (effect). If there is a cause, there will be an effect; this is determinism. If there is a possibility of an effect given a cause, this is probability. If an event has a consistent probability of occurrence, it is called randomness. If a system's input–output and influential factors contain such elements of randomness, or if the system's structure itself exhibits a certain probability, it is referred to as a stochastic system. Luhmann once said that the possibilities of a system exceed its realizations, and this makes the system complicated [3]. Clearly, the presence of randomness and probability in stochastic systems can lead to a perception that these systems are "complicated." However, in the real world, due to various reasons, almost all real systems have possibilities that outweigh their realizations. As a result, almost all systems in reality can be considered stochastic systems. Therefore, using randomness as an attribute of "complicated systems" lacks depth in abstraction and refinement to a large extent.

2. System structure

Since system structure has a significant impact on the overall functionality of a system, can we directly distill the attributes of "complicated systems" from the characteristics of system structure?

An example is hierarchical systems. If a system exhibits multiple levels in its structure, it may indeed introduce complex phenomena between these levels, particularly the emergence of complex phenomena in systems with micro and macro levels. However, even systems with a single level can exhibit complex phenomena. Furthermore, the concept of multiple levels represents an external observation of the system by individuals and still relies on intuitive perception. Therefore, it is challenging to use the concept of multiple levels to exhibit the system's inherent mechanisms.

Another example is multi-scale systems [4]. If a system's element exhibits variation based on a specific dimension, whether measured quantitatively or qualitatively, that property displays multi-scale dynamic changes along that dimension. In such cases, even if the system undergoes no structural changes, it can still show variations in the overall system form at different scales. For instance, when transitioning from one scale to another, the system's overall form may exhibit complex phenomena such as abrupt changes or bifurcations. However, whether the multi-scale phenomena exist and how scales are defined strongly rely on human's subjective cognition and judgment. Therefore, multi-scale is not a completely objective complex attribute of the system itself.

Another example is self-organizing systems. Apart from purely natural systems, a large number of artificial systems are designed, created, and organized by humans. A system, especially a socio-economic system, resembles a living organic entity. It often spontaneously establishes certain new structures, forms a certain order, or exhibits new functions based on a specific purpose and relevant rules in the absence of direct external influence. This phenomenon is known as the self-organization of the system. Clearly, if a system demonstrates self-organization, observers from an external standpoint might perceive that the system lacks awareness of when or why it will spontaneously exhibit new characteristics at the overall level. This perception would naturally be considered a complex phenomenon by people. However, perception alone does not provide a comprehensive understanding of the essence, particularly because self-organization is just one manifestation of complex phenomena in a system, not its entirety.

In summary:

1. In order to study complicated systems at the level of rational thinking, people have attempted to extract the attributes of such systems based on the characteristics of system components and define them using specific scientific terms.

2. The approaches discussed earlier have their own rationale since these characteristics of complicated systems indeed lead to the emergence or provide conditions for certain complex phenomena. However, these approaches also reflect their limitations, mainly because they have not achieved a comprehensive characterization and distillation of the essential attributes of "complicated" systems.

Since attributes are the most fundamental qualities of a thing in a philosophical sense, these qualities determine why something is what it is and not something else. Therefore, we need to explore the philosophical essence of "complicated systems" rather than rely on the induction and extraction based on the specific material form or physical correlations of the complicated systems. Therefore, we are required to engage in deeper thinking, transcending the superficial-level induction and extraction of intuitive characteristics in system components. Furthermore, we should engage in abstraction and refinement at the level of rational understanding of complicated systems.

3.3 Two Paths of Thinking About the Attributes of Complicated Systems

In Chap. 1, we introduced the definition of a system and identified wholeness as its fundamental attribute, known as systemicity. This is straightforward because wholeness is the essence of the system concept, and, thus, it can be directly considered as the essential attribute of a system. Following this logic, can we consider a certain characteristic as the attribute of complicated systems and call it complexity? Unfortunately, we cannot because we haven't provided a concept of complicated systems with a clear essence like the concept of a system. Therefore, we cannot directly take a certain essence as the essential attribute of complicated systems. Even if we refer to the essential attribute of complicated systems as "complexity," we don't know what "complexity" means.

As discussed later, scientists from various fields spent several decades gradually clarifying the essence of "complexity" as an attribute of complicated systems and applying it in the field of management. The main reason is that the "complexity" of "complicated systems" is too complex. It is not enough to create a new term and provide a definition for it. Any valuable concept must be an abstraction and refinement of the reality in the relevant field, and. it must be coherent and compatible with other concepts.

The specific content will be introduced below.

3.3.1 Complexity and Complex Systems Based on Complexity Terms

3.3.1.1 Complexity Jungle

Since the early twentieth century, scientists in fields such as physics and chemistry have gradually discovered some new phenomena that contradict traditional knowledge in natural sciences. These phenomena give a sense of strangeness and complexity. For example, during his doctoral research, French scientist Bénard added some liquid to a transparent dish and heated it on a stove. The liquid exhibited a temperature difference from bottom to top in the vertical direction, and when this temperature difference exceeded a certain threshold, convection occurred. The hotter liquid from the lower layer flowed into the cooler upper part. If the liquid had a free surface, under the combined effects of buoyancy, heat transfer, and viscosity, the liquid formed a large-scale hexagonal honeycomb-like grid structure. The scale of this honeycomb structure is approximately one hundred million times the molecular distance. To maintain the stability of this macroscopic structure, countless molecules interacted and coordinated with each other. When a flat plate is placed on top of the liquid, the hexagonal grid structure can turn into a rotating cylindrical shape or even concentric circles. All these phenomena make it seem as if the previously lifeless molecules are now following the instructions of a conductor, demonstrating remarkable complexity and incredibility [5].

In the 1960s, American meteorologist Lorenz, while using a computer to simulate weather prediction, discovered that deterministic systems exhibited random phenomena and complex regularities behind the phenomena. He wrote this discovery in a paper titled "*Can a Butterfly's Wing Flap Cause a Tornado in Texas?*" The meaning of the title is that a small change in initial conditions in a system can lead to significantly different long-term outcomes for that system. It's just like the Chinese proverb says: "*A difference of a fraction leads to a difference of a thousand miles.*" He creatively referred to this phenomenon as the "butterfly effect," and, today, almost everyone is familiar with this term.

In the 1970s, physicist Prigogine, while studying far-from-equilibrium physical phenomena, discovered dissipative structures. Haken, while researching laser theory, discovered the complex laws governing the formation and evolution of material system structures. He proposed the use of synergetics to explore the fundamental principles behind the diverse structures that emerge in nature. Additionally, various complex phenomena, such as self-organization and the emergence of order from disorder, were discovered in chemical reaction experiments and mathematical iterative algorithms. These phenomena observed in the inanimate physical world, along with similar phenomena discovered in other scientific disciplines, were undoubtedly considered novel at the time. Scientists collectively referred to these complex phenomena as "complexity." Of course, when comparing them to the even more astonishing complex phenomena exhibited by living systems, such as biological and social systems, scientists believed that these phenomena represented the "minimal complexity" evolved by nature.

Scientists have mainly created new definitions within their respective disciplines to describe the complexity of these complex phenomena. The fundamental approach involves quantitatively and formally measuring the information contained within these novel, complex phenomena. Subsequently, we delve into providing a "unified" explanation for these intricate complexities.

After some time, scientists realized that the complex phenomena discovered in various scientific fields were deeply rooted in the specific theoretical frameworks of those fields. However, the theoretical frameworks of different disciplines varied greatly, making it impossible to formulate a "unified" concept that could cover complex phenomena in multiple fields. As a result, they could only use "complexity" as a general term for complex phenomena.

Since the path of unifying complex phenomena encountered difficulties and obstacles, scientists had to resort to alternative approaches and make efforts to characterize the underlying features of complex phenomena within their respective scientific fields using their domain knowledge.

These efforts broadly followed two paths:

1. Based on the composition and characteristics of systems, describe complex phenomena using concepts such as nesting, nonlinearity, multi-level, heterogeneity of elements, and entanglement of correlated forms.
2. Utilizing imaginative metaphors such as chaos, self-organization, symmetry breaking, unpredictability, or even referring to "complexity" as "singularity."

During this exploratory process, many new concepts (terms) were developed to describe or characterize specific types of complex phenomena. Commonly used terms include information entropy, fractal dimension, self-similarity, adaptability, chaos, and many others. There are no fewer than 40 different "complexity terms," forming what is metaphorically referred to as the "complexity jungle." Scientists discovered that the complex phenomena referred to as complexity in various fields were indeed "too complex." It was challenging to extract common attributes of all complex phenomena, let alone cover them in a comprehensive manner.

At the same time, the scientific community recognized that the use of complexity terms drew heavily on knowledge from various disciplinary fields, creating many new definitions to describe different aspects of "complexity." The main objective was to characterize the uniqueness of complexity by quantitatively characterizing the information features of systems. This approach strongly reflects the inherent properties of the physical objects and the diversity of subjective cognition, which is still far from the common essential characteristics of complex systems. Therefore, it lacks the academic quality that should be represented as an essential attribute.

3.3.1.2 Complex Systems Derived from Complexity Terms

With the development of science, we know that since the inherent and diverse complexity of everything in the world is a universal reality, it is inevitably difficult to distill from the rich and colorful complexity terms a single concept that can cover all the core characteristics.

In 1979, the founder of dissipative structures theory, Prigogine, integrated the scientific research related to "complexity terms" in his work "The New Alliance," collectively referred to as "complexity science" in response to the research boom on complexity at that time. He hoped to capture the essence of complexity by "packaging" various complex phenomena in the natural and social fields and naming them as complex systems. However, the work demonstrates that since complexity terms heavily rely on specific scientific knowledge in different fields, "complexity science" also struggles to possess profound academic inclusiveness [6].

Scientists continued to advance on the path of complexity science but changed their approach. Instead of horizontally expanding the complexity terms, they started to think vertically about where complexity comes from. They hypothesized the existence of systems that gave rise to complex phenomena and considered them as the source and substance of complexity, naming these systems complex systems. It should be emphasized that complex systems here are speculative concepts based on complexity terms and built upon the foundation of complex phenomena, rather than referring to a specific type of tangible reality.

The association of complexity with the concept of systems had a significant influence on two groups of scientists. One group consisted of European scientists led by Prigogine, the founder of dissipative structures theory, and Ervin Laszlo, the founder of synergetics. Prigogine's view of complexity as a product of self-organization aligns closely with the connotation of systems, where simplicity gives rise to complexity through self-organization. Laszlo, on the other hand, takes a significant leap forward in systems theory. He straightforwardly states that synergetics emphasizes the cooperative actions of numerous subsystems to generate macro-scale structures and functions, considering his synergetics as a branch of systems science.

The other influential group of scientists was the Santa Fe Institute (SFI), established in the United States in 1984. The primary purpose of SFI was to study complexity science, and it gathered a group of renowned scientists from various fields, gradually forming the Santa Fe school. The emblematic academic ideas of this school include studying complexity through the lens of emergence, such as how complexity emerges from adaptation, and the intrinsic cause of complexity generation is the adaptation of systems or entities to maintain endurance and seek development, leading to the emergence of complexity through adaptation. The most famous concept proposed by this school is complex adaptive systems, which refers to systems composed of a large number of different components with close interconnections, capable of learning in the environment, adapting to environmental changes by altering their own behavior, and becoming more vital and sustainable. The Santa Fe school associated complex adaptive systems with complexity, establishing a connection between complexity and systems science.

In April 1999, the journal Science published a special issue on complexity science titled *Complex Systems*. The editor invited scientists from physics, chemistry, biology, economics, ecology, and other fields to introduce the state of complexity research in their respective domains. They straightforwardly pointed out that if we are studying complexity or complexity science, why use the term "complex systems?" The main reason is to avoid using the term "complexity," which is ambiguous and unclear. The aim is to add the term "complex systems" alongside complexity, using a word that implies "complexity" to study specific complex phenomena in certain fields, while using "complex systems" to study complex phenomena for a more abstract level.

It must be said that scientists have made sincere efforts in exploring complexity. They have deeply felt that the connotation of "complexity term" is not unified, and it is challenging to raise "complexity" from the concrete to the abstract in different fields. However, by introducing the concept of a "complex system," they can, at least to some extent, transcend specific fields and take a step closer to a unified understanding of complexity.

This "roundabout" thinking is fully reflected in the book titled *Complexity and Postmodernism: Understanding Complex Systems* by Paul Cilliers, a South African philosopher of science (1998). The author straightforwardly states that complex systems need to be understood from the perspective of complexity. In the first chapter, "Toward Complexity," he mentions the hope of providing some working definition of what "complexity" means. Unfortunately, the concept remains elusive both qualitatively and quantitatively [7]. Luhman (1985) proposed a useful description, suggesting that complexity in a system implies the existence of more possibilities than actual realizability. … We can attempt to analyze the characteristics of complicated systems to establish a general description without trying to confine it to a specific a priori.

He further explains that when dealing with things we do not yet understand, we need new ways of thinking. In this sense, philosophy plays an important role, not as a meta-description of what happens in science and technology but as an integral part of scientific and technological practice. This postmodern perspective has particular value in studying complexity, as it offers a way to approach complex systems [8].

Here, Paul Cilliers explicitly presents the notion of complex systems within the framework of postmodernist theory, aiming to use this concept to study complexity, which is a collective term for complex phenomena perceived in various fields.

In that case, given the diverse nature of complexity, complex systems are also bound to be rich and varied. Paul Cilliers indeed meticulously compiled a list describing various characteristics of complex systems, encompassing a total of ten items. After listing these characteristics, he breathed a sigh of relief, stating, "These ten features proposed here aid us in discussing complexity in a qualitative manner."

Clearly, since the fundamental purpose of introducing complex systems is to describe and characterize complex phenomena (complexity) using the relative physicality of systems, there is no dedicated theory of so-called complex systems [9] from a logical perspective. If we insist on mentioning it, it is merely an understanding of various complex phenomena, such as metaphors about complex phenomena and philosophical reflections on the ontology and cognition of complex systems. Some even simply refer to it as "complexity theory, also known as complex systems theory" (Complexity theory, otherwise known as Complex systems theory). In particular, if the individual behaviors of complex systems represent complexity itself, then complex systems, like complexity, are not "one-size-fits-all" and cannot be "homogenized." Instead, they can only be classified based on specific characteristics of complexity. Therefore, complex systems operate independently and autonomously.

In fields dominated by natural sciences, such reductionist thinking principles are beneficial for discovering and addressing new problems at interdisciplinary boundaries, playing a significant role in advancing modern science. For example, significant academic developments during this period, such as nonlinear system theory, dissipative structure theory, synergetics, bifurcation theory, fractal theory, chaos theory, adaptive system theory, and complex networks, have become important milestones and valuable treasures in the development of human science in the latter half of the last century. The introduction of genetic algorithms, evolutionary algorithms, the development of the swarm software platform, agent-based system modeling, artificial life described by agents, artificial societies, and more have greatly enhanced and deepened our understanding and analytical capabilities of complex phenomena in natural and social systems [10].

For decades, this profound path of scientific understanding has continued to play a tremendous role. One of the most iconic examples is when the 2021 Nobel Prize in Physics was awarded to three scientists from the United States, Germany, and Italy for their contributions to "*physically modeling the Earth's climate, quantifying its variability, and reliably predicting global warming*" and for "discovering the interaction of disorder and fluctuations in physical systems from the atomic to the planetary scale;" it further helped us deepen our understanding of the nature and evolution of complex physical systems.

3.3.2 Complexity and Complex Systems Based on Methodology

The trajectory of scientific exploration has never followed a single path. While scientists in Europe and America delved deep into the exploration of complexity and complex systems, Chinese systems scientist Hsue-Shen Tsien, with his rich logical thinking in system engineering practice, Chinese philosophical ideas, and autonomous theoretical logic in systems science, suggested another new methodology to understand the attributes of complexity and complex systems.

3.3.2.1 Hsue-Shen Tsien's Philosophical Methodology on Complex Systems Attribute

Hsue-Shen Tsien is a renowned contemporary scientist who has made pioneering and groundbreaking contributions not only in control science, aerospace science, systems science, and management science but also in the fields of scientific philosophy and scientific thought.

It is well known that aerospace engineering technology is a complex technical system, aerospace manufacturing engineering is a complex manufacturing system, and the corresponding organizational management is a highly typical complex management system. The combination of these three elements inevitably forms a complex composite system. For decades, as the chief technical leader of China's aerospace industry, Hsue-Shen Tsien has successfully applied systems thinking and systems engineering technology to the practice of China's "*Two Bombs, One Satellite*" aerospace project. These important practices and experiences have laid a solid practical foundation for Hsue-Shen Tsien's ideas on complexity and complex systems.

In the late 1980s, on the basis of the characteristics of open, multi-subsystem, and multi-type subsystems proposed by Hsue-Shen Tsien for complex mega-systems, he introduced the feature of multi-level structure, reflecting his deepened understanding of the connotation of complicated systems. For instance, he utilized the latest achievements in contemporary natural sciences to interpret complex system phenomena from a hierarchical perspective.

The following are the basic viewpoints he proposed on this issue in an article published in 1989, excerpted as follows:

> At the beginning of the nineteenth century, scientist Laplace believed that everything in the world was determined by mathematical theories and equations. However, by the end of the nineteenth century, Boltzmann introduced non-deterministic statistical mechanics and molecular kinetic theory to explain thermodynamic laws, which led to the contradiction between deterministic Newtonian mechanics and non-deterministic molecular motion. This contradiction was not resolved until the rise of chaos theory in the 1960s. 'Chaos,' a seemingly non-deterministic disorder, is actually deterministic. Chaotic disorder is caused by deterministic laws, but it can be dealt with using non-deterministic statistical mechanics. This segment of the history of science illustrates how deterministic Newtonian mechanics evolved into non-deterministic statistical mechanics and how deterministic chaos was then introduced to explain the non-determinism within statistical mechanics. It appears that the objective world is deterministic, but human perception of the world has limitations, which necessitate the introduction of non-determinism. However, this process does not come to a halt; cognition continues to advance [11].

Similarly, how do we address the indeterminism in quantum mechanics today? D. Bohm believes that the world is deterministic, and there is a "hidden order" in quantum mechanics that has not been observed. Hsue-Shen Tsien said that this "hidden order" cannot only be sought in the microscopic world; it is concealed at a "macroscopic" level deeper than the material world. How small is the "macroscopic" level? In physics, there is a Planck length, which is the smallest unit of distance in physics, where gravity and spacetime cease to exist. The Planck length is determined by the relative values of the gravitational constant G, the speed of light c, and the Planck constant h, expressed as $\sqrt{\frac{h}{2\pi}\frac{G}{c^3}}$, approximately equal to 1.6×10^{-35} m, or 1.6×10^{-33} cm. This length is approximately 10^{-34} cm, which coincides with the length of "superstrings" in the "superstring theory" in physics. It is 19 orders of magnitude smaller than the basic particles, such as neutrons and protons, in the microscopic world.

In addition, the world of superstrings is not a 4-dimensional spacetime but a 10-dimensional spacetime, with an additional 6-dimensional spacetime that is invisible in the higher level microscopic world. Hsue-Shen Tsien said, "*This leads me to speculate that the indeterminism exhibited in quantum mechanics at the microscopic level is actually the deterministic chaos formed by the motion of 10-dimensional spacetime at the macroscopic level. In other words, this is because the macroscopic world of superstrings is a 19-dimensional spacetime, with 6 dimensions that are invisible and unknown in the microscopic world, thus leaving out 6 factors. It is this cognitive 'omission' that turns the originally deterministic objective world into something that appears to be indeterministic. The 'hidden order' of things is actually hidden in the macroscopic order.*"

According to the difference of 10–19 orders of magnitude from the macroscopic to the microscopic, let us assume that the difference from the microscopic world to the macroscopic world familiar to us is also 19 orders of magnitude. Therefore, the typical length of the macroscopic world is 10–15 cm $\times$ $10^{19} = 10^2$ m.

From the macroscopic world, if we move up to the cosmic scale, it would be $10^2 \times 10^{19}$ m $= 10^{21}$ m$\approx$105 light-years, which is the size of the Milky Way. Hsue-Shen Tsien also proposed that, beyond the cosmic scale, the typical scale would be 10^{16} billion light-years, referred to as the "expanding view," and so on.

The above content reflects Hsue-Shen Tsien's compilation of experimental results and theoretical conclusions from the latest natural science research conducted by contemporary humanity. These findings are used as a basis to explore the principles of complex system properties. This approach is consistent with that of many natural scientists who emphasize scientific experimentation and empirical methods on this issue.

It is worth noting that in the 1970s and 1980s, Hsue-Shen Tsien formulated a series of important independent academic thoughts and theoretical viewpoints on the systems science framework, system complexity, and complex systems. These contributions had transformative significance for the study of complexity and the properties of complex systems. Some of the more important aspects of his work include the following.

1. Hsue-Shen Tsien proposed a new system classification system, dividing systems into simple systems, simple mega-systems, complex systems, complex mega-systems [12], and special complex mega-systems. Examples of complex mega-systems include biological systems, human body systems, human brain systems, social systems, geographical systems, and galaxy systems. Among them, the social system is the most complex and is also referred to as a special complex mega-system. These systems are all open systems that exchange matter, energy,

and information with the external environment. Hence, they are called open complex mega-systems. Existing methodologies and methods cannot adequately handle complex systems, complex mega-systems, and social systems, necessitating the use of new methodologies and methods [11].

2. In the 1980s and 1990s, Hsue-Shen Tsien proposed, "The issue of complexity needs to be given special attention now. The development of a nation and the development of society all involve complex problems. By addressing this issue, there will be significant advancement in science and technology. We need to break free from the limitations of scientific research methods that originated centuries ago." The complexity referred to by Hsue-Shen Tsien here is no longer limited to the "minimum complexity" of the physical world but includes the emergence of "advanced" complexity in human socio-economic mega-systems. It reflects Hsue-Shen Tsien's contemplation on the development of a new scientific system through the study of complexity and complex systems, where "there will be significant advancement in science and technology."
3. In understanding the complexity and attributes of socio-economic systems, Hsue-Shen Tsien advocates not treating complexity solely as a human-made formalized definition. The definition of complex systems is not merely a speculative expedient proposed to alleviate the abundance and complexity of terminology related to complexity. His basic principles of thinking are as follows: Firstly, elevate the study of complexity from individual disciplines to a more general and broad systemic category. Secondly, based on the systems science system, he autonomously established that such systems belong to a higher level category called complex systems. Thirdly, complexity should be regarded as an essential attribute of social and economic system entities. Systems possessing such essential attributes are referred to as complex systems in the socio-economic field. The "soul" of complex systems is complexity itself, and complexity should have its own recognizable and identifiable essential attributes. Fourthly, there should be dynamic mechanisms behind these attributes that give rise to complexity. To understand complexity and realize its practical value, it is necessary to study these mechanisms and gain a clear understanding of them.

3.3.2.2 Irreversibility of Reductionism: A New Window

As a Chinese scientist, Hsue-Shen Tsien naturally follows the attributes and characteristics of complicated system entities based on Chinese philosophical thinking. Therefore, Chinese philosophy's understanding of existence and the essence of existence, specifically ontological understanding, is closely related to Hsue-Shen Tsien's thoughts and viewpoints in this regard. The following points provide some interpretations:

1. In reality, complicated systems are diverse and colorful phenomena. Any complicated system manifests as a specific phenomenon entity in the real world that can be distinguished from other entities.

2. In a general sense, the reason that the real existence of a complicated system exists and can be distinguished from other entities is that it possesses specific properties that are unique to its own existence. When we consider self-cause and fundamentals of all complicated systems existence, it becomes not just the properties of specific complicated systems but rather the essential properties of all complicated systems. Essential properties are unique attributes that have a decisive significance in the existence of complicated system entities, representing the fundamental quality and inherent determinism of complicated system existence.
3. Philosophically, the study of the fundamental and essential aspects of entity existence is ontology. Ontology represents the fundamental and essential aspects of entity existence. Therefore, studying the fundamental and essential aspects of the existence of complicated systems falls within the realm of philosophical ontology. If complexity is the fundamental and essential aspect of complicated system existence, even if we have not yet determined the specific connotation of complexity, we can conclude that complex systems with the essential attribute of complexity are the ontology of complicated system entity existence.
4. It is evident that studying the ontology of complicated systems requires starting from the entity phenomena, abstracting the phenomena through the process of understanding, and forming an understanding of the structure and mechanisms of complexity and the existence of complex systems. Only through this can we clarify the true nature of complex systems, which is of greater practical value. The thinking and process of understanding ontology is called epistemology. It is clear that the quality of epistemology in exploring the essence of complicated systems or exploring the essence itself is of utmost importance.
5. Chinese philosophy believes that the understanding of real existence and the essence of existence should not separate the ontological world from the phenomenal world. The cognition of ontology should not depart from the way humans cognize ontology in the real world. Therefore, Chinese philosophical ontology emphasizes heaven-man oneness and knowledge-action unity. In particular, humans are the subjects of the social and economic fields, and they possess the most complex thinking and behavior. Therefore, the understanding and exploration of the essential properties of complex social and economic systems should not consider humans as external "observers" or "outsiders" of the system. It should not solely focus on contemplating the "physical nature" (natural laws of matter) in the complex phenomena of social and economic systems. Instead, it should fully recognize that complex social and economic systems in reality encompass not only complex "physical nature" but also complex "human nature" (the essence of human existence). These systems represent the integration of complex "physical nature" and "human nature" in social and economic systems. Therefore, when contemplating the essential properties of complex social and economic systems, it is advisable to integrate the inherent nature of humans, who exist between heaven and earth (human nature), the inherent nature of human cognition in understanding the world (mind nature), and the inherent nature of the objective natural laws followed by systems (physical nature). This integration achieves the unity of the essential nature of humans as well as the essential

nature of matter, known as the "heaven-man oneness." It enables the realization of a comprehensive understanding of the essence of the complex real world of social and economic systems, revealing the profound interconnections among individuals, between individuals and objects, and among objects within complex entities, elucidating the fundamental originality and roots that are established based on these connections.

As is well known, Chinese philosophy emphasizes the understanding of ontology through ethical principles, such as Confucianism's "仁(Ren)" (benevolence), Lao Tzu's "道(Taoism)" (usual principles), and Buddhism's "性(Xing)" (nature). Here, ontology is not only physical and material but also encompasses the rationality of human beings beyond the physical and material realm. Departing from the core of Chinese traditional culture, Hsue-Shen Tsien saw not only the physical and material ontology but also the conceptual ontology of "inner nature" in human beings. Based on the cognition of "complicated systems" that encompass the physical, material, and social aspects, He broke free from the constraints of ontological and reductionist thinking in the field of complexity. He integrated the multiple connotations of complexity with the ontology of complex systems. While not entirely governed by natural laws, he formed an epistemology and methodology. This approach addresses the essential attributes of social and economic systems that adhere to natural laws while also respecting human "humanity" and "inner nature." Therefore, the conceptual overview of ontology in traditional Chinese philosophy becomes the thinking principle of Hsue-Shen Tsien's ontology epistemology and methodology for understanding and managing complicated systems in the field of socio-economic systems, and it also serves as the logical starting point for constructing the academic system of complex systems management in this book.

Starting from this principle of thought, Hsue-Shen Tsien proposes that any problem that cannot or should not be addressed with reductionist methods and instead requires or is better suited to new scientific methods is a type of complexity issue. Such issues are present in complex mega-systems [13]. Hsue-Shen Tsien believes that instead of broadly discussing complexity, it is more practical to begin with the study of specific complex systems with complexity issues and to seek out the specific forms and mechanisms of complexity within these systems. On this accumulating foundation, a new theoretical system can be established.

Why does Hsue-Shen Tsien use the applicability of reductionism as a principle to judge whether a system is complex or not? This reflects his methodological principle of thinking about the fundamental attributes of systems. As aforementioned, people have discovered two types of systemic integrity from the overall form of real systems: one type is simple (additive) integrity, and the other type is complex (non-additive) integrity. Hsue-Shen Tsien takes the essential distinction of systemic integrity as the logical starting point for understanding the attributes of ontology to judge the complexity of systems and the essential attributes of complex systems [14, 15].

Specifically, according to general logic, the opposite of "complex" is "simple," and the opposite of "whole" is "part." If a system is complex in terms of holistic perception, we can break down the system into several parts, and, relatively speaking, each part will be simpler, making it more convenient to study. If the study of the parts is still unclear, further decomposition and research can be conducted, and each part will become simpler. This is the basic law of "gradual decrease in complexity from the whole to the parts" in a system [16]. This law tells us that if we start from the lowest level parts that have been clarified and gradually assemble the knowledge and solutions of each part, it can naturally become a cognition and solution for the entire system. Therefore, we can consider that the so-called "complicated system" is not "essentially" complex. The above-mentioned simple (additive) whole systems are such systems because the simple (additive) wholeness of the system can be deciphered using reductionist methodology to reveal that the system is complex. However, the above-mentioned complex (non-additive) whole systems, because of the system's complex (non-additive) wholeness, generally cannot be deciphered as complex using reductionist methodology. Such systems are essentially complex.

In the paradigm of scientific research, reductionism belongs to the category of methodology. The above discussion provides us with a methodological criterion for determining whether a complex system is "essentially" complex.

This opens a new window for Hsue-Shen Tsien, who used the principle of the reversibility of reductionism to test and identify whether complex systems have essential complex attributes.

In reality, the following situations often involve problems that cannot be handled by reductionism or are not suitable to be handled by reductionism, indicating that the corresponding complicated systems possess essential complexity [17]:

(1) The environment is complex and closely related to the system, profoundly affecting it. Therefore, the new composite system is formed by the environment, and the system exhibits unpredictable and irreducible complexity.
(2) The form and process of the system generally exhibit scenes and flows that are relatively independent yet coherent and holistic. Due to the lack of subjective knowledge, information asymmetry, and incompleteness, the use of reductionism can easily lead to fragmentation and rupture of scenes and flows, thereby damaging the complex wholeness of the system.
(3) If the system exhibits complex wholeness behaviors such as mutations, emergence, submergence, and evolution, it indicates that the system has complex relationships and structures, unclear information transmission mechanisms, or is significantly affected by "severe" uncertainties. In any of these cases, reductionism is not suitable.
(4) Describing, analyzing, or modeling the system requires interdisciplinary knowledge, techniques, and methods. Reductionism is generally unsuitable for addressing these types of problems.

The analysis above informs us that there is indeed a category of systems in reality that, in the reductionist sense, are "indivisible" or "inappropriate to divide." These systems inherently possess an inherent complexity that cannot or should not be reduced gradually through reductionism to ultimately become "simple."

3.3.2.3 Attributes of Complex Systems Based on Irreducibility

In summary, for a category of systems in reality that are "indivisible" or "should not be divided," it is impossible for people to fully reduce the system's complex wholeness to a simple entirety without damaging the system's wholeness. This experience involves methodology. Hence, it is not just a matter of people's perceptual understanding, but also a rational thinking of "complicated systems." This rational thinking becomes a fundamental distillation and abstraction of the existence of such systems. This is the key step in the second stage of our understanding of "complex systems" (the first stage is described in Sect. 2.3 of this book).

For a long time, people have mainly followed the path of reductionism to think about and solve the overall problems of systems. However, once they encounter situations where the system as a whole is "indivisible" or "not suitable for division," this approach becomes ineffective. We need to change our methodology to address this problem. Irreducibility based on reductionism has given birth to a new challenge, allowing us to find a new approach to abstract and condense the attributes of "complicated systems."

As early as the early twentieth century, when people first conceived the concept of systems, they realized that there were such complex attributes inherent in the integrity of systems. However, they had not been able to find a "starting point" to address this problem in depth. Now, we can use the "irreducibility of reductionism" as a ruler to judge whether "complicated systems" possess inherent complex attributes.

Following this logic, as long as a system possesses the attribute of "irreducibility of reductionism," it fundamentally differs from systems that may seem complex but can ultimately be simplified using reductionism. This signifies that such systems have an inherent essence of their own existence and cannot be transformed into simplicity. Based on this, we refer to systems with the attribute of "irreducibility of reductionism" as complex systems and emphasize the following:

1. "Complex systems" are symbol representations that result from the essential condensation and abstraction of "complicated systems" based on the principle of "irreversibility of reductionism." It can be said that "complicated systems" represent the entities of phenomena, while "complex systems" represent the essence of these systems. "Complicated systems" are expressions of subjective perceptual cognition using natural language and colloquial terms, while "complex systems" are expressions of rational cognition using scientific terminology and concepts.
2. Therefore, the "irreversibility of reductionism" is the primordial and essential nature of complex systems, reflecting the inherent self-originating nature of complex systems. If we define it as complexity, then complex systems and complexity share the same essence in terms of theoretical learning. They both represent the "self-cause" of the existence of "complicated systems" in the socio-economic domain. They encompass both the entity of phenomena and the essence of thinking.
3. Section 2.2.2 previously pointed out that the non-additive wholeness (complex wholeness) and the "irreducibility of reductionism" have the same connotation.

> Thus, the irreversibility of reductionism, non-additive wholeness, and complex wholeness can be seen as different expressions of the attributes of complex systems. In other words, these concepts and complexity are synonymous.

Chapter 2 of this book introduces the concept of "complicated systems," and this chapter introduces the concept of "complex systems." Here, we integrate these two concepts for comprehensive discussion:

When we begin to understand that "a system is complicated," it is initially based on our perceptual cognition of the system's external and surface characteristics, leading to the perception of a "complicated system." This is a colloquial expression in natural language and represents the first stage of understanding, not yet a scientific concept. Then, we enter the second stage of understanding, which involves the internal and essential rational cognition of the "complicated system," abstracting and condensing the attributes of this system. At this point, scientific terminology is used to define the system's attributes. Through analysis, this chapter extracts that the irreversibility of reductionism (complex wholeness) is the essential attribute of the existence of complicated systems and defines it as complexity. Therefore, "complex systems" can be seen as another symbol representing complexity. Hence, terms such as irreversibility of reductionism, non-additive wholeness, complex wholeness, and complexity are synonymous.

Concepts play a crucial role in elevating thinking from concrete to abstract, serving as tools for transmitting and communicating ideas within the scientific community. They form the "building blocks"[1] of theoretical scientific discourse and are essential for constructing the theory of "complex systems management." Good concepts not only enhance the systematic and logical nature of new theories and promote theoretical development but also help us observe and understand the essence of complicated systems with a sharper and deeper perspective, grasping the fundamental significance of complicated system attributes within the framework of complex systems management theory. To achieve this, methodology naturally plays a crucial role.

While new concepts can be helpful in describing and abstracting new scientific problems, we should exercise caution and refrain from proposing concepts that lack a profound practical background or have little scientific content. Good concepts should not only have descriptive functions but also help us capture the connotation of phenomena and possess theoretical logic.

For example, as mentioned earlier, if we use concepts such as nonlinearity and emergence to characterize complex systems or complexity, we can only capture the phenomenological features of complex systems. However, it becomes challenging to characterize the essential attributes of complex systems based on reductionist methodology.

[1] Foundational elements.

3.4 Review of Two Complex Systems Views

As mentioned above, for a century, humans have conducted extensive explorations around the concepts of complexity and complex systems, reflecting their perseverance and continuous innovation in the path of scientific progress. The different paths in the history of understanding complexity and complex systems have enriched our knowledge base on the one hand but have also caused confusion and ambiguity in certain contexts. Therefore, we have dedicated a significant portion of the preceding section to introduce and compare two influential yet significantly different cognitive paths.

3.4.1 Two Methodologies for Complex Systems

One methodology for complex systems is to build upon a plethora of complex phenomena; primarily in material systems, "complexity" is employed as an overarching concept for summarizing these complex phenomena. In conjunction with the specific characteristics of various disciplines, numerous "complexity terminologies" have been created to describe them. With an abundance of complexity terminology, condensing their shared essence becomes challenging. Thus, the conceptual idea of "complex systems" is used as the common "framework" from which all complexities arise. However, due to the diversity of complexity terms, this results in a variety of complex system types. This methodology is referred to as "complex systems methodology A (CSMA, hereafter)."

Another methodology for complex systems is based on constructing a hierarchical systems science framework. Starting from the levels of systems, it extracts systems with the irreversibility of reductionism. This property is equivalent to the theoretical concept of non-additive wholeness or complexity within systems. Termed as complexity, the complexity of systems fully reflects that the "complex" in such systems is intrinsic, substantive, and fundamental. Systems possessing these attributes are referred to as complex systems. This approach is denoted as "complex systems methodology B (CSMB, hereafter)."

3.4.2 Discussion of Two Complex Systems Methodologies

3.4.2.1 Introduction of CSMA

For several decades, CSMA has not only been successfully applied in fields such as natural sciences and engineering, achieving a series of groundbreaking results, but also significantly propelled scientific development in various domains for over half a century. Simultaneously, this research paradigm has been applied to issues within social and economic systems, yielding numerous novel outcomes.

In recent years, as complexity science research has deepened, scientists have further expanded the connotations of complexity and constructed new frameworks to study broader aspects of complexity and complex systems. They particularly aim to unveil the profound relationships between physics, biology, society, and life and discover common laws and principles for interdisciplinary scientific problems. An outstanding representative of this field is SCALE, The Universal Laws of Growth, Innovation, Sustainability, and the Pace of Life in Organisms, Cities, Economies, and Companies (2018), written by Geoffrey West, a theoretical physicist from the United Kingdom. In this book, Geoffrey West has extensively researched the inherent growth rules governing life forms, cities, and companies, exploring whether there are common internal growth principles that determine life and death, urban expansion, and corporate lifespan. The book introduces the scaling laws, which elucidate how growth and decline in various systems, from living organisms to cities and from economies to companies, are constrained by their own scale and exhibit certain proportional relationships, following unified principles (see the book's introduction). The novelty of the conclusions drawn in SCALE arises from Geoffrey West's background as a physicist. He emphasizes the quantitative, formalized approach and the use of data and statistical analysis, which are characteristic of traditional complexity science. Furthermore, he transcends the constraints of material systems' "minimum complexity" and introduces complexity thinking related to life, death, energy, and matter metabolism [18].

While we appreciate the aforementioned research achievements and the wisdom of the researchers, we can generally identify three limitations when applying CSMA to study complexity in social and economic systems:

1. The complexity and manifestations of social and economic systems, in reality, are far more intricate than the complexity arising from physical experiments or the connotations of complexity terms. For example, the complex forms and dynamics of social and economic environmental systems are much more multifaceted than the behaviors of molecules and atoms in material systems. Many complex phenomena in social and economic systems are difficult to replicate in physics experiments. In other words, it is challenging to find the complex scenarios and real dynamic mechanisms of social and economic systems within the basic thinking framework of complexity science. Consequently, it is even more difficult to provide in-depth explanations or specific solutions to real social and economic problems through the paradigm of natural science.
2. CSMA requires the adoption of established paradigms, such as using mathematical language as a tool for modeling systems and applying fundamental concepts and principles from relevant disciplines to explain complex social and economic phenomena. These concepts, principles, and models are primarily reflections of the physical laws specific to material systems. However, achieving a high degree of alignment and interpretability in social and economic systems is challenging.

In particular, the presence of numerous semi-structured phenomena and principles in social and economic systems, as well as unstructured human psychological, cultural, and value-related phenomena and principles, diverges significantly from the CSMA framework. It is not reasonable to expect social reality to conform to the same particle, molecule, atom, and mechanical force patterns as found in physics and chemistry.

3. In CSMA, it is generally necessary to elucidate the essence of complex phenomena in a given field using complexity terminology. However, the terminology of complexity primarily consists of information and data from fields such as physics, chemistry, and meteorology. In social and economic systems, such states often do not exist, resulting in a lack of corresponding information and data. Even if one attempts to forcibly create such complexity terminology, it is challenging to find genuine counterparts in the real scenarios of social and economic systems. In summary, simply and directly applying the complexity terminology from CSMA often struggles to represent and reveal the mechanistic aspects of the real situations and issues within social and economic systems. Instead, it can only serve as metaphorical, data-driven descriptions at the surface level of phenomena.

In conclusion, when using CSMA to study socio-economic system problems, it can provide some interpretation and metaphor from the perspective of complexity science, allowing us to make associations and gain insights. However, real-world socio-economic systems are much more complex than natural systems like physics and chemistry. The "minimum complexity" encompassed by complex terminology is insufficient to capture the profound complexity of socio-economic systems in reality. It also lacks the guidance to address management scientific problems in socio-economic systems, accurately capture information and data that reflect the essence of the problems, reveal underlying patterns, and generate new complex systems thinking and academic propositions.

3.4.2.2 Introduction of CSMB

CSMB is based on a complex systems viewpoint rooted in the reductionism of irreversibility or complex wholeness. Its notable characteristic lies in revealing the essential attributes of complex systems at a determined philosophical level, which is particularly important for the field of complex systems management as a new management theory domain.

The reason why a new theoretical domain needs a determined philosophical basis is that any theoretical system in a field must have a common logical starting point. It is necessary to establish an understanding of the essential attributes of the domain's problems at the epistemological and methodological level, which is accomplished at the philosophical level. To achieve this, it is generally insufficient to think within the domain, within its local scope, or within the specific technical and methodological levels of the domain. Instead, one must think at a higher level and larger scale that is closely related to the domain. This delves into the realm of philosophical thinking, elucidating the essential attributes of the domain's issues and establishing the foundational paradigm for academic research. Furthermore, philosophical thinking demands the unity of epistemology and methodology, thus establishing a methodological and method system for theoretical research in the domain.

Civilization is the accumulation of human intellectual activities. It is under the guidance of philosophical thinking that humanity creates management theory civilization in each era. Based on different management practices, thinking approaches, understandings, and explanations of knowledge, people have correspondingly defined the boundaries and characteristics of the connotation of management theory, forming various management theoretical doctrines. The dialectical, critical, and practice-oriented nature of philosophical thinking provides a correct rational thinking paradigm for creating and conducting new management theory research [19].

CSMB fully embodies this point because "irreversibility of reductionism" or "complex wholeness" is a rational cognitive understanding of the essence and objective attributes of human management practice. It represents a refined and abstract understanding of "complexity" in its essential sense and uses "complexity" to represent the uniformity, universality, and regularity of the essential attributes of these complicated systems. By achieving this, it becomes possible to study the fundamental laws of management activities and behaviors from the perspective of complex systems, with "complexity" as the logical starting point. In fact, this solidifies the theoretical foundation of complex systems management.

In a sense, management is the implementation and practice based on "designed knowledge" by people. This means that management theory not only directly provides cognitive guidance and methodological support for human management practices but also provides a thinking approach and behavioral rules for management entities. In both cases, we need to be guided by clear philosophical thinking to achieve and excel in using existing knowledge to solve problems when conditions are ripe, as well as to create new knowledge to solve problems when conditions are not yet established.

Compared to CSMA, CSMB not only avoids the fragmentation caused by the diversity of complexity terms but also avoids the lack of consistency in the theoretical concept of complex systems. Specifically, concepts such as "irreversibility of reductionism" or "complex wholeness" reside at the cognitive level of philosophical thought, providing a foundational framework for CSMB to explore diverse facets of complex management. This exploration encompasses foundational theories, specific theoretical inquiries, techniques, and methodologies. This is precisely what is required to enrich CSMB with profound insights that encapsulate the distinctive nature of intricate management practices. Conversely, CSMA, owing to the prescriptive essence of its complexity terms, risks draining the vibrancy from vivid, real-world management scenarios.

Overall, both CSMA and CSMB both have their own theoretical and application domain advantages. CSMA is more rooted and applied in natural science systems, and complexity terms are closer to the natural principles of the corresponding natural disciplines. Therefore, it has already demonstrated and will continue to demonstrate strong scientific power in revealing the research on the complexity of specific disciplines in the natural sciences. At the same time, precisely because of this, CSMA demonstrates a clear orientation toward complexity terms and dependence on specialized topics but lacks comprehensive coverage. For example, it is difficult for it to cover the extensive domain of socio-economic management theories. On the other hand, CSMB, at the philosophical thinking level, condenses the essential attributes of system ontology and essence. By applying CSMB in the field of social-economic management, we can establish the "root" of complexity in a new class of management practices, achieve a deep integration of complex systems management, and leave us with ample space to expand the scenarios to fill management activities with rich and indispensable social and humanistic content. Without these contents, management will be "lifeless" and "lack warmth." It is precisely in this important aspect that CSMB possesses the desired qualities of management thinking.

Because CSMA and CSMB have their own principles and application domains, they should not be compared as opposing ideologies, and one should not be used to negate the other. On the contrary, since both paths have their initial intentions regarding complexity and complex systems, we should distinguish the larger context and overarching problems and then make use of the strengths of both paths, allowing them to complement each other.

References

1. Sheng Z, Xue X, An S. Construct a theoretical system and discourse system for the management of major projects with Chinese characteristics. Manage World. 2019; 35(04):2–16+51+195.
2. Sheng Z. Fundamental theories of mega infrastructure construction management. Cham: Springer;2018.
3. Ding D. Luhmann and his "Social Systems Theory." World Philos. 2005;05:34–8.
4. Ye Q. On the self-organization mechanism of the system: the interpretation of the dissipative structure mechanism diagram. J Syst Dial. 1994;02:57–63.
5. Yang L, Liu Y, Zheng Y, et al. Cholesteric liquid crystal box Bernard effect experiment. Chin J Liq Cryst Displays. 2012;27(03):288–91.
6. Kenhua Zhan XS. Prigogine and dissipative structure theory. Shaanxi Science and Technology Press;1982.
7. Paul C. Complexity and postmodernism: understanding complex systems. Taylor and Francis;2002.
8. Celias. Complexity and postmodernism. Shanghai Science and Technology Education Press;2006.
9. Rotmans J, Loorbach D. Complexity and Transition Management. J Indus Ecol. 2009; 13:184–196. https://doi.org/10.1111/j.1530-9290.2009.00116.x
10. Sheng Z. Management: from systemic to complex. J Manage Sci. 2019;22(03):2–14.
11. Qian X. Basic scientific research should be guided by Marxist philosophy. Philos Stud. 1989;10:3–8.
12. Qian X. A new field of science—open complex megasystems and their methodology. J Univ Shanghai Sci Technol. 2011;33(06):526–32.
13. Yu J. Wisdom from the convergence of wisdom—Qian Xuesen's systematic scientific achievements and contributions. Spacecraft Eng. 2011;20(03):1–11.

14. Yu J. Qian Xuesen's achievements and contributions in systems science and systems engineering: innovation from system thinking to system practice (part I). Aerospace China. 2021;12:15–23.
15. Yu J. Qian Xuesen's achievements and contributions in systems science and systems engineering: innovation from system thinking to system practice (part II). Aerosp China. 2022;(01):41–5.
16. Yu J. Development and application of systems science and systems engineering. Sci Decis Making. 2017;12:1–18.
17. Yu J. Qian Xuesen's systems engineering thought and system theory. Civil-Mil Integr Cyberspace. 2021;12:9–10.
18. West J. Scale: deconstructing the growth logic of everything and reshaping the framework of thinking for understanding the world. CITIC Press;2018.
19. Sheng Z. Management theory: the epochality and modernization of character. J Manage Sci. 2019; 22(04):1–10.

Chapter 4
Paradigm Shift in Complex Systems Thinking

Chapter 1 of this book demonstrated that complicated management activities exist along with complicated management issues within them. Chapter 2 informs us that any management activity is a type of system, and the property of management is systemic. Complicated management activities are a type of complicated system. Chapter 3 elaborates on the property of complicated systems being the irreversibility of reductionism, referred to as complexity. Systems possessing the attribute of complexity are termed complex systems. Therefore, complicated management attributes are complex systems. Furthermore, by adopting the CSMB as the guiding principle and logical starting point, a new discipline in management studies focusing on addressing complex holistic management issues in fields such as socio-economics will emerge.

Since this is a new field in contemporary management studies, particularly with the shift in the characteristics of the management objects in this field, the corresponding epistemology and methodology must also adapt to undergo a comprehensive transformation to address and solve the scientific problems arising from this essential change. Conversely, if we depend only on traditional experiences or skills without the necessary transformation of thinking principles, epistemology, and methodology, we will not only be unable to cope with the challenges of analyzing and solving various new problems in this field but will also lack the ability to construct an academic and theoretical system.

Therefore, when establishing the new concepts and thinking principles of complexity, complex systems, and complex wholeness, we must carry out a cognitive transformation of this new field of management at a philosophical level. This is referred to as a thinking paradigm shift in the realm of scientific philosophy. A thinking paradigm shift is the manifestation of the transfer of scientific paradigms at the level of thinking and cognitive stages.

Z. Sheng, *Outline of Complex Systems Management Theory—Based on Irreversibility of Reductionism Thinking*, International Series in Operations Research & Management Science 360, https://doi.org/10.1007/978-981-97-3361-3_4

4.1 Overview of Paradigms and Paradigm Shifts

In a scientific domain or field with a certain history and accumulation, apart from having a domain of discourse and significant scientific problems with high relevance, there are also academic thoughts, theories, methods, and discourse systems that are widely agreed upon by scholars in the field. These systems have abundant successful examples to prove their correctness and effectiveness, ensuring the continuous academic vitality of the domain or field [1]. There are two profound implications here: firstly, there is a shared normative thinking and behavioral agreement within the domain or field, and secondly, based on this thinking and behavioral agreement, there are sufficient examples to gain trust. In this case, we consider that a scientific research paradigm has formed in that domain or field.

As a scientific concept, the paradigm was initially proposed by the American philosopher of science, Kuhn, in his book *The Structure of Scientific Revolutions* [2]. The basic principles of a paradigm can be manifested at three levels: ontology, epistemology, and methodology, which respectively address the questions of the existence of things, the relationship between the knower and the known, and the system of research methods. These theories and principles play a guiding and normative role in a specific scientific community, coordinating their views on a domain's problems and their behaviors. Academic paradigms mainly refer to the ways and perspectives of viewing research objects, such as how to perceive objects, what to perceive and see in objects, how to handle relationships with objects, and so on.

Additionally, there is a concept closely related to paradigms called pattern. A pattern refers to summarizing and generalizing the methods and means of solving a certain type of problem into methodological principles, which vary according to the fields. Patterns define the core methodology for solving problems in a particular domain so that we do not need to repeat the same work of thinking about how to approach a new problem. In this sense, a paradigm is a recognized pattern.

If a scientific domain faces new circumstances, such as substantial changes in the characteristics of the research subjects or a significant shift in perspectives toward the subjects, resulting in the original established paradigm no longer being able to provide problem-solving capabilities or its capabilities weakening, and simultaneously, a new adaptive paradigm emerges, this is known as a **paradigm shift**. Kuhn believes that almost every breakthrough in scientific research is achieved by first breaking the old mindset and then succeeding. Therefore, "paradigm shift" refers to a situation in a field where new challenges or non-incidental academic achievements break the existing consensus or rules, leading to significant revisions in the basic thinking, theories, and methods of that discipline. The concept of paradigm shift reveals important turning points caused by the inherent laws of the development of things, and accurately predicting and actively adapting to a new paradigm has significant guiding implications for clarifying the new development path of a field and grasping its future direction [3].

The basic connotations of paradigms and paradigm shifts are not only applicable in the scientific domain but also manifest in other practical domains of human activities.

For example, Kodak, as the pioneer of digital cameras, failed to actively explore new product functional paradigms when the digital revolution emerged, sticking to their traditional film-based paradigm. This failure to adapt to the new paradigm eventually led to their decline and bankruptcy. In contrast, in the same year, a small company called Instagram, with just a dozen employees, was acquired by Facebook for $500 million, relying on a photo-sharing software app.

Taking music as another example, there are Chinese traditional instruments and Western instruments. A paradigm shift can be seen when using Western instruments to perform traditional Chinese music. For instance, accompanying Chinese Peking Opera with a piano involves playing the melodies of Chinese Peking Opera using Western instruments. It combines the original scores of Chinese Peking Opera with the adaptation for the piano, reflecting a paradigm shift.

4.2 Thinking Paradigm Shift in Complex Systems

4.2.1 Overview of Thinking Paradigm Shift

Thinking primarily refers to the process of generalizing and responding to things, phenomena, or problems in the human brain. Thinking is based on perception but goes beyond its limits. Due to differences in individuals' worldviews and cognitive principles, different stable and repeatedly used thinking paths and methods are formed during the thinking process. These paths and methods are considered effective norms and are referred to as **thinking paradigms**. For example, the way in which problems are studied, such as how to conduct analysis and synthesis, comparison and classification, abstraction, and generalization, and the starting point for these thinking activities, reflects different thinking paradigms.

In various social practices, as people deepen their understanding of the objective world or encounter new things and phenomena, they realize that the effectiveness of the existing thinking paradigms has decreased or even become basically ineffective. In such situations, people reflect on more and more similar cases and summarize occasional successful cases, thus forming and solidifying new thinking paradigms. With the continuous improvement of these new paradigms and the increasing number of successful applications, people gradually accept and replace the original thinking paradigms, giving rise to what is known as **thinking paradigm shift**. The thinking paradigm shift is the most fundamental and comprehensive guiding factor in a paradigm shift.

According to the dialectical relationship between objective things and subjective thinking, the appearance of things and phenomena with new essential attributes is the fundamental reason for the thinking paradigm shift. Therefore, when a type of management problem emerges in today's activities that possess new attributes of complex wholeness, it is inevitable to undergo a thinking paradigm shift in order to effectively, persistently, and stably respond to and solve such problems. This means that a thinking paradigm shift will inevitably occur in the ontology, epistemology, and methodology of complex and holistic problem management.

In summary, the so-called thinking paradigm shift in this chapter is essentially a cognitive transformation concerning complex wholeness management problems. As this transformation takes place within the framework of thinking in complex systems, it is referred to as **a thinking paradigm shift in complex systems. Its fundamental purpose is to utilize the principles of complex systems thinking, knowledge frameworks, and discourse systems to reveal and analyze the fundamental characteristics, laws, and principles of managing complex holistic problems. It provides corresponding method systems and enhances the ability to solve complex holistic problems.**

Since it is a "shift" in thinking paradigms, there must be a pre-existing thinking paradigm. From the perspective of the development of human science, the Cartesian paradigm should have had the greatest influence on scientific thinking in recent times. René Descartes (1596–1650), a French philosopher, mathematician, and physicist, made significant contributions to the development of modern mathematics and is considered the father of analytical geometry for formalizing the coordinate system. He was also one of the founders of Western modern philosophical thought and the precursor of modern materialism, advocating "universal doubt." His philosophical ideas deeply influenced subsequent generations of Europeans and laid the foundation for European rationalistic philosophy.

One of Descartes' significant contributions to the scientific paradigm with a close relationship to thinking paradigms is his seminal work published in 1637 titled "Discourse on the Method of Rightly Conducting One's Reason and of Seeking Truth in the Sciences," commonly referred to as the "Discourse on the Method."

In the "Discourse on the Method," Descartes outlines a method for investigating problems, which involves four steps [4]:

1. Never accept any truth as evident until you clearly and distinctly understand it as such, leaving no room for doubt, solely relying on your own judgment.
2. Break down complex problems into simpler ones, attempting to solve them one by one.
3. Arrange these smaller problems in order from simple to complex and start with the ones that are easier to solve.
4. After solving all the problems, combine the solutions and examine whether they are complete and if the issues have been thoroughly resolved.

Clearly, Descartes' proposed methodology is a typical reductionist paradigm. For hundreds of years, both Western and Eastern scientific communities have, including their research methods, largely followed Descartes' paradigm. It is important to acknowledge that Descartes' method has significantly contributed to the development of modern science in the East and the West. However, when facing increasingly common complex problems in modern times, including those in natural and social sciences, the emergence of properties related to complex wholeness has made it increasingly difficult to solve problems solely through reductionism. Therefore, to address the irreversibility of reductionism, new methods must be created. Thus, transforming the Cartesian reductionist thinking paradigm becomes the starting point for the thinking paradigm shift in complex systems.

To illustrate the above issues concretely, let's consider a universally applicable example.

It is well known that the conservation of natural ecology, especially the environmental management and protection of densely populated and socio-economically developed water basins, has gained significant attention worldwide. The Taihu Lake in eastern China is the country's third-largest freshwater lake and a crucial water source. Today, the Taihu Lake Basin has become one of the most densely populated, economically developed, and vibrant regions in China. Over the years, the water environment of Taihu Lake has played a significant role in driving and supporting the socio-economic development of the Yangtze River Delta region. In 2019, the total population of the Taihu Lake Basin reached 61.64 million, accounting for 4.4% of the national population. Its GDP amounted to 9,684.7 billion yuan, contributing to 9.8% of the national GDP. The per capita GDP of the region was 157,000 yuan, which was 2.2 times the national average. However, with the rapid socio-economic development and the rise of urban clusters, the deterioration of the water environment in the Taihu Lake Basin has become an increasingly severe problem, making it a "heavily affected area" by eutrophication. Therefore, the country has undertaken large-scale water environmental management in the Taihu Lake Basin.

In the current complex and rapidly changing context, the water environment governance in the Taihu Lake Basin faces a series of new and profound challenges. Therefore, it is imperative that we confront and effectively address this significant real-world issue. The foremost step is to establish a paradigm shift in the thinking of complex systems for Taihu Lake governance. This effort aims to enhance the modernization of the water environment governance system and capabilities in the Taihu Lake Basin, building upon a comprehensive review and reflection on the history and current state of Taihu Lake water environment governance.

The paradigm shift in Taihu Lake water environment governance thinking primarily refers to the epistemology of how people perceive the essence of Taihu Lake's water environment and its governance actions. It involves questions such as: What are the fundamental properties of Taihu Lake's water environment? How can we solve Taihu Lake's water environmental problems in a way that aligns with our understanding and reality, ensuring effective and sustainable governance? Given the current state of Taihu Lake's environmental governance and the new challenges it faces, the fundamental transformation in thinking, which we make in terms of thinking principles, is the paradigm shift in Taihu Lake's governance thinking.

While analyzing numerous factors connected to the water environment of Taihu Lake and its related environmental challenges, including causal mechanisms and evolutionary processes, several novel and transformative principles of thinking have emerged, laying the foundation for a paradigm shift in the governance philosophy of Taihu Lake:

1. The geographical scope of the Taihu Lake Basin is extensive, covering a wide spatial area, and the interconnection between the water environment and regional

socio-economic systems is intricate and complex. They mutually support and constrain each other. The water environment conditions not only exhibit dynamic changes but may also undergo qualitative evolution and complex phenomena. Many issues arising in the Taihu Lake water environment exhibit characteristics such as uncertainty, hiddenness, ambiguity, and difficulty in precise determination.

2. The stakeholders in Taihu Lake water environment governance include governments, social organizations, enterprises, the general public, and more. While this diversity is necessary and important, it also complicates the formation of consensus and objectives among governance stakeholders. It can lead to conflicts and negotiations, particularly concerning issues involving conflicting interests. Each of these elements enhances the level of complexity inherent in the behaviors of governance stakeholders and the structural models of governance.
3. Governance solutions refer to the approaches, plans, means, and methods proposed to address Taihu Lake water environmental problems. For highly complicated problems, the pathway for developing solutions inevitably involves fundamental methods and paths related to solving complicated management, including complexity analysis and decomposition, adaptive selection, iterative generation, and more.
4. Understanding the complex and holistic issue of Taihu Lake water environment governance is inherently a process that evolves from ignorance to knowledge, from limited knowledge to more extensive knowledge, from partial knowledge to comprehensive knowledge, and from superficial knowledge to profound understanding.

In conclusion, irrespective of the approach to analyzing the water environmental governance of Taihu Lake, there have been substantial shifts in the ontological, epistemological, and methodological viewpoints concerning the lake's water environment and governance actions. These shifts signify a paradigm transformation in the governance thinking related to Taihu Lake.

The core of this thinking paradigm shift is based on the cognitive understanding of the complex systems of Taihu Lake's water environmental governance, particularly the coupling of the aquatic ecological complex systems with the complex systems of human socio-economics (systems of complex systems). This explicitly indicates that to achieve further improvements in Taihu Lake's water environmental governance, it is necessary to adopt a thinking paradigm based on complex systems to guide the overall process of governance work.

4.2.2 *Basic Goals of Thinking Paradigm Shift*

The thinking paradigm shift is not an end but a prerequisite for performing well and solving problems under new thinking principles. In other words, the thinking paradigm shift has its specific objectives. For example, in practice, the basic objectives of thinking paradigm shift for complex systems are as follows:

1. **Deciphering Complex Mechanisms** Irreversibility of reductionism manifests as the difficulty of clearly and comprehensively explaining complex phenomena at the macroscopic level by means of information from lower level local forms and behaviors, i.e., there is a rupture in the mechanistic understanding at the intermediate level or scale between the microscopic and macroscopic. This, to some extent, illustrates the existence of complex components in the macroscopic mechanistic entirety that cannot be derived from or simply aggregated from the microscopic level. The thinking paradigm shift is needed to assist in deciphering such complexities.
2. **Filling Cognitive Gaps** In the emergent processes from the lower to upper spatial levels or from the past (present) to the future on the temporal scale, due to the deep uncertainties [5], potential correlations, and hidden causalities inherent in complex interactions, there is a lack of cognitive understanding regarding the formation, propagation, and transformation of system emergent processes. The thinking paradigm shift provides new pathways to fill these cognitive gaps.
3. **Increasing Tool Supply** Irreversibility of reductionism leads to phenomena such as information asymmetry and incompleteness. In such situations, some traditional and mature descriptive and analytical tools may experience reduced functionality or even fail. Under the guidance of the thinking paradigm shift, it is necessary to increase new channels for tool supply and construct new paths for knowledge acquisition, particularly in the synthesis of academic ideas and methodological innovation, by utilizing interdisciplinary knowledge to obtain solutions to complex scientific problems.

The above three points analyze the epistemological and methodological difficulties caused by the abstract attribute of "irreversibility of reductionism" and propose objectives to address these challenges. This segment is beneficial for us to clarify the goals and starting points for solving difficulties during the paradigm shift in complex systems thinking.

4.2.3 Fundamental Elements of Thinking Paradigm Shift

Based on the aforementioned basic goals of the thinking paradigm shift, what are the fundamental elements of the thinking paradigm shift for complex systems? This can be summarized as how to perceive and approach complexity issues.

1. Complexity entities (activities, phenomena, problems, etc.) not only possess simple wholeness, additive wholeness, or reversibility of reductionism but also have complex wholeness, non-additive wholeness, or irreversibility of reductionism. It is indeed these specific characteristics endowed with regulatory properties that prevent merely applying reductionism to fragment a complex issue,

investigating each segment exhaustively, and thereby fully comprehending the whole entity. However, this does not mean that reductionism should be completely denied or rejected in the process of solving complex problems. Instead, based on reductionism, other means and methods should be employed to overcome the challenges posed by the irreversibility of reductionism according to practical circumstances.

2. Complexity issues make it difficult for individuals to clearly discern and interpret the internal dynamics from bottom to top, from local to global. In short, individuals may be able to observe complex phenomena or shapes at the holistic level but lack knowledge about the relationships between these phenomena or shapes and the behaviors at the mesoscopic and microscopic levels, as well as the complete functioning and evolutionary paths of such relationships. When we consider the macroscopic, mesoscopic, and microscopic levels as a whole and understand the emergence of macroscopic phenomena by tracing the internal dynamics from the microscopic individuals, we can not only observe the holistic emergence of the system but also understand the principles and pathways that lead to the emergence. In that case, we will have a comprehensive understanding of the entire emergence mechanism across different scales.
3. Complexity issues may exhibit complex phenomena on the whole that are mechanistically unclear and of uncertain origin. Especially, the spatial structures, temporal evolutions, and inherent regularities and orderliness manifested by such phenomena are not directed by external forces but are the result of self-coordination, where the original order is changed, and a new order is formed. This is commonly referred to as self-organization. Therefore, when studying complex problems, in addition to investigating how they form new holistic orders under external influences, it is also necessary to explore how they form self-organized new orders through their own internal synergistic interactions without direct external influences.
4. For a long time in the scientific community, alongside the influential reductionist paradigm, there has been another influential empirical research paradigm. Empirical research is premised on the fundamental cognition that there exists an objective world. It seeks to approach this objective world through continuous study. Consequently, empirical research emphasizes that investigations must be based on observable and empirically proven facts. It aims to make knowledge about the objective world "visible" and "tangible" through observed data and experimental results. This places the research focus on understanding "why things are the way they are" and uncovering "deterministic causality." This approach leads to the pursuit of empirical evidence, the construction of clear causal models, and methodologies that establish deterministic input–output relationships to reveal general conclusions. Moreover, these conclusions should be verifiable and replicable under the same conditions. However, the nature of complex problems largely challenges these traditional thinking paradigms. For instance, the complexity of these problems not only makes determinism, structural stability, and causal relationships within the problems rare but also transforms the traditionally "manageable" uncertainty into "severe" uncertainty. Due to the complex

mechanisms arising from such uncertainties, conventional tools and methods for handling uncertainty are no longer applicable. In the context of complex problems, the expectation of verifiability and repeatability of regularities, which is typical in empirical research paradigms, is no longer taken for granted. Instead, due to the influence of emergence, self-organization, and "severe" uncertainty, these complex patterns become either extremely rare or exceedingly difficult to reproduce.

5. In the general systems thinking paradigm, people's basic worldview is characterized by systemicity or "slightly" complexity. For example, the relationships among system elements are mostly linear, uncertainty is relatively regular, and human behavior is mostly stimulus/response-oriented. Consequently, people's understanding of reality and problems tends to be dominated by balance, stability, linearity, negative feedback, causality, and the complete decomposability of holistic problems. However, for complexity problems, these modes of thinking largely lose their fundamental assumptions. People can no longer pursue the necessary causal laws and why they are necessary within the system; instead, they focus on "what is possible" and "why it is possible." This expands the Cartesian analytical and Newtonian empirical paradigms.

This will give rise to a new cognitive process paradigm for complex problems. It involves a spiral approach of practical exploration, cognition, re-practice, and recognition when dealing with complex phenomena and issues. This cognitive process can be generally described as follows: it transforms cognitive and programmatic systems that are relatively disordered, one-sided, vague, and lacking in logic into systems that are relatively ordered, comprehensive, clear, and logically coherent. This process requires multiple repetitions and revisions and manifests as a process of comparison, optimization, approximation, and convergence. It is not a traditional one-time modeling and one-time optimization "all-at-once" process. Through this spiral approach, people gradually reduce the vagueness and uncertainty in their understanding of complexity, enhancing the clarity and precision of their understanding of complexity [3].

6. The complexity of a problem stems partly from the operation of various complex dynamic mechanisms within the problem and partly from the manifestation of a series of unique characteristic features. Regarding dynamic mechanisms, we can start with regularity and use various structural modeling methods to understand and analyze them, providing us with architectural and mechanistic cognition of complexity as much as possible. However, logically speaking, the reason why a problem is complex lies in its general regularities being few and its uniqueness being abundant. Furthermore, uniqueness often originates from contingency, and contingency is often embedded in the details of the problem. Given that "details determine the outcome," when addressing complexity, we should not overlook the minutiae but also contemplate how to interpret complexity through the analysis of the roles of such details. This important insight requires us to conduct in-depth investigations and studies of the complexity problems in practical activities, avoiding the use of preconceived concepts or detached hypotheses to analyze and

solve specific problems. This thinking paradigm often manifests as a focus on adopting "real-world research methods" or comprehensive analysis and decision-making methods that consider all elements and perspectives when dealing with complexity problems.

7. Regarding the complexity of the problem, people need to use different knowledge to understand, explain, analyze, and acquire relevant new knowledge based on the different characteristics and attributes of complexity. This indicates that in the process of transferring the paradigm of complex systems thinking, new ways and approaches to acquiring knowledge have emerged. For example, complexity exhibits macroscopic emergence. Initially, the cognitive thinking of reductionism led to a cognitive gap between the micro and macro levels, and emergence was considered a complex phenomenon. However, scientific research should not only observe phenomena but also explore the underlying laws and principles. Therefore, by employing comprehensive tools such as computers, interdisciplinary theories, and multi-scale modeling, a spatiotemporal scale model has been constructed between the original macro and micro levels, obtaining new knowledge through new means and filling the cognitive gap between the micro and holistic levels. This reveals the mechanistic nature of emergence. This indicates that the acquisition and formation path of knowledge under the paradigm shift of complex systems thinking follows a new paradigm that is inconsistent with tradition.
8. Human beings are the subjects of all practical activities, and they are the most vibrant and complex elements in all practical activities. Especially in the current era, the assumptions about "humanity" or the "nature of humans" have a significant influence on how to establish a thinking paradigm for complexity in management activities. According to the fundamental principles of management philosophy, in the process of transferring to complex systems thinking paradigm, it is necessary and appropriate to establish that humans are "complex beings" based on the presumption of complexity. Conversely, any form of thinking that presumes humans are "simple beings" is unrealistic and cognitively lagging. The core essence of "complex beings" includes the diversity of human nature and the adaptability of human behavior. Building upon this foundation, it extends to various dimensions of human behavior in practical activities. This will lead to a series of significant transformations in traditional assumptions about the principles, values, behavioral preferences, and habitual goals of "humanity."

Taking human decision-making behavior in socio-economic systems as an example, the famous American management theorist Douglas McGregor once said, "*Behind every management decision or measure, there must be certain assumptions about human nature and human behavior.*" He also said, "*These assumptions about human nature and human behavior, to some extent, determine the starting point, process, and outcome of management*" [6]. In fact, in order to achieve their own goals, decision-makers adaptively adjust their behavior according to environmental changes. It is this adaptability that gives rise to complexity [7]. For example, decision-makers tend to repeat behaviors that have previously yielded satisfactory results, refuse to use behaviors that have led to unfavorable outcomes, or attempt to solve new problems based on their memory and knowledge base. In this context, humans exhibit complex patterns of inheritance, exchange, and mutation, which are characteristic of the broader evolutionary process in systems. They have developed a complete self-evolutionary mechanism. To fully describe the decision-making psychology and behavior of "complex individuals," it is necessary to introduce elements of human nature such as the subject's storage elements (subject memory, preferences, knowledge), recognition elements (subject perception, cognition, judgment), adaptation elements (subject learning, replication, change), action elements (subject behavior after decision-making), and mind (subject physiology, psychology, culture). These elements form the virtual entity of "complex individuals" and establish new theoretical concepts and principles for the science of decision-making behavior. All of these require a thinking paradigm shift to be pursued.

9. In the current era, modern information technologies such as the Internet, Internet of Things (IoT), cloud computing, big data, artificial intelligence, and blockchain have rapidly developed and become powerful forces driving social progress. Within a short period of 20–30 years, this force has not only brought about significant changes in human production, work, and life styles but has also profoundly transformed human cognitive concepts, thinking habits, behavioral patterns, interpersonal relationships, self-adaptation, consciousness, and emotional experiences and expressions.

Furthermore, in the past, humans were the only intelligent entities in management activities. However, today, in the age of artificial intelligence, a new type of non-human intelligent entity, namely robots, has emerged in the management of complex problems. Unlike machines invented by humans in the past, these robots are not entirely controlled or manipulated by humans. They possess a certain level of autonomous intelligence that continues to improve over time. This opens up the possibility of humans and non-human intelligent entities merging into a non-fully human composite intelligent entity in a specific practical scenario under certain norms and guidelines. This greatly deepens the traditional understanding of the concept of "complex individuals." For example, they may form partnerships and become "integrated" subjects or objects in new practical activities, challenging other entities as more powerful, intelligent opponents. Today, we can already see that artificial intelligence technologies based on deep learning can use big data to solve small problems, while human intelligence can solve big problems with limited data. As a result, composite intelligent entities can demonstrate superior leadership, adaptability, and complex decision-making capabilities by combining their wisdom and spirituality. This trend will undoubtedly have a significant and profound impact on the formation, evolution, and behavioral choices of decision-making activities in complex management scenarios.

As a fundamental consideration, we have presented the nine key points related to the thinking paradigm shift in complex systems mentioned above. Certainly, considering the intricate and diverse nature of complexity, along with the impact of uniqueness, it's inevitable that the specific patterns of paradigm shift in thinking within complex systems will exhibit a wide range of variations. The nine key points mentioned earlier serve as a foundational framework and offer inspiration for further exploration and expansion across various levels and dimensions.

Having established the fundamental points of the complex systems thinking paradigm shift, it is possible to further supplement new details based on the unique scenarios or contexts of specific systems. This is one of the core abilities for analyzing the complexity in real-world management activities and mastering such complexity. Possessing this ability enables us to closely integrate complex systems thinking with concrete practices, resulting in the emergence of two new concepts in the field of management: "complexity (management) issues" and "complexity management (entities)." These concepts can be further integrated into the core concept of "complexity management," which will be the fundamental content of Chap. 5 of this book.

References

1. Zhaohan S, Tao S, Zeng E, et al. The evolution of system thinking and the paradigm shift of complex systems in Taihu Lake environmental governance engineering. Manag World. 2023;39(02):208–24.
2. Kuhn BL, Shuli JI. The structure of the scientific revolution. Shanghai Science and Technology Press;1980.
3. Bai J, Yan X, Zhi Y. Research on complex decision-making in the prevention and control of the novel coronavirus epidemic: based on the practice of cloud deployment of nucleic acid testing system for all employees in Jiangsu province. Foreign Econ Manag. 2023;45(01):55–66.
4. Descartes TW. Let's talk about the method. The Commercial Press;2000.
5. Flyvbjerg B, Bruzelius N, Rothengatter W. Megaprojects and risk: an anatomy of ambition. Cambridge University Press;2003.
6. Stoyan S, Monique D. The human side of enterprise. Macat Library; Taylor and Francis;1960.
7. Holland JH. Hidden order—adaptability creates complexity. Shanghai Science and Technology Education Press;2000.

Chapter 5
From Systemicity to Complexity in Management

In this chapter, we will connect the core points from the previous four chapters in logical order. Overall, we can establish two parallel yet coherent theoretical chains.

Firstly, centered around production and management practices, people have developed a cognitive path as follows. Within human production and creative activities, there is a category of activities known as management. Integrality and purposefulness (functionality) are the two fundamental qualities of any management activity. As human productive and creative activities have expanded in scope and scale, people have developed distinct perceptions and impressions of "simple production and creation" versus "complex production and creation." At the same time, individuals have acquired intuitive insights into both "simple management" and "complex management." The term "complex management" pertains to the management of intricate production and creative processes. Within complex management activities, "complex management issues" have emerged, abbreviated as complex issues. Complex management issues serve as the core representation of complex management activities.

Secondly, regarding the composition and universal connections of objective world phenomena, people have formed another cognitive path. A system is the most basic concept that reflects and summarizes the universal connections of objective world phenomena, forming a whole. Wholeness and functionality are the basic attributes of a system, referred to as systemicity. When people observe, listen to, or sense that a system comprises numerous distinct components with intricate interconnections, they perceive it as a "complicated system." Through in-depth analysis, people abstract and condense the inherent and essential attributes of "complicated systems." In the field of management, an important foundational viewpoint is that if a type of system exhibits "irreversibility of reductionism" (complex wholeness), it signifies that the system is fundamentally complicated. This type of system is called a complex system. The irreversibility of reductionism (complex wholeness) is the

Z. Sheng, *Outline of Complex Systems Management Theory—Based on Irreversibility of Reductionism Thinking*, International Series in Operations Research & Management Science 360, https://doi.org/10.1007/978-981-97-3361-3_5

attribute of a complicated system called complexity. In complicated systems, problems that possess the attribute of complex wholeness are called complexity problems. Analyzing and solving such problems requires a shift from traditional thinking paradigms to complex systems thinking paradigms.

Now, let's integrate the two cognitive routes mentioned above and, within a new thinking paradigm, develop a new management model to deal with complexity problems.

5.1 Systemicity of Management

Next, let's explore the theoretical equivalence between general management attributes and systemic attributes.

As mentioned earlier, any management activity is both a whole and a complete process. In terms of its integrality, any management activity is composed of fundamental elements such as the management environment, management subject, management object, management objectives, management organization, management issues, and management plans. In terms of its process, although management activities can be divided into multiple relatively independent stages, these stages are inevitably interrelated and orderly. Ultimately, they must manifest as a continuous process from beginning to end to demonstrate the functional role of management. In this way, any management activity is essentially a human-made system that serves a specific production and creative practice.

Therefore, by comparing the basic concepts of a system, we can clearly delineate the systemic connotations of the management concept:

(1) Any management is a whole that consists of various elements and parts.
(2) These elements and parts are indispensable and interact and depend on each other within the management whole.
(3) The significance of management lies in its specific functionality.
(4) Any management is both a complete whole and a complete process.

The aforementioned "functionality" and "wholeness" of management are precisely the core attributes of a system. This indicates that "systemicity" is the fundamental attribute of all management. In this way, systemicity not only organizes the fragmented understanding of management concepts into a logical path but also becomes a thinking paradigm for designing, constructing, implementing, and executing management activities. The management model based on systemicity as a thinking principle is called **systemic management**.

The core idea of systematic management is to understand, analyze, and solve management problems based on the concepts, principles, and methods of systems. It involves considering the management service recipient as a complete system and using systematic element analysis, correlation analysis, function analysis, and organizational behavior analysis to plan, design, and organize management activities comprehensively. At the specific technical level, systematic management emphasizes setting clear goals, rigorous analysis, a focus on quantification, and procedural approaches to planning, designing, and coordinating management activities at the operational level to achieve overall management objectives and excellent comprehensive results. In summary, systematic management is about maintaining and ensuring the unity of integrity, correlation, and dynamics in management activities and processes.

This core idea is highly important because it tells us that due to the fundamental equivalence between systems and management attributes, there exists a theoretical connection between them. Not only can systemic principles provide planning, configuration, and organizational patterns for management activities in general management practices, but with the development of systems science and management science, new models can also be expanded. Complex systems management, introduced in this book, is an example of such expansion [1].

5.2 Complexity Problems

In Sect. 1.3 of this book, it is pointed out that the most important perception of "complicated management" is the emergence of "complicated management problems." It briefly summarizes some external connections and surface characteristics of these problems. For example, the problem boundaries are often fuzzy and unclear, it is difficult to describe these problems completely using structured methods (models), and solving the problems involves multidisciplinary and multi-domain knowledge. These are intuitive perceptions of the external connections and superficial characteristics of complicated management problems akin to "plain language" in natural language discourse.

Furthermore, we can infer some characteristics of these complex problems from an objective attribute perspective based on intuitive perceptions:

(1) These problems generally have a very close relationship with the management environment. Various changes in the environment profoundly impact the problems. In particular, the complex forms and formation mechanisms of the problems often result from the interaction between the problems and the environment. Therefore, if we sever the connection between the problems and the environment, we cannot fully understand and analyze why these problems are complex.

(2) These problems typically exist within a specific management activity and process. Observing any specific management activity and process is like watching a relatively independent and coherent story with people, objects, events, relationships, causality, and changes unfolding in temporal and spatial order. Every story has a background, plot, and plot development, that is, a complete scenario. The more complex the problem, the more complex the scenario, and the closer the relationship between the problem and the scenario. It

requires us to consider the problem within the context, think about the problem, analyze the problem, and find solutions by placing the problem in the given scenario. This demands that we analyze and synthesize the scenario from top-down and bottom-up perspectives within its entirety and process to solve the problem rather than dissecting the scenario and fragmenting it or completely separating the problem from the scenario.

(3) These problems generally exhibit various forms of complexity dynamics, such as sudden changes, emergent phenomena, disappearance, evolution, and so on. The mechanisms behind these dynamic changes are highly complex, primarily due to the close and complex explicit or implicit connections between the elements of the problem. Therefore, whether we break these complex connections at the physical level, the system level, or the level of thinking, the complexity dynamics of the problem will be greatly compromised.

From the above, if we apply reductionism to decompose complex problems with numerous constitutive elements, complex interconnections, and close interactions with the environment in complicated management activities and then study each part individually, it will inevitably sever the complex interrelationships between the problem parts, thus disrupting the inherent complexity mechanisms of the problem. Even if we thoroughly study each part, it often does not lead to the resolution of the holistic problem. This indicates the existence of a new attribute of complex problems that is incompatible with reductionist thinking. It signifies that our understanding of complicated management problems has reached the stage of rational thinking.

Furthermore, based on the description in Chap. 4 of this book, the irreversibility of reductionism (complex wholeness) of CSMB serves as the starting point for establishing a new thinking paradigm shift of complex systems in the field of management. In the 1970s and 1980s, Hsue-Shen Tsien coined the term "complexity problems" to describe these intricate challenges characterized by the irreversibility of reductionism. It is precisely this irreversibility of reductionism that inherently renders these problems complex.

Hsue-Shen Tsien said, "*Complexity problems deserve special attention now because social and economic development all involve complexity problems. Solving these problems will lead to significant advancements in science and technology. We need to break free from the limitations of scientific research methods that have been in place for several centuries.*" He further emphasized from the perspective of systems methodology that any problem that cannot be addressed or is not suitable to be addressed through reductionism should be treated as a complexity problem that requires new scientific methods.

Thus, the "complex problems" in complicated management activities and the "complexity problems" proposed by Hsue-Shen Tsien share the same essential attribute of the irreversibility of reductionism or complex wholeness [2]. This suggests that one can discern the nature of complex management problems based on the irreversibility of reductionism. From this, we gain the following two insights:

(1) The complex problems in complex management activities not only possess "complex" concreteness but also have "complexity" abstraction. The term "complexity" here does not refer to the vocabulary used in complexity science but rather represents the characteristic of irreversibility (complex wholeness) inherent in the field of management. **Management problems that display this level of complexity, characterized by complex wholeness, are commonly referred to as complexity management problems or simply complexity problems.**
(2) The application of Hsue-Shen Tsien's methodological principles to assess complex problems in management has led to the important concept of complexity (complex wholeness) problems. This concept represents a deep integration of complex management problems with complex systems thinking and holds significant theoretical value.

The irreversibility of reductionism in complexity problems provides strong guidance on how to correctly perceive and address these problems in practice.

Firstly, the boundaries of complexity problems are often fuzzy and not fully clear. In addition to definite input/output relationships among the elements within the problem, there are also incomplete or even highly uncertain associative relationships. Apart from explicit and knowable relationships, there are implicit and difficult-to-know relationships. Moreover, certain relationships or associated elements that we identify may be subject to variation due to other factors in the actual process. All of these factors contribute to a cognitive understanding of the problem that is often fuzzy, uncertain, or even unknown.

Secondly, these complexity problems are generally difficult to be fully described by a clear and structured approach (model). In fact, complexity problems often involve elements from science and technology, socio-economics, human behavior, and cultural values. Scientific and technological elements are primarily governed by natural science and technical principles, which can be described using a structured approach. Socio-economic elements are mainly influenced by social or economic laws and can be described using a semi-structured approach. On the other hand, human behavior and cultural value elements often require a non-structured approach to description. As a result, complexity problems as a whole must be described using structured, semi-structured, and even non-structured approaches to capture their complete essence. This not only significantly increases the difficulty of describing the problem but also adds complexity in integrating different types of expression.

Lastly, complexity problems involve knowledge from multiple disciplines and fields, requiring experts from various domains to apply interdisciplinary knowledge to solve them. However, based on the principles of human cognition, the understanding of these problems by cognitive subjects is inevitably a process of transitioning from ignorance to knowledge, from limited knowledge to more extensive knowledge, from partial knowledge to comprehensive knowledge, and from superficial knowledge to profound knowledge. Therefore, the generation and formation of solutions for such problems manifest as an adaptive "trial and error" process that involves continuous exploration. In this process, solutions are typically not optimized in a single attempt but are gradually refined through multiple comparisons, adjustments, and improvements of alternative management approaches. Overall, this is a dynamic, iterative process where intermediate solutions progress along an ordered path from relative vagueness to comparably clear, from partiality to comprehensiveness, and from lower quality to higher quality, converging toward the formation of a final solution.

5.3 Management Complexity

In the preceding section, we conceptualized the complex problems within complex management as complexity problems, delving into the theoretical connection between the irreversibility of reductionism and the concept of complexity wholeness. Although complexity problems are only a part of the challenges in complex management, the impact of the complexity of problems on complex management activities is profound and comprehensive.

Firstly, Hsue-Shen Tsien revealed the complexity (complex wholeness) attribute of problems in management activities at a higher level of epistemology and methodology, transcending specific fields and expertise. Any management activity can be seen as a type of system. Based on this thinking principle, people can discern the characteristics of real management systems for a system whose attributes can be described as reducible, signifying that the entirety of the system can be expressed as the sum of its individual components (subsystems). For another type of system, its attributes can be summarized as irreducibility, meaning the wholeness of the system is not the sum of the partialities of all its components (subsystems). Clearly, management activities belonging to the former can be addressed using reducibility, and thus, they are essentially not complex. On the other hand, management activities belonging to the latter cannot be solved using reducibility and thus possess inherent complexity. Therefore, by introducing complexity thinking beyond specific fields and expertise, we can unify complexity with the essential characteristics of systems and classify system types, once again highlighting the important guiding role of philosophical thinking in CSMB of complex systems.

Secondly, as human productive and constructive activities become increasingly complex, the accompanying complex management activities and problems have gained more weight and deeper implications in various types of management. Therefore, when we examine the changes in the attributes of management activities from a complex systems thinking perspective, we deeply experience the trend from systematicity to complexity in the field of management.

In conclusion, by using irreversibility as the criterion for determining complexity (wholeness), the complexity of management and its complex problems are revealed to possess complexity (complex wholeness) attributes. In short, complexity (complex wholeness) is a universal attribute of complex management (activities, problems), referred to as **management complexity**.

Let's take the example of water environmental governance in the Taihu Lake Basin. Since water environmental governance in Taihu Lake is a complex problem, it inevitably exhibits the characteristics of complexity, namely governance complexity. By analyzing the complexity of governance, we can realize the thinking paradigm shift of complex systems and establish the principles of governance complexity. Specifically, the key points of governance complexity in Taihu Lake are as follows:

(1) The approach to water environmental governance in Taihu Lake has evolved in its conceptualization, progressing from simple systems to complex systems and ultimately to social-ecological complex systems. Specifically, we should recognize that the lake's water environment is a social-natural ecological complex system composed of human beings, nature, and society. The essential attribute of this complex system is social-natural ecological complexity, which involves social-natural symbiosis, decentralization, shared values, and orderly evolution. Therefore, humans themselves need to adopt social-natural ecological complexity thinking to address the complexity of water environmental issues in lakes.
(2) "Human and Natural Life Community" is the core concept guiding the paradigm shift in Taihu Lake governance. The term "governance" implies the subject and the object, but the true meaning of water environmental governance in Taihu Lake should not be simply understood as human beings having the power to suppress and deprive the water environment at will. Instead, we need to deeply understand that first, the water environment of Taihu Lake itself is a community of life; second, the human society in the Taihu Lake Basin is also a community of life; third, the combination of society and the water environment forms a larger community of life. Therefore, a multi-level, nested, complex community system is formed. Fourth, each community and the interactions between communities should be interdependent, mutually beneficial, and evolve in synergy. Human beings can only receive generous returns from the environment by treating it well, utilizing it rationally, and protecting it. Conversely, disorderly development and unrestrained exploitation will inevitably harm the sustainable development of human society, ultimately leading to the collapse or decline of the entire system.
(3) On the surface, "pollutants" in the water environment are, in fact, a type of "negative" product from the entire watershed's socio-economic system, transported to the lake's water body through specific supply chains and logistics networks. They have complex natural, ecological, social, economic, and cultural implications. Consequently, water environment management not only needs to consider traditional natural mechanisms like water quantity, water quality, biology, chemistry, meteorology, and more but also has close logical connections with "pollutant" manufacturers, suppliers, logistics networks, water resource trading, value conversion, and the ultimate water body management. As such, the issue of the water environment is not only an "ecological problem" but also a "social problem." In addition to allocating administrative authority and legal governance, it requires an element-based market allocation model that

conforms to market mechanisms and supply chain management rules. This could include mechanisms for cost and compensation calculation in water environment management, water environment data resource ownership rights, efficient flow, and trading, among other factors. This requires the establishment of an open, shared, fair, and trusted circulation platform in the governance system. Otherwise, the water environment cannot form a value network, and modern technologies like the Internet of Things, blockchain, and big data cannot be effectively socialized. Thus, the metaphor of lake water quality based on the supply chain network of water pollutants will deepen our understanding of the attributes of watershed water environment management. This means transitioning from "physical" and "engineering" management of lake water environments to comprehensive management that includes "ecological," "social," "economic," and "integrated" dimensions. To achieve this, governance thinking and mechanisms involving markets, symbiosis, trust, valuation, trading, and alliances need to be integrated into the governance process.

(4) The transformation of Taihu Lake governance is a core issue in the paradigm shift because it answers how to perceive and discover new things about the research object under a new thinking paradigm. The transformation of Taihu Lake governance mainly refers to addressing the deep-seated governance problems in the complex giant socio-ecological system of the Taihu Lake basin. Obviously, if these problems can be decomposed into individual subproblems using reductionist methods and solved one by one, they would no longer be "deep-seated" problems. Therefore, the governance of such problems is inevitably complex and irreducible, representing complex holistic problems. Although general systems thinking is effective for complex holistic problems, it often only partially works or lacks strong effectiveness. Therefore, to comprehend and solve these problems, a paradigm shift based on complex systems thinking is required, employing multi-level epistemology and methodology to grasp and decipher this special attribute. For example, the forms of governance problems related to the water environment involve the heterogeneity of solid, liquid, and gaseous physical forms. The elements associated with the problems involve nature, society, economy, and humanities. The formation and evolution of the problems involve potential associations and large-scale evolutionary laws that are difficult to comprehend, ultimately representing a comprehensive integration of all these aspects. This determines that addressing these problems cannot be achieved by limited decomposition but requires facing the complex holistic scenario that contains all the elements of the problems and various dynamic mechanisms and adopting comprehensive, integrated approaches to obtain solutions [3].

(5) Lake governance involves not only managing the lake itself but also managing the people who interact with it. It is necessary to consider not only the physical "pollution discharge pipes" but also the socio-economic "supply chains" of symbiotic pollutants. Therefore, the essence of water governance transformation lies in the comprehensive transformation and reconstruction of human

behavior, values, preferences, interests, and social development patterns. Therefore, governance strategies include not only engineering projects for lake water protection and improvement at the physical level but also new conceptual changes in governance subjects themselves, behavioral changes of governance subjects, and the reorganization of the lake water environment symbiotic systems at the social level. Therefore, improving the biochemical indicators of lake water bodies should not be approached solely with administrative, engineering, and technical thinking. Instead, it should be based on the comprehensive panoramic scenario of the ecological, social, and economic symbiosis behind the "lake disease" and construct a new understanding of water environment scenario governance transformation and corresponding methodologies.

5.4 Complexity Management

Every management activity or research should be oriented by management issues. Hence, with the emergence of complexity (or complex wholeness) as a new type of problem, it is foreseeable that significant, even fundamental, changes will occur in managers' thinking, objectives, management activities, and management processes. These alterations primarily aim to enhance adaptability and control over complexity (or complex wholeness), inevitably leading to transformations in thought processes, management models, and methodologies.

How do we comprehend this new theoretical viewpoint? Firstly, the "complexity" in this section, as a new attribute of management issues, not only encapsulates the new quality of complex management activities and scenarios but also exhibits important consistency and extensibility. For instance, within specific contexts, combining "complexity" with other discourses can express multiple rich connotations, such as **"complexity issues" refer to management problems that cannot or should not be handled with reductionism, "complex thinking" represents the complex systems thinking paradigm adopted by subjects in management, "management complexity" denotes the various forms of complexity (or complex wholeness) inherent in management activities, and "complexity management" implies a type of management model based on the paradigm shift of complex systems thinking.**

Secondly, the semantic expressions above can more precisely and comprehensively convey the implications of management activities and behaviors oriented toward "complexity" through logical associations. For example, **complexity problems** give a certain connotation to **management complexity**, prompting managers to establish **complexity thinking** and form **complexity management** models. Furthermore, **complexity management** is a comprehensive concept that integrates **complexity problems, complexity thinking, and management complexity**. Its main implication is that managers first develop a complex, intuitive sense for practical problems in management that are difficult to articulate clearly, analyze thoroughly, and predict accurately, and thus challenging to identify causes, make decisions, devise solutions, and propose plans. Based on this, by extracting the irreversibility of reductionism as the essential attribute of such problems, **complex problems** are abstracted as **complexity problems** at the level of rational cognition. The complex systems thinking paradigm adopted by managers throughout the process is referred to as **complexity thinking**. Consequently, the various forms of complexity (complex wholeness) inherent in corresponding management activities due to the existence of complexity problems are defined as **management complexity**, and the corresponding management models based on complex thinking are referred to as **complexity management**. The **complex systems management** introduced in this book is a type of **complexity management** with specific specifications.

Regarding the aforementioned example of the governance of Lake Taihu mentioned above, there is actually the concept of "complexity management-based governance of Lake Taihu's water environment," abbreviated as complexity governance of Lake Taihu. Its basic theory is that the governance of the water environment in the Lake Taihu basin is both a creative activity of constructing an artificial complex system and a complex system engineering. It should adhere to the basic thinking and principles of complexity management, achieving the unity of theoretical thinking and practical thinking. The key points are as follows:

(1) The governance of the water environment in the Lake Taihu basin is a human-made complex system creation activity and its complexity governance engineering, which occurs at the convergence point of the complexity attributes of the basin's water environment and the complexity of various related behaviors by agents.
(2) The core objective of Lake Taihu governance is to study the functional requirements of governance based on the analysis of the water environment problems of Lake Taihu and the socio-ecological complexity of the water environment. The focus is on how to emerge new governance functions through transformative reform and optimization of the governance system. For example, how to improve the ability to control the complexity of water environment governance, robustness, resilience, etc.
(3) Lake Taihu governance itself is also the construction of a new artificial complex system with its own system elements, relationships, and structures. It requires appropriate top-level design, realizing the path of multi-level, multi-scale goals of Lake Taihu governance through the Lake Taihu governance organization platform, as well as the market-oriented allocation of digital assets in governance and the Lake Taihu water environment ecological compensation mechanism based on Gross Ecosystem Product (GEP).

In summary, it can be observed that the scientific concept of "complexity" is not equivalent to the everyday usage of "complex" in common communication. The latter only represents the surface and external expression of things, while the former is the condensed abstract cognition of the attributes of things, presenting an abstract concept using scientific terminology. Therefore, when discussing "complexity" in the field of management, it must be "immersed" in the context of management activities. That is the reason we "tirelessly" combine "complexity" with other discourses to form compound terms or provide interpretations in specific contexts.

Furthermore, with the emergence of complex forms in human production and construction activities, there is a "corresponding" increase in complexity management issues in the field of management. These issues have become more numerous, significant, diverse, and profound. The complexity of human production and construction activities continually triggers and generates practices and academic research on management complexity and complexity management. When we view all of this from the perspective of complex systems, it represents the evolution trend of management from (simple) systematicity to complex (systematicity).

Following such a trend, if we take the methodology of analyzing and solving management problems as the mainline and review the development process of management models, we can use a broader perspective to understand the significance of management methodologies for the evolution of management models.

Based on the dialectical relationship between epistemology and methodology, the following table summarizes the correlation between a certain management model and its corresponding methodology over a period of time, in accordance with the evolution of people's perceptions of the attributes of management objects and problems. Today, this relationship has been further enriched by the proposal of a new model of complexity management, as shown in Table 5.1.

Among them:

(1) Experience management is a management model that involves summarizing knowledge and principles about management from specific cases or a few examples and "replicating" them to other management objects and issues.
(2) Scientific management is a management model that places management under scientific systems and norms through clear rules and standards.
(3) System management is a management model that views both the management object and the managers themselves as systems and organizes and controls management activities through systems thinking and principles.

Table. 5.1 Development of methodology-based management model

Management mode	Generalized managed object	Key management techniques	Management Methodology
Experience management	Individual	Generalization	Duplication
Scientific management	Subsystem	Generic extraction	Standardization
System management	Simple system	System analysis	System management and control
Complexity management	Complex system	Complexity analysis and synthesis	Comprehensive integration [4]

(4) Complexity management is a management model that situates both the subjects of management and the managers themselves within the framework formed by a paradigm shift to complex systems thinking, principally employing methodologies designed to address issues of complexity.

The above analysis reveals that as people's understanding of the essential attributes of management continues to deepen, management thinking, methodologies, and models are constantly evolving and enriching. This development aims to enhance the ability to analyze and control the complexity of management objects. It is a fundamental phenomenon and law in the cognitive development process of human management.

References

1. Sheng Z. Management: from systemic to complex [J]. J Manag Sci. 2019;22(3):2–14.
2. Yan Y, Ren H, Fan G. System complexity analysis and complexity management of large-scale engineering projects [J]. Sci Technol Manag Res. 2009;29(06):303–5.
3. Sheng Z, Wang h, Hu Z. Supply chain resilience: adapting to complexity—based on a complex systems management perspective [J]. Chinese Manag Sci. 2022;30(11): 1–7.
4. Ruwei Dai LC. Development of a comprehensive integrated seminar hall [J]. J Manag Sci. 2002;(3):10–6.

Chapter 6
Introduction to Complex Systems Management

Combining the discussions in the previous chapters about the development of management studies and the deepening of systemic thinking, we can draw two important conclusions:

1. Management has evolved from a systemic approach to a trend of complexity on the whole.
2. A new paradigm of complexity management has emerged based on complex systems thinking.

In the past few decades, the academic field of management has been aligning with the above development trends, contemplating and exploring how to further materialize and substantiate the complexity management paradigm. Efforts have been directed toward ensuring that in terms of connotation, extension, logicality, and maturity, this paradigm possesses more standardized norms and operability. Additionally, there has been a continuous expansion and completion of its academic propositions, frameworks, and content, as well as its discourse system, which has become a new domain within contemporary management studies.

From this chapter, we will follow a standardized theoretical framework to introduce the fundamental content of complex systems management theoretical framework.

6.1 Definition of Complex Systems Management

In the field of management, the term "complex systems management" is formed by the synthesis of the words "complex systems" and "management." Its meaning is akin to cell fusion in the field of life sciences. The two cells, "complex systems" and "management," permeate each other and combine to form a new cell called "complex systems management." This new cell is neither the original "complex systems" nor

Z. Sheng, *Outline of Complex Systems Management Theory—Based on Irreversibility of Reductionism Thinking*, International Series in Operations Research & Management Science 360, https://doi.org/10.1007/978-981-97-3361-3_6

the traditional "management." It has undergone essential changes and has generated a new quality known as "complex systems management."

The term "complex systems management" can be interpreted in various ways, depending on how complex systems are managed. Currently, the most representative interpretations are twofold. One is the research on complex social-economic systems management by Western scholars based on complex phenomena, complexity vocabulary, and complexity science. The other is the research on the management of "complex wholeness" issues in complex socio-economic systems based on Hsue-Shen Tsien's complex systems thinking. They represent two different logical thinkers that explore complex systems management. However, if we consider complex systems management not only as a skill, technique, or experience but as a significant thinking paradigm shift in epistemology and methodology about management, it becomes a new field of management studies with its unique attributes, theoretical framework, and discourse system, we must propose and construct corresponding academic propositions and knowledge transformation regarding complex systems management.

In such a new field of management studies with significant academic frontiers and an interdisciplinary nature, it is evident that we cannot simply adopt the academic ideas and discourse systems of other disciplines or blindly imitate and replicate theories from other disciplines for extension. It requires us to establish new logical and theoretical thinking principles through a change in mindset. Furthermore, many scientific problems in complex systems management have profound new connotations. We cannot solely rely on traditional management thinking, theories, and methods. Instead, we must fully utilize academic imagination, including thinking about the complex wholeness of problems and the deep uncertainty, indivisibility, irreversibility, and unpredictability of large temporal and spatial evolutionary scenarios. Creativity is synonymous with innovation, and compared to any other time in the past, creativity has never been as crucial as it is now. Although it is highly challenging, without creativity, it is difficult to achieve theoretical progress and academic development in complex systems management.

Now, let's unleash our creativity around the theoretical framework of complex systems management. Firstly, the complex management activities correspond to the complex production and construction activities of human beings. Within these complex management activities, there exist complex problems characterized by the irreversibility of reductionism (complex wholeness). This is consistent with Hsue-Shen Tsien's ideas on complexity, and thus, Hsue-Shen Tsien's academic thoughts on complexity and complex systems can be applied to the activities of complexity management.

Specifically, the complex systems management proposed in this book is a complexity management model based primarily on Hsue-Shen Tsien's complex systems thinking. It also incorporates other academic ideas related to complex systems. In terms of management philosophy, Hsue-Shen Tsien's viewpoint on complex systems forms the core of complex systems management. In practice, it primarily deals with management activities related to "complex wholeness" issues in complex social-economic systems. Academically, it represents a paradigm within the complexity management field driven by new methodologies.

It is noted that the following points are particularly important:

1. Firstly, it is important not to interpret "complex systems management" literally as the management of complicated systems or managing complicated systems. It is not "complex management systems" or any similar interpretations.
2. The complex systems in complex systems management are not based on complexity terms or complexity science's CSMA. Complex systems management is a new field in management studies, and as mentioned earlier, CSMA is not suitable for analyzing and solving the management of "advanced complexity" issues in complex socio-economic systems.
3. CSMB does not require predefined artificial definitions or the creation of numerous new terms. Instead, it introduces complexity through the irreversibility of reductionism methodology and achieves the integration of complex systems and management through the homogeneity of the essential attributes of complexity problems and complex wholeness problems. This integration provides a theoretical basis for complex systems management.
4. In summary, based on the previous third point, the basic essence of complex systems management is the application of complexity thinking to manage the problems of complex wholeness.

"Complex systems" originally belonged to the field of systems science. It appears in the concept of "complex systems management" as a holistic concept, primarily providing philosophical and methodological principles and logical starting points for understanding and managing complex wholeness problems. In this sense, we can consider the idea of "complex systems" as a philosophical perspective and a navigator for exploring the paradigm of complexity management. When faced with various challenges of complexity and a lack of capability to handle them, there is a need for new, higher level philosophical thinking to provide precise guidance and support. Complex systems thinking offers a valuable source of inspiration and empowers paradigm innovation in management, thereby forming the new field of complex systems management.

Therefore, from a philosophical perspective, complex systems management can be understood not as an intersection between complex systems and management but as a comprehensive reconstruction and sublimation of management paradigms under the guidance of complex systems philosophy when traditional management thinking encounters significant difficulties in dealing with complex wholeness problems. It serves as an assistance to the transfer of management paradigms and demonstrates the power of management philosophy. This deep thinking logic and inherent nature exemplify the profoundness of complex systems + management = complex systems management.

6.2 Introduction to Complex Systems Management Theory

Generally, a theoretical framework is not a simple collection of unrelated and illogical theoretical viewpoints. A theoretical framework should not only have consistent theoretical thinking principles but also encompass core concepts, fundamental principles, and scientific topics that deeply explain the essential attributes of the field, as well as corresponding methodologies and methodological frameworks. Compared to studying individual theoretical issues, constructing a theoretical framework for complex systems management is much more complex and challenging.

Given this circumstance, developing a theoretical framework for managing complex systems is considerably more intricate and challenging than studying individual management questions that have a well-established theoretical foundation.

The theoretical framework of complex systems management encompasses disciplines such as philosophy, cognitive science, systems science, management science, information science, mathematics, and more. Therefore, in-depth exploration and construction of this framework should not be limited to the level of individual theoretical issues, nor should it be limited to technical and methodological considerations. Instead, we need to adopt a proper level of thinking to clarify our scientific understanding of the theoretical framework and systematically complete the tasks of each stage according to the logical chain of constructing the theoretical framework.

The Chinese philosopher Xu Changfu has offered profound academic insights while examining the challenges and issues stemming from the intersection of theoretical disciplines and engineering disciplines in research within the humanities and social sciences. His scholarly ideas serve as a valuable source of inspiration for our comprehension and development of the theoretical framework for managing complex systems. Here, we reference and cite several of his academic perspectives.

6.2.1 Theoretical and Practical Thinking in Complex Systems Management

People perceive and transform the world through sensory organs such as ears, nose, eyes, and hands and use their brains to think through language. In the practice of complex systems management activities, how do people engage in basic thinking activities?

Initially, people in complex systems management practice need to experience many phenomena and deal with various problems. Through senses and experiences, they gain relevant insights and understandings, forming their intuitive understanding and knowledge of things and phenomena. Then, through thinking and abstraction in the brain, they achieve deeper understanding and knowledge. This type of understanding and knowledge is more universal and provides practical guidance for complex systems management. In this way, they gradually surpass individual-specific management phenomena and problem entities, forming "principles" of management that have universal significance. These principles can correctly reveal the general laws and principles of complex systems management.

We refer to the type of thinking and form of reasoning employed by individuals in the practice of complex systems management, where the fundamental objective is to "uncover the true nature of the subject" and to understand the general principles of management as theoretical thinking within complex systems management. It results in a knowledge system concerning the principles of complex systems management at the level of attributes [1].

Theoretical thinking is crucial in the practice of complex systems management. Firstly, theoretical thinking focuses on the "generality" of management activities and phenomena, surpassing the specific "uniqueness" of individual activities and phenomena. The form of theoretical thinking mainly involves abstract generalization and logical deduction while minimizing cognitive biases caused by personal value preferences. This enhances its objectivity, universality, and scientificity, allowing it to reflect the essential laws of complex systems management as deeply as possible.

However, theoretical thinking also has its limitations, mainly due to the following reasons:

(1) Theoretical thinking mainly provides principles derived from logical thinking. However, in practical complex systems management activities, besides logical thinking, there are also principles derived from non-logical thinking and other forms of thinking. Therefore, it is not enough to rely solely on theoretical thinking to achieve a comprehensive understanding of complex systems management activities.
(2) Theoretical thinking mainly obtains "general" principles of complex systems management. However, for a specific management activity, in addition to following general principles, it is also necessary to respect the "unique" principles specific to the activity itself.
(3) Human theoretical thinking, as a cognitive activity, also has its limitations. Therefore, the principles derived from it are always partial and incomplete, and their scientific nature is relative.

Therefore, although the principles obtained from theoretical thinking have a guiding role in practice, one should not think that relying solely on theoretical thinking can solve all practical problems in complex systems management.

In addition, we need to note that any specific complex system management activity is individual, concrete, and unique, ultimately aiming to achieve specific management objectives. In this regard, a specific management activity requires both theoretical thinking to provide general principles as guidance and intuitive, non-logical thinking to gain an understanding of the uniqueness and reality of the activity through human intuition and various non-logical thinking processes. Based on this understanding, general principles are transformed into the intentions, plans, and methods of specific entity activities. It involves the planning of turning the blueprint of "virtual activities" into complete "actual activities," including specific planning, processes, methods, and skills. Furthermore, it is essential to implement these plans at the operational level.

Only then can complex system management activity have ultimate practical significance. The thinking mode that focuses on "planning" as the main task in management activities, aiming to transform "virtual activities" into "actual activities," is called practical thinking, which is another important thinking mode in complex system management activities, distinct from theoretical thinking.

Evidently, practical thinking is primarily aimed at understanding how to construct management activity practices rather than understanding the principles. Practical thinking values the implementation of planning in practice, and the planning itself is guided by the intentions and values of the subject [2]. Practical thinking can be applied to the planning of a specific activity practice or multiple activity practices. However, practical thinking inevitably involves direct engagement with complex system management practices, thus incorporating human preferences and value orientations.

From this, two points can be observed:

(1) For a specific complex system management activity, management activities encompass both theoretical thinking and practical thinking.
(2) In complex system management activities, theoretical thinking primarily involves setting goals, intentions, and overall understanding of management activities, while practical thinking proposes corresponding plans, methods, and processes. The former aims to clarify the principles as much as possible, while the latter aims to transform virtual activities into actual activities based on those principles. The former provides regulations, while the latter ensures operability. The former guides the latter, and the latter serves the realization of activities. During the construction phase of actual activities, practical thinking is the dominant thinking mode. However, overall, cognition, planning, and operation are unified, and theoretical thinking and practical thinking are also unified [3].

Distinguishing between theoretical thinking and practical thinking in complex systems management activities holds significant importance. Specifically, theoretical thinking involves clarifying the general principles of complex systems management in a "general sense." It requires, to some extent, setting aside the details, uniqueness, and differences of individual management activities and extracting the commonalities and essence behind the phenomena, abstracting them to obtain fundamental knowledge with universal significance. Therefore, this knowledge should also apply to certain individual activities. In essence, this process involves recognizing the "what" and "why" of management from numerous individual complex systems management activities, thereby highlighting the guiding role of theoretical thinking in complex systems management practices. Therefore, theoretical thinking primarily represents a type of thinking that clarifies "what" and "why" generates complex systems management theories and establishes a theoretical framework for complex systems management.

On the other hand, practical thinking is a different mode of thinking that serves specific complex system management activities. Once management activities are concretized, not only the general intentions of individuals need to be realized through the integrity and entity of the management activities, but also the uniqueness of specific activities must be fully respected. This allows for the specific planning and methods of complex system management activities to be achieved through a combination of intuitive, logical, and non-logical thinking. In fact, this combines the principles of complex system management with the practical execution and details, designing complete operational methods, skills, and processes for complex system management. Therefore, practical thinking primarily involves designing management systems, processes, and execution procedures, as well as specifying and planning the required operational methods and management steps. It essentially involves knowing "what to do" and "how to do it" in the specific operations and implementation of complex system management.

However, we should view these two modes of thinking as interconnected rather than isolated. In fact, theoretical thinking is based on an understanding of the general principles and commonalities in complex systems management, and this understanding needs to be refined, abstracted, summarized, and elevated through observations of numerous specific and individual management phenomena. In other words, theoretical thinking in complex systems management derives from practical thinking in individual management activities. Conversely, practical thinking, although focused on individual management activities, requires guidance and direction from overarching general principles. It often employs "non-individual" tools and methods to address "individual" problems. This means that practical thinking in individual management activities needs to be guided by the general principles of theoretical thinking. Furthermore, the practical application of these general principles in specific management activities demonstrates the practical significance and effectiveness of these principles. Thus, while there are important distinctions between theoretical and practical thinking in complex systems management, they are closely integrated and inseparable in the practice of complex systems management activities.

Based on the above, theoretical thinking and practical thinking are the two most important and basic types of thinking in the management of complex systems. The value of theoretical thinking lies in its ability to categorize and systematically analyze the attributes of activities and problems, as well as the relationships between these attributes. In contrast, the value of practical thinking lies in its ability to synthesize the attributes within different relationships into a cohesive reality of management activity. The former strives for universal and persuasive reasoning, while the latter aims for actionable and executable behavior and activities [3].

In other words, the outcome of theoretical thinking is the formation of theories, while the outcome of practical thinking is the formation of tangible management activities. Understanding the connotations and differences between the two is of great guiding significance for our subsequent understanding and construction of complex systems management theories.

6.2.2 *Experience, Knowledge, and Theory in Complex Systems Management*

This section mainly provides an interpretation of the concept of complex systems management theory.

Firstly, people gain insights and experiences in complex management practices. Over time, these insights and experiences naturally evolve into practical knowledge for certain management activities. With more occurrences, these practical knowledge and experiences become richer and deeper. In subsequent similar situations, people consciously or unconsciously apply these insights and experiences to complex systems management practices, further expanding and improving them based on multiple successful attempts. The knowledge gained through experience or observation in complex systems management activities is referred to as management experience. Experience can be a form of cognition, skill, or technique.

In specific complex systems management activities at different times and locations, people often generate similar management experiences repeatedly. Such experiences are more easily recognized and solidified. Solidified experiences become more reliable and valuable through repeated testing and provide stronger guidance for complex systems management practices.

The experiences generated from complex systems management practices, especially the solidified ones, become knowledge about complex systems management. It represents the systematic understanding of general principles regarding complex systems management activities. "Knowledge" is a broad concept that includes facts, information, descriptions, skills, techniques, etc. Any complex systems management knowledge must be accurate (i.e., based on facts), validated through practice, and believed and accepted by people. Individual pieces of knowledge may be independent and scattered, but they can have certain correlations, and knowledge generally improves throughout its formation process.

Under certain guiding principles of thinking, after identifying the essential attributes of complex systems management activities, individual knowledge pieces that are relatively independent and scattered can be integrated according to certain rules. This integration is further systematized and logical, resulting in a knowledge system with logical correlations, referred to as a "knowledge framework," or is referred to as the formulation of complex systems management theory. The theory reveals the universal principles of complex systems management activities in a more profound way and can guide the construction of the complex systems management knowledge system. In summary, experience is a "primary" form of knowledge, while theory is an "advanced" form. As theory is not only fundamentally based on knowledge but also represents the systematization and logical organization of knowledge, it inherently possesses a systematic nature in its entirety—meaning that the overall theory in a certain field is a theoretical system. For instance, the holistic theoretical representation in the field of complex systems management should be the complex systems management theoretical system.

So, where does theory surpass knowledge in terms of being "advanced"?

(1) Theory is the systematization of knowledge.

In general, knowledge sets in various fields are often independent or weakly related. Especially, knowledge related to different problems often originates from different types of practical activities, resulting in their relative independence. Due to different backgrounds, prerequisites, and conditions, the disparities in knowledge among different types are significant.

However, the correlations between "knowledge" within complex systems management theory are relatively strong. First and foremost, during the formation process, they must have a consistent understanding of the attributes of complex systems management activities. This consistency reflects the unity of theoretical thinking principles and, based on this unity, demonstrates the generality of the theory. It ensures a relatively comprehensive coverage of complex systems management activities and their various problems, which can only be achieved through systematic top-level design. Furthermore, because the knowledge within the theory needs to have a strong "fusion" and "extension" among each other, i.e., exhibiting close correlations, new theoretical elements can grow within a highly systematized framework. Lastly, a high-quality theory system should provide guidance and rules for the corresponding methodology, further enhancing the overall functionality of the theory. Therefore, theory is inevitably a systematic form of knowledge and cannot be a simple collection of scattered and isolated perspectives.

(2) Theory is the logicalization of knowledge.

Complex systems management theory, in order to reveal the essential attributes and inherent laws of management activities, must be able to abstract and contemplate based on language that reflects these unique attributes. Language carries its own semantics and context, as well as connections between words and connections between words and objects. This requires adherence to people's language habits and language logic. Although different countries and ethnic groups around the world use multiple languages with different language logic and habits, complex systems management theory represents the collective understanding of the essential attributes of human complex systems management activities by people who use different languages. Therefore, it should have a "higher level" scientific discourse system and its corresponding logical framework that goes beyond the general logical relationships of natural language. Furthermore, theory needs to generate new insights and knowledge through logical thinking processes such as reasoning and judgment in order to explain and interpret the phenomena of complex systems management activities, reveal the objective laws within them, and guide practice. This requires the theory to have its own specific logic and rules. To ensure the theory embodies fundamental qualities, it must exhibit normativity, certainty, organization, and consistency. Specifically, complex systems management theory must have its own core concepts, basic principles derived from these concepts, and further refined universal scientific issues. All of this must be done within the logical framework of a scientific discourse system. Thus, a theory is inevitably a logical system with knowledge as its foundational element.

Thus, complex systems management experience, knowledge, and theory share a common origin and consistency. However, theory, in its formation process, scientific content, and quality, inevitably exhibits increasing systematization and logical progression, as well as maturing.

6.3 Basic Connotations of Complex Systems Management Theory

In this way, according to the general definition of theory and the content of Chap. 5, complex systems management theory refers to the systematization and localization of knowledge related to complexity management issues.

In general, complex systems management knowledge not only needs to consolidate knowledge from multiple fields but also, in many cases, requires the integration of knowledge from different fields to form new knowledge. For example, when making decisions regarding complex holistic problems, we need to integrate scientific theories, human experiences, knowledge, wisdom, computer technology, and data science to develop new decision analysis and judgment capabilities. This indicates that complex systems management knowledge includes unitary knowledge, the integration of knowledge units, and the fusion between knowledge units, reflecting the systematization of complex systems management knowledge.

Furthermore, the elements of complex systems management knowledge need to establish various logical relationships such as subordination, inclusion, parallelism, connection, and feedback among themselves. Through deduction, judgment, and inference, these relationships help us understand the phenomena of complex systems management, analyze the complexity of problems as a whole, and generate and expand new knowledge within a systematic knowledge group and knowledge chain. Only through the logical structuring of knowledge can the knowledge system of complex systems management transform into a "dynamic and evolving" system imbued with vitality.

For example, the essential attribute of complex systems management is complex wholeness. Specifically, this wholeness is formed by the complexity of the management entity, management subject behavior, management environment, and the close interrelationships and couplings between them. Therefore, it is necessary to logically and holistically generate and express knowledge about complex wholeness by integrating various aspects of complexity knowledge and using it as the core for studying issues related to complex wholeness.

In summary, complex systems management theory formed through the systematization and formalization of knowledge can guide us in understanding the essential characteristics of complex system management. It enables us to study complex wholeness management problems that reflect the fundamental attributes of complex system management through standardized thinking and logical deduction.

In summary, complex systems management theory is a systematic and logical system established by knowledge as a fundamental element through the practice and thinking activities of complex systems management. With the support of this system, people can describe and understand various phenomena in complex systems management practice in a more organized manner. It also facilitates a deeper understanding of the essential characteristics and general laws of management problems and activities because the system has been endowed with the qualities of systematized and logical research into the essential attributes of complexity.

As a nascent field of management, complex systems management theory naturally requires the construction of a comprehensive theoretical framework comprising "thinking principles-core concepts-basic principles-scientific issues-methodology and method systems." The thinking principles have been established through complex systems thinking paradigm shift, and what needs to be gradually completed are the links of "core concepts-basic principles-scientific issues-method systems," which will be introduced in subsequent chapters of this book.

6.3.1 Construction of the Academic Discourse System in Complex Systems Management

The academic discourse system refers to the discourse platform through which the academic community engages in academic discourse activities and achieves discourse functions. This platform provides the basic environment and conditions for academic discourse exchange and dissemination in a specific field and maintains the stable growth and development of academic discourse content through certain mechanisms, ensuring the lasting and orderly functioning of academic discourse functions. As a new field in management studies, when considering the functions and roles of the discourse system in this field together with the disciplinary system and academic system, it is recognized that these three components are interrelated and mutually beneficial. The disciplinary system serves as the foundation, the academic system as the core, and the discourse system as the expressive carrier.

In a disciplinary field or within an academic research community, the vigorous and enduring vitality of theoretical innovation can be attributed to several positive factors. The fundamental reason lies in the mutual promotion and impetus among the disciplinary system, academic system, and discourse system within that field. The academic discourse system, in particular, plays a crucial role in providing strong support for the continuous enrichment and growth of academia. Conversely, if academia lacks autonomous innovation in its academic activities, it is often due to weak academic discourse power or the absence of an autonomous academic discourse system. Therefore, to expand, enhance, and innovate the academic development of complex systems management, an important task is to strengthen and improve the construction of the academic discourse system of complex systems management [4].

The key principles and basic paths for constructing the academic discourse system in complex systems management are as follows.

6.3.1.1 Grasping the Principles and Key Points of Constructing the Discourse System

The complex systems management proposed in this book is a type of complexity management paradigm based on Hsue-Shen Tsien's complex systems thinking, enriched by other academic ideas related to complex systems. It possesses its own originality and guiding principles. Therefore, it is not suitable to use certain complex terminology or rely on discourse elements from other disciplinary fields to construct its own discourse system. Instead, it should be built upon the following two main directions:

(1) Discourse innovation based on epistemological transformation: drawing from Chap. 3 on CSMB and Chap. 4 on Complex Systems Thinking Paradigm Shift, it is necessary to deeply explore and expand the connotations related to complexity and wholeness and to study how to use a new discourse system to describe and characterize them. Even if such discourse has appeared in other domains, here, it should be given a new interpretation within the context of complexity.
(2) Discourse innovation based on methodological transformation: currently, modern information technologies such as the Internet, big data, and artificial intelligence are deeply integrated with management activities, giving rise to a series of new management techniques and methods. The academic discourse system of complex systems management must reflect and express these important technologies and methods adequately. Corresponding discourse innovation is not only of direct significance for the development and implementation of new methods in complex systems management but also contributes to providing a "scaffold" for grounding methodological innovation through discourse innovation based on epistemological transformation. The metaphorical discourse introduced in Chaps. 12 and 13 of this book, regarding scenario cultivation, scenario slicing, slicing flows, and other big data-driven complex systems scenario modeling techniques, offers exploratory examples for us to construct a new discourse system based on methodological transformation.

6.3.1.2 Grasping the Basic Paths for Constructing the Discourse System

After clarifying the principles and key points mentioned above, it's necessary to determine the basic paths for actually constructing a new discourse system. These paths represent the operational "entry points" for building the discourse system. The following are examples of these paths: refining distinctive new concepts, expanding the use of new language, and achieving a large cycle in the theoretical chain.

(1) Refining Iconic New Concepts

The expression of discourse in complex systems management should be based on specialized scientific language. This foundation is comprised of concepts. Concepts facilitate the transition from concrete thinking to abstract thinking in academic research and serve as the "building blocks" of academic discourse for communication and dissemination among members of the scientific community.

Academic research in complex systems management requires conceptualization as a first step in understanding and analyzing management activities, phenomena, and problems. It involves proposing concepts that reflect essential attributes, providing the basic components for expressing scientific questions, achieving consensus within the scientific community, and facilitating communication and dissemination on recognized platforms. Therefore, autonomous academic research in complex systems management necessitates the independent proposal of core new concepts within the theoretical framework to reflect innovative academic thoughts.

The more a concept embodies the essential attributes of complex systems management, the more fundamental it becomes within the theoretical framework. Therefore, concepts of this nature need to be "carefully selected" and become the "masterpieces" in the conceptual framework, generally referred to as core concepts or identifiable concepts. Iconic concepts are fundamental language symbols unique to a discipline. The extraction of identifiable concepts can involve creating a completely "new discourse" or giving existing or traditional concepts new connotations. Both approaches are feasible and effective. However, regardless of the approach, the concepts must contain profound new ideas and theoretical implications and have the ability to encompass, expand, associate, and elevate within the theory chain. If the proposed concepts lack originality, they merely repeat others' ideas, and the significance of the research work often remains limited to reconfirming the correctness of others' thoughts.

It is essential to recognize that academic discourse reflects our cognition and understanding of the world, and concepts, especially identifiable concepts, condense the core of our academic thoughts. As Albert Einstein said, "Without the use of defining categories and general concepts, thought is like breathing in a vacuum, impossible."

Clearly, extracting such concepts to create "new discourse" is not an easy task. Generally, as people deepen their understanding of things, their mode of expression gradually shifts from "plain language" to academic or analytical language. Although this transition may make it less accessible to many people, it represents an important advancement within the academic community. During this transition, concepts acquire new normative, essential, and extensible aspects, making them a "universal language" within academic exchanges. However, the realization of this transition primarily depends not on language skills but on the injection of rich complexity thinking and theoretical knowledge. The deeper the thinking and theory, the more it can be expressed using the most profound, concise, and essential language.

As an example, in Chap. 7 of this book, we mentioned that when researching managerial problems, we restructured the traditional concept of scenario, which was relatively simple, infusing it with new implications and making it an iconic concept. The introduction of this new foundational concept allows us, within the theory of complex systems management, to shift from problem-oriented to scenario-oriented approaches. This signifies a deepening of theoretical research from general wholeness to complex wholeness, recognizing academically that the theoretical problems originally perceived are embedded within scenarios. To ensure that it doesn't lose its contextual reality and uniqueness, it is necessary to perceive the truthfulness of theoretical research conclusions in the original ecological scenario as much as possible.

This requires researchers to not only observe the scenario of the problem as a "bystander" during the research process but also to immerse themselves as a "participant" within the scenarios where the phenomena and attributes of the problem are most significant, and even to act as a "doer," creating new scenarios directly. Furthermore, using the scenario as an iconic new concept leads to the basic principles of scenario robustness regarding deep uncertainty decision-making and forms a new scientific problem of deep uncertainty decision-making in the theory of complex systems management. Subsequently, it designs a key technology in complex system management: big data-driven complex system scenario modeling technology. Every step of this not only embodies the foundational discursive function of the iconic concept but also reflects its promotive role in constructing a new theoretical system.

When we acknowledge the importance of concepts, particularly innovative or redefined distinctive concepts, in developing the academic discourse system for complex systems management, we should also understand that when addressing real-world problems and embarking on research related to these problems, we should not immediately start with a pre-established set of concepts as a discourse framework. Instead, when we genuinely extract concepts that reflect the essential attributes of the problem from real-world issues, autonomous academic discourse naturally emerges.

By consistently applying new principles of thought and expression to address these issues, autonomous discourse content is formed. Throughout this process, if we fail to extract concepts with profound, essential attributes or, if the concepts we extract tend to become homogeneous, or if new concepts cannot be elevated from perception to analyzable scientific concepts, it becomes challenging to propose fundamental assumptions and prerequisites based on concepts in theoretical research, and it is difficult to construct an autonomous academic discourse system.

(2) Expand the Use of New Languages

The basic tool of discourse is language. Initially, oral language was a symbolic system consisting of speech as a material form composed of vocabulary and grammar. Later, the expression of language was defined as the form in which people use language, and after the invention of writing, the main form of language use became written language. In a broader sense, written language also includes graphics, mathematical symbols, and so on. Thus, the core elements of a discourse system based on "language use" can be further abstracted and understood as rules, logic, instructions, procedural processing, and ways of thinking. This expanded cognition makes it easy for us to understand the consistency or high similarity between the paradigm of academic discourse and the "algorithm" as a recent universal philosophical thinking method.

Israeli historian Yuval Noah Harari, in his book *Homo Deus: A Brief History of Humankind*, regards "algorithm" as a universal philosophical thinking method of our time: "An algorithm is a methodical set of steps that can be used to make calculations, resolve problems, and make decisions." Therefore, any set of paths (rules, processes, steps) that can be put into a logical program and processed using methods including computers is an "algorithm." Thus, in addition to traditional natural language, the paradigms of experimental science, theoretical derivation, computer simulation, and data-intensive science are all discourse modes in the sense of "algorithm."

This conclusion provides us with the insight that enhancing the ability to describe, interpret, and solve complexity management problems through the further deepening and expansion of the connotation of discourse is expected. In this sense, in addition to improving research capabilities through innovation in complex system management thinking principles and theoretical systems, we can also use advanced information technologies such as big data and artificial intelligence to promote it. Therefore, the understanding of the discourse system can also be further expanded from traditional natural language, scientific language, and general computer functionality to include broad algorithms, new symbol systems, data-driven approaches, and artificial intelligence.

It is not difficult to understand that the evolution of complex system management from systematicity to complexity in essential attributes and the extensive use of modern information technology will inevitably give rise to new discourse demands for managing complexity. At the same time, potential conflicts and imbalances between technology and management, technology and ethics, and technology and human behavior will arise. Additionally, with the continuous emergence of interdisciplinary and composite disciplines resulting from the integration of management science, natural science, and engineering technology, numerous new problems will arise regarding the rationality, compatibility, and expansibility of fundamental theories. The study of these new problems will result in a lack or even a blank space in the discourse system in terms of thinking, expression, dissemination, and communication. All these situations indicate that advanced information technology in contemporary times not only provides us with powerful capabilities but also forms a significant "blank area" in the discourse that urgently needs to be filled, and new technology itself is one of the main actors participating in filling this void. It can be imagined that once there is a breakthrough in this aspect, it will greatly enhance the ability to construct an academic discourse system for complex system management.

In terms of promoting the construction of an academic discourse system through methodological innovation, the comprehensive integration methodological system proposed by Hsue-Shen Tsien is a classic example. The comprehensive integration methodological system is a holistic approach proposed by Hsue-Shen Tsien based on major project management, integrating multidisciplinary and multi-domain technologies and methods to understand, analyze, and solve complex management problems. Essentially, the complexity of such problems mainly stems from the insufficient cognitive abilities of the subjects, the inherent complexity of the objects, and the deep uncertainty of the environment. Thus, the characteristics and principles of comprehensive integration methodology are consistent with the complexity problem and its resolution. It should be noted that the comprehensive integration methodological system is a holistic design of methods for solving complex problems under the guidance of complex systems thinking. It can ensure the scientificity of methodology in complex system management and enable us to describe and abstract the research process in a new discourse manner.

(3) Achieve a Large Cycle of Theoretical Chains

Scenario-oriented thinking and theoretical innovation are the academic foundations of innovation in complex systems discourse. Based on this foundation and following a "grand path," we can systematically innovate the connotations of complex system management discourse.

Currently, in academic research on complex system management, there exists a series of fundamental theoretical problems, such as paradigm shifts, knowledge acquisition patterns, and theoretical system construction. These problems cannot be solved through a few scattered theoretical issues or empirical summaries. Instead, it is essential to establish a complete theoretical chain in a general sense, consisting of philosophical thinking, core concepts, basic principles, scientific issues, and methodology and method systems. In each stage of the theoretical chain, there should be specific content that reflects the essential attributes of complexity problems, and at each turning point of the theoretical chain, there should be reasonable, logical transitions. Such a theoretical research path ultimately forms not a specific and localized theoretical outcome but a general foundational theoretical system in the field. It is evident that this kind of theoretical system is an academically valuable new proposition in the field of complex system management. Therefore, in terms of influence, it is actually a fundamental, complete, and emblematic path of academic discourse innovation.

Of course, the construction of such a systematic discourse approach needs to adhere to the basic laws and forming paradigms of theoretical systems. Its construction process involves both innovation of complex systems management theoretical architecture and innovation of discourse framework.

In the above discussion, we have devoted considerable space to the "entry points" related to the content of discourse in academic discourse architecture, emphasizing that the most fundamental aspect of discourse architecture lies in the basic characteristics possessed by management thinking and the theory itself, as well as the truthfulness and universality of academic discourse content. This is the embodiment of the universal value of human management civilization.

6.3.2 Glossary of Complex Systems Management Terms

At the beginning of this chapter, an explanation of complex systems management is provided, focusing on the interpretation of the scientific content of complex systems management. It points out that as a new management model, in practice, it primarily involves the management activities of a type of "complex wholeness" issues within socio-economic systems. Academically, it is a paradigm within the category of complexity management, driven by a new methodology.

Given this, why use the term "complex systems management" instead of simply "complexity management"? This relates to the logical nature of scientific conceptual discourse and the characteristics of path dependence in the formation of terms.

Generally, terms are labels used in specific academic fields to denote concepts. Unlike mere nouns in grammar, terms are conventions of language symbols expressed through speech or writing to convey or delimit the conceptual language of scientific thought and communication, requiring clear connotations.

According to the theoretical system of this book, "complexity" precedes complex systems with complexity attributes. Logically, "complexity" is a concept broader than "complex systems." Complex systems management is a complexity management paradigm driven by Tsien Hsue-Shen's complex systems thinking. Its core is the irreversibility of reductionism. Therefore, naming it "complex systems management" helps distinguish it from other reflective types in the category of complexity management. Additionally, the usage of a particular term often reflects the trajectory of scientific development, which we will elaborate on further below.

As human society develops, complexity has emerged in what was once simple management. Therefore, it is essential to understand what complex management issues look like, whether and how we can uncover the essence of complex management problems, and what kind of methodology should be employed to systematically solve these complex management issues. Further, this leads to the need to clarify, comprehend, and profoundly explore whether there is a knowledge system that captures the essence of complex management and whether it is possible to construct such a system.

All of these questions go beyond specific management problems and delve into questions involving philosophical thinking, scientific paradigms, theoretical logic, knowledge formation paths, and major issues. It is evident that this is a significant and challenging problem in the field of management. The fundamental reason is that "complex" is not just a phenomenon or characteristic on a specific domain, region, or level; rather, it is a global, universal entity and intrinsic attribute of contemporary human management activities. Therefore, we need to think and reveal the intrinsic attributes of "complex" at a higher level, on a larger scale, and across multiple dimensions. This requires us to construct a methodological framework that is "cross-cutting" across multiple scientific fields to adapt to the "complex" inherent in different disciplines. Otherwise, even if there is progress in a specific case or aspect, it will be challenging to address this major issue comprehensively.

Starting from this holistic perspective, the developments and achievements in the field of systems science over the past century have provided important inspiration and assistance in solving this problem.

Firstly, regardless of the characteristics of specific domains and specific issues studied by natural sciences, social sciences, etc., systems science treats them as abstract "systems." The cross-disciplinary nature of systems science is consistent with the cross-cutting nature of "complex." This determines that many ideas, concepts, and principles in systems science have higher level and deeper generalization and interpretability when describing and abstracting universally "complex" phenomena and problems across various vertical disciplines.

In fact, since the early twentieth century, scientists in the field of natural sciences discovered diverse, complex phenomena in physics and chemistry and used the term "complexity" as a collective term for these phenomena. This marked the beginning of modern academic research on "complex" and the labeling of this research as complexity science. Later, the scientific community introduced the concept of "complex systems," hoping to use a certain "complexity" vocabulary when studying specific complex phenomena and adopting more abstract "complex systems" when studying complex phenomena above specific fields. This indicates that "complex systems" are conceptualized as the "matrix" giving rise to complex phenomena. This view of complexity and complex systems has played a crucial role in advancing modern scientific development, representing a remarkable achievement of humanity. People have also applied this thinking on complexity and complex systems to study complex management issues in the socio-economic field, offering many new perspectives and insights. However, overall, the complexity vocabulary and the concept of complex systems derived from natural sciences struggle to align directly with social phenomena, organizational behavior, and individual human behavior in the socio-economic field, and they face challenges in delving into the mechanisms behind complex socio-economic management phenomena. This suggests that, on the one hand, the attempt to use the concept of systems to address the "complex" cross-cutting nature has been successful, while on the other hand, the complex system perspective derived from the "complexity vocabulary" struggles to serve as the theoretical foundation for studying "complex management."

To fundamentally resolve this dilemma, it is necessary to endow complex systems with the "complex" intrinsic attributes in the socio-economic field. Tsien Hsue-Shen proposed the concept of "complexity" with "irreversibility of reductionism" as its core as the intrinsic attribute of "complex" in the socio-economic field. If a system possesses the attribute of "irreversibility of reductionism," it signifies that such systems have the fundamental nature and essence of "complex." This is referred to as a complex system. Clearly, at the theoretical level, "complex systems" can serve as an academic symbol representing the intrinsic attributes of complex systems. At the practical level of thought, "complex systems" can serve as the tangible embodiment of the attribute of "complexity." Consequently, the term "complex systems management" signifies a paradigm regarding the essential nature of complex management in the socio-economic domain.

Moreover, since the essence of complex management is characterized by complex systems, a profound understanding of the essence of complex management and the construction of a theoretical framework for complex system management inevitably require the maturity of complex systems thinking in the field of systems science. Only then can the necessary theoretical support capabilities be established to address the intrinsic nature of complex management. The profound intrinsic connection between systems science and management science illustrates the characteristics of the current era, where different disciplines mutually intersect and promote "big science" integration. This understanding provides a deeper insight into and appreciation for the independent yet mutually reinforcing development of systems science and management science over the past few decades, as well as their joint response to the myriad of complex real-world management issues and the underlying mechanisms.

This, in essence, is the interpretation of the term "complex systems management."

References

1. Ding Y, Sheng Z. Make good use of complex system management thinking to improve problem-solving skills. Masses. 2021;14:54–6.
2. Xu C. Ways of thinking: transgression and demarcation: a critique of theoretical thinking and engineering thinking in the humanities and social sciences. Academia Bimestris. 2001; 01:5–14+207.
3. Xu C. On the encroachment of theoretical thinking and engineering thinking in the humanities and social sciences. Tianjin Soc Sci. 2001;02:25–31.
4. Sheng Z. Discourse system: speak the "Chinese Dialect" of management academic innovation. J Manage Sci 2019; 22(06):1–14.

Chapter 7
Core Concepts of Complex Systems Management

Based on the practice and thinking principles of complex systems management, this chapter attempts to put forward several core concepts. Each section introduces a concept, and, as the theoretical framework enriches, these core concepts will become more profound and comprehensive.

7.1 Complex Systems Management-Environmental Composite System

Any complex systems management activity empowers a complex, productive, constructive activity and constructs a new artificial complex system within the original environment. This indicates that once complex systems management activities are completed, a new system is formed by the artificial complex system and the original environmental system. Because this new system is composed of a combination of complex systems management activities and the original environment, it is referred to as the "complex systems management-environmental composite system."

1. Generally, any management activity can generate such a composite system, but it does not attract special attention. This is because the artificially constructed systems in general management activities are relatively small in scale, and the combination with the environment is often insufficient to form new systems with a major impact. However, artificially constructed systems in complex systems management activities tend to be large in scale, have a profound impact on the environment, and have a long-lasting effect. They play a crucial role in the original environmental system. Examples include the planning and approving of major projects, important socio-economic development strategies, and the design of social governance policies. All these fall under the scope of complex

Z. Sheng, *Outline of Complex Systems Management Theory—Based on Irreversibility of Reductionism Thinking*, International Series in Operations Research & Management Science 360, https://doi.org/10.1007/978-981-97-3361-3_7

systems management and have a tremendous and far-reaching impact on the original socio-economic environment. At this point, it is necessary to introduce the concept of the "complex systems management-environmental composite system" and pay attention to the important theoretical implications of this concept. From the perspective of system functionality, the new composite system may generate new functions that were not present before, especially unforeseen and undesirable negative functions. This poses an implicit management risk for the artificially designed complex systems in complex systems management. Due to the deep uncertainty or the characteristics of large spatiotemporal scales often associated with such risks, sufficient attention needs to be given to stakeholders.

The above analysis indicates that, in comparison to general management activities, research in complex systems management should pay full attention to the interactions and feedback between the originally conceived artificial systems of management and the complex systems mentioned above. Specifically, due to the complexity features such as uncertainty, multiple scales, and adaptability, the interweaving of these interactions and feedback results in new systems exhibiting even more intricate, delayed, and variable behaviors. These complexities far exceed the scope of people's general management experience and capabilities, making them challenging to anticipate and mitigate in practical management.

7.2 Complex Wholeness

Whether considering complex systems management as a field of management studies, a unique type of practice, or a theoretical framework, complex wholeness is the most fundamental concept within its scope. This is because the essence of complex systems management, people's epistemology, and methodology regarding complex systems management are all developed from the perspective of complex wholeness.

In different fields, "complex wholeness" has different backgrounds and connotations. This is mainly because "complex wholeness" in a specific context can often be interpreted based on the cognitive needs of the subject, leading to specific explanations. For example, if the focus is on "complexity," nonlinearity, randomness, and chaos can be considered forms of complex wholeness. If the focus is on "wholeness," then all associations, functionalities, and emergent properties where $1 + 1 > 2$ are considered complex wholeness. Therefore, we cannot simply "borrow" the "complex wholeness" from other disciplines to define the complex wholeness of complex systems management. Otherwise, this concept will be difficult to align closely with the attributes of complex systems management.

From the discussions in the previous chapters of this book, the essence of "complex wholeness" as a concept in complex systems management inevitably refers to the "irreversibility of reductionism" concerning complex systems or complexity issues. It is an abstraction and condensation of the attributes of complex systems management in terms of the overall meaning and process. Therefore, it can be the physical complex wholeness of complex systems management practice, the systemic complex wholeness in attribute cognition, or the management complex wholeness in management activities. It can also be a synthesis of these different aspects of complex wholeness.

While complex wholeness serves as a representation of the irreducibility of reductionism, to better understand it in a concrete and clear manner, we will provide a decomposed interpretation of this concept from different perspectives below.

Firstly, the basic connotation of complex wholeness can be understood as follows: In the formation and characteristics of phenomena, scenarios, and problems in complex systems management activities, the cognition and behavior of the subject not only involve various complex forms but also these complex forms are closely interconnected and mutually influential in multiple ways. This makes it inappropriate or impossible to divide a complex management activity, scenario, or problem, as a whole, from top to bottom into several relatively independent and relatively simple parts. After solving each different part separately, we can then integrate them to form a complete solution to the original problem. In short, complex wholeness represents that the phenomena, scenarios, and problems in complex systems management are not only complex but also cannot be "reversible" divided into parts.

The above basic connotation has the following more detailed practical contents:

(1) Complexity of human mind and behavior: both the subjects and objects of complex systems management involve highly intelligent individuals. Humans, at the core of complex systems management activities, exhibit various mental characteristics such as rationality, irrationality, agency, passivity, subjectivity, objectivity, initiative, and adaptability. They also exhibit behavioral attributes as social beings, economic actors, and natural entities. Humans actively change their behaviors to adapt to changing environments and problems through means like learning, imitation, trial and error, inheritance, and mutation. Adaptability contributes to complexity. The higher the level of human cognition, the greater the adaptability of their behavior. Therefore, in complex systems management activities, humans demonstrate profound complexity, continuously creating complexity through adaptability and manifesting an integration and ordering of mind and behavior. Moreover, individuals in complex systems management are generally part of various groups, and the interconnections and mutual influences between these groups, as well as between groups and other management elements, form an inseparable whole. All of these ultimately result in the emergence of macro-level functions based on the complexity of human minds and behaviors.
(2) The complex wholeness nature of system interrelationships. In general, complex systems management involves various forms of interrelationships among its constituent elements. These relationships include both explicit and implicit associations, with impacts that can be real-time or lagging. The nature of these relationships exhibits both deterministic causality and uncertain correlations. The transmission and transformation of information through these relationships can be both complete and symmetric, as well as incomplete and

asymmetric. The interplay of these relationships is not only observed within the same level but also across different hierarchical levels. The emergence of overall functionality in complex systems management is precisely the manifestation of the "irreducibility" of these interconnected effects. This statement contains two layers of meaning: firstly, the holistic functionality and qualities of complex systems management, such as adaptability and robustness, are a result of the simultaneous existence and mutual coupling of the aforementioned forms of interrelationships. Secondly, the orderly complexity and holistic nature of complex systems management cannot be attributed to mere chance. It can only be explained by acknowledging that removing any part of the complex interrelationships within the system would result in the cessation of the driving force behind the overall functionality of complex systems management. This expresses the complex wholeness of system interrelationships in complex systems management.

(3) The complex wholeness of system structure. Complex systems management activities are dynamic integrated subjects composed of elements such as humans, objects, and information. The integrated subject consists of physical and structural parts as well as relational, logical, semi-structural, and non-structural parts. Each part plays an irreplaceable and indispensable role in the overall functionality and behavior of the integrated subject. Therefore, for complex systems management, the wholeness of its structure is symmetrical to the wholeness of its functional behavior. We cannot decompose the integrated subject into multiple parts at the physical, relational, or logical level and construct the wholeness of its functional behavior using only one part. Conversely, the mechanism of the overall functionality and behavior of any integrated subject cannot be driven solely by the mechanisms and relational connections of one part of the structure. This understanding of complex wholeness comes from the functional connotation and emergent connotation of the system as a whole.

(4) The complex wholeness of the environment. As mentioned earlier, for a specific complex systems management activity, the environment has multiple complex influences on the management activity and the composite system formed by the management activity and the environment. Some of these influences are direct effects of the original environment, and some are indirect effects of the composite system. However, they all originate from the wholeness of various relationships and couplings formed between complex systems management activities and the environment. Therefore, it is necessary to carefully consider the complexity of the environment itself and the possible emergent properties of the whole.

Based on the above four aspects, it can be seen that.

1. Complex wholeness has become an essential characteristic of complex systems management and complexity issues in an all-around manner.
2. The complexity of human minds and behaviors is closely intertwined with the complexity, indivisibility, irreversibility, and unpredictability found in complex systems management activities, phenomena, and issues. They are interconnected

and mutually causal, forming the fundamental dynamics of complex systems management through various feedback mechanisms and transmission pathways.
3. In complex systems management, the complexity of the whole manifests differently across various levels and dimensions. This requires us to consider the different forms and characteristics of complex wholeness in complex systems management at different levels and dimensions, as well as their various impacts on management activities. This can be achieved by analyzing the complexity of individual behaviors at the micro level, emergent behaviors at the macro level, and multi-level, cross-scale transmission and transformation mechanisms to fully uncover the profound laws governing complexity.

In summary, the significance of introducing the concept of complex wholeness in complex systems management theory lies in the following aspects:

1. Through detailed and subtle analyses of the complex wholeness of complex systems management activities, phenomena, and problems, we can ensure the accuracy of understanding and controlling the complexity of problems and address new challenges effectively.
2. Analyzing and summarizing different types of complex wholeness helps abstract and distill common attributes and principles in the study of complex systems management theory, leading to the development of new theories.

7.3 Essence of Uncertainty

In typical management activities, people generally prefer to describe management phenomena and solve management problems with "clear, definite, and certain" causality. If this were possible, research would be easier, and conclusions would be definitive. However, the fact is increasingly evident that in various aspects of management cognition, decision-making, prediction, and solution optimization, people are confronted with situations where phenomena and problems are "not certain, indeterminable, difficult to decide, or hard to predict." This is what we intuitively perceive as "uncertainty." Although we do not like "uncertainty," we have to face it. Moreover, we find that different phenomena and problems exhibit different patterns of "uncertainty," and people's perception of "uncertainty" also varies. This indicates that "uncertainty" itself possesses certain properties or qualities that can be considered as attributes, and this is what we call "uncertainty."

From a general theoretical perspective, it is easy to understand the close relationship between uncertainty and complexity or complex wholeness. Certainty always gives people a sense of "simplicity," while uncertainty always makes people feel "complex." Correspondingly, the basic perception of "complex wholeness" also exists. This is why "uncertainty" is a core concept in complex systems management theory.

Initially, people understood "uncertainty" through the process of thinking. For example, to accurately, comprehensively, and profoundly understand management phenomena and problems, one needs to master the corresponding management knowledge. However, due to incomplete or insufficient knowledge, it becomes impossible to make definite and certain descriptions, predictions, and judgments about the status, trends, and future scenarios of a fact or phenomenon. This is generally referred to as subjective uncertainty. Additionally, management problems and phenomena themselves may have objective mechanisms that give rise to various possibilities in terms of the states, paths of change, and outcomes of phenomena, which is generally referred to as objective uncertainty. Objective uncertainty is independent of human subjective cognitive abilities; it is an objective attribute of things or phenomena. This kind of uncertainty only changes when the structure and mechanisms of things or phenomena themselves change.

In general, uncertainty in management activities includes both subjective uncertainty and objective uncertainty. In many cases, these two types of uncertainty coexist and integrate with each other. Furthermore, through careful observation, people have recognized that the causes and degrees of "uncertainty" can be divided. For example, one type of "uncertainty" follows "certain" rules, thereby reducing the degree of inherent uncertainty. This type of uncertainty is relatively "simple." On the other hand, people have also discovered that some uncertainties do not follow clear and stable rules or no longer fit traditional approaches for handling regular uncertainties. These uncertainties are referred to as "deep" uncertainties or "serious" uncertainties.

It is not difficult to understand that in management research, based on determinism and gradually introducing uncertainty, simple uncertainty, deep uncertainty, or serious uncertainty, management theories and theoretical frameworks have progressively adapted to the increasing complexity of management reality and problem complexity. However, in order to transform this adaptability into a powerful ability to manage complexity in practice, we need to further analyze the underlying causes and evolutionary forms of uncertainty at a finer granularity and find the entry points to unravel complex wholeness.

In light of the paradigm shift towards complex systems thinking, we expect that management activities and issues will inevitably exhibit various uncertainties, many of which may have subjective origins. However, we should pay more attention to uncertainties originating from its inherent attributes, namely the essential uncertainties resulting from the irreversibility of reductionism. Obviously, these uncertainties cannot be simply described as "deep" or "severe" using natural language. Instead, we should start with the irreversibility of reductionism as the logical starting point to reveal the corresponding characteristics and connotations of uncertainty.

Irreversibility of reductionism indicates that complex systems, through multiple different micro and partial levels, form a system as a whole that cannot be determined. This statement includes the following:

1. The emergence or self-organizing phenomena of the system as a whole do not exist or are difficult to fully discover in the lower level micro and partial elements.
2. Even if there are sporadic pieces of information about macro states in the lower level micro and partial elements, the complete mechanism of transformation from micro to macro is broken, fragmented, and ambiguous in people's cognition.

3. Consequently, traditional methods, even so-called comprehensive mathematical analytical algorithms, cannot guarantee complete predictions of the system's macro future.

In this way, the so-called "certain" future disappears. The future you "see" may be surprising: the future cannot be extrapolated from past and present laws and rules; the future that may occur in the sense of probability and statistics is not necessarily encountered; the same deterministic rules cannot guarantee identical future results in two separate instances; and the formation paths of all complex holistic phenomena are uncertain. In short, the future is unknown in the deterministic sense.

Based on the above cognitive logic, we find that at the micro and macro levels of complex systems, the initial position of the micro, initial information characteristics, and slight biases formed by algorithms all contribute to some contingencies or changes in details. As mentioned earlier, contingencies and details are important causes of complexity, which means that we cannot use the traditional "hypothesis-driven" approach to set assumptions between micro individuals and macro wholes and obtain holistic responses based on certain rules. Instead, we can only break through the irreversibility of the reductionism dilemma between the micro and macro levels through exploration, trial and error, iteration, approximation, and convergence.

This implies that although our research object and target are complexity as a whole, we need to adopt a horizontal "fragmented" and vertical "segmented" cognitive and analytical approach. This fully reflects that solving complexity management problems generally cannot be achieved using direct or simple linear thinking but requires a "crossing the river by feeling the stones" approach.

Furthermore, the above cognitive logic tells us that in complex systems management, different levels, especially the macro whole level, exhibit complex conditions where structured, semi-structured, and unstructured elements coexist. Even if we know all the rules, we cannot guarantee that applying these rules will definitely lead to that "certain" complex whole. We can only know what it is after all the rules have been fully executed, which is referred to as "computational irreducibility."

To summarize the above academic ideas, for complex systems management, the irreversibility of reductionism manifests distinct characteristics in two dimensions: the space of management activities and the time of the process. These characteristics are as follows:

1. In the time dimension, the future overall form of the system is unknown. This mainly implies that the future is no longer or, to a large extent, cannot be extrapolated from past and present conditions and rules. People can use traditional knowledge and experience to understand and predict a part of the future but not the entirety. Understanding the future and the path toward it in complex systems management requires a "segmented" iterative and approximating process.
2. In the space dimension, the whole system cannot be simply "assembled" from micro and local elements. At the same time, every element, local component, and their correlations and interactions are indispensable for the formation of the system as a whole. If any part is missing, the integrity and functionality of the entire system will be compromised (Fig. 7.1).

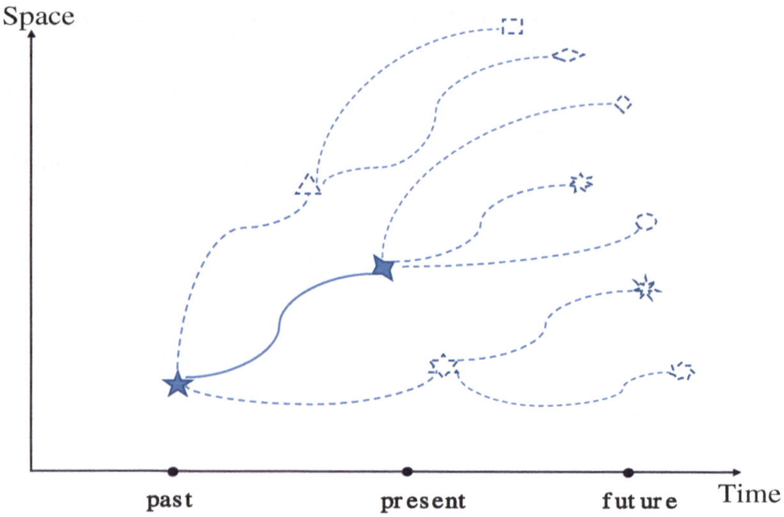

Fig. 7.1 The future of complex systems is unknown

3. Based on the two points mentioned above, it can be argued that the irreversibility of reductionism in the management of complex systems introduces a "non-traversal" thinking regarding the understanding of uncertainty. In simple terms, "traversal" implies that all possible events will eventually occur, meaning that the mathematical expectation in both space and time is the same. However, in reality, this is not always the case. For example, let's consider 100 individuals in different locations engaging in uncertain transactions of the same type with a certain cost, each achieving different performances. The average performance is denoted as "a." Can we assume that if someone performs this action 100 times, the final performance will definitely be "a?" It is not accurate to think so because each person actually only performs the action once, and the outcome of winning or losing is determined in a single instance. "a" represents a spatial-mathematical expectation, whereas achieving an average performance of "a" requires someone to perform the action 100 times, resulting in a temporal-mathematical expectation. However, in this process, the spatial-mathematical expectation and the temporal-mathematical expectation are not equal because an individual who performs the action continuously 100 times may lose all their capital in one of the instances, thereby terminating subsequent actions. This is known as "non-traversal."

The analysis and interpretation from various aspects mentioned above inform us that in the real world, the irreversibility of reductionism leads to the "future," the "whole," and the path toward the "future" and the "whole" in complex systems always being unknown and uncertain until they materialize. This uncertainty encompasses not only the uncertainty in realizing future paths in the temporal dimension but also the uncertainty in the formation of overall structure through emergence, evolution, and self-organization in the spatial dimension. Furthermore, it is impossible to eliminate such uncertainty from both of these dimensions. We refer to this type of uncertainty as essential uncertainty.

For complex systems management, essential uncertainty holds profound, comprehensive, and fundamental significance. The concepts presented in this book, such as "multi-scale," "adaptability," and "iterative," can all be extended or expanded based on the characteristic of essential uncertainty. Therefore, it can be argued that essential uncertainty fully embodies the essential characteristics of complex systems management activities and holds significant theoretical importance:

Firstly, essential uncertainty reveals the primary root of the complexity and wholeness of complex systems management activities in many aspects. This is beneficial for designing approaches and technical strategies to unravel the complexity wholeness of complex systems management problems based on the concept of essential uncertainty.

Secondly, the concept of essential uncertainty has close logical correlations with the concepts of scenarios, multi-scale, and adaptability in complex systems management theory, as well as with the principles of resolving complex wholeness, adaptive selection, multi-scale management, and "iterative" generation. This further strengthens the intrinsic systematicity and logical nature of the complex systems management theoretical framework.

7.4 Multi-scale

Multi-scale is a core concept in the theory of complex systems management, primarily because it serves as a logical starting point and regulatory principle for the attribute of complex wholeness.

To facilitate the understanding of the concept of multi-scale, we first analyze the complex wholeness of complex systems from two perspectives: complexity and wholeness. However, this does not imply that the complex wholeness of complex systems can be arbitrarily separated into these two relatively independent dimensions.

7.4.1 Multi-scale from the Perspective of Complexity

Complex systems management activities, phenomena, and problems are composed of various elements that form complex systems through various correlations and structures. However, "complexity" does not equate to "disorder." Within the "complexity" of a system, there exist various explicit and implicit stable orders. Whether establishing or maintaining order, corresponding criteria, measures, and guidelines based on the attributes and representation methods of system elements or states are required, collectively known as "scales."

In this section, scales mainly refer to the division of a thing or its characteristics into relative orders based on certain regulations. They serve as fundamental criteria and standards for observing complex systems management phenomena or understanding the complexity of problems. For example, when considering a system, how to perceive the relationship between the whole and the environment, the whole and other wholes, the whole and parts, parts and parts, parts and elements, and elements and elements requires a sequence of division standards. Otherwise, we cannot distinguish the differences in the nature of the whole, locality, and elements. These division standards are scales.

In specific problems, the division of scales often follows the changes in the nature of the characteristics. For instance, to differentiate the categories of a feature, a categorical scale is established. To distinguish the orderliness of features, an ordinal scale is defined. To differentiate the hierarchy of features, a hierarchical scale is established. To measure the quantity of features, a quantitative scale is defined.

In general, the concept of multi-scale in complex systems management abstracts the category, order, hierarchy, and quantity of the same feature, performance, or attribute. Clearly, the prerequisite for the concept of scale is the existence of the same feature, performance, or attribute. For the sake of simplicity, we can collectively refer to this prerequisite as the same dimension.

Intuitively speaking, scale is a concept based on the same dimension. Suppose it is necessary to finely distinguish the features, performance, or attributes represented by the same dimension; a certain regulatory criterion is used as the scale to distinguish the categorical, ordinal, hierarchical, and quantitative aspects within the same dimension. The way in which these features, performance, or attributes are described is not important.

If multiple distinct and distinguishable differentiated results appear within the same scale for the features, performance, or attributes of the same dimension, this is called a multi-scale phenomenon. It is not difficult to understand that the multi-scale phenomenon is an important representation of complexity and a certain "root cause" of complexity. Multiple scales are a common objective phenomenon in the management activities of complex systems. For example:

(1) **Temporal scale**. In the management activities, phenomena, and problems of complex systems, it is often found that certain features not only have the same temporal dimension but also have different scale connotations. This requires us to consider the holistic problem from a temporal multi-scale perspective. For example, global climate change is a multi-temporal scale problem that spans decades, centuries, thousands of years, or even longer. This problem is closely related to human carbon emissions, which are also multi-scale. For an individual, a car trip may only last a few days or hours, while the carbon emissions of a company occur within its lifespan of several decades. The development and large-scale promotion of technology to reduce carbon emissions may take several decades. Therefore, various carbon emission activities of different time scales collectively contribute to global climate change. Thus, addressing and solving the problem of global climate change is a multi-temporal scale problem.

(2) **Spatial scale**. This concept mainly involves the characteristics and attributes of the spatial scale and boundary of management activities, phenomena, and problems in complex systems. It can be broadly divided into two aspects: first, the physical subject scale of the management activities, phenomena, and problems of complex systems is large. For example, global climate change is a large-scale spatial problem on a global scale. The global geography is complex, and the roles and effects of different geographical regions, such as land, oceans, forests, and deserts, in global climate change are different. Therefore, it is necessary to divide the global geographical regions into different spatial scales, resulting in a multi-spatial scale problem on a holistic level. Second, the functional impact of complex systems has a large spatial scale. Within a large-scale range, the form and strength of functional effects vary with changes in spatial scale. Therefore, it is necessary to differentiate the impact space of large scales into different scales to clarify the detailed conditions of different scale spaces in the complex wholeness problem. For example, the extensive marine pollution caused by a nuclear power plant accident belongs to a large spatial scale problem. To accurately analyze and respond to the impact of the accident, it is necessary to evaluate the pollution impact of different regions through multi-scale spatial divisions based on factors such as distance from the accident site, ocean currents, climate, and meteorology.

(3) **Hierarchical scale**. This generally refers to the division of a certain feature or attribute in the management of complex systems based on vertical standards, such as macro, meso, and micro; higher level, middle level, and lower level; and significant relevance, general relevance, and irrelevance. This division forms a hierarchical multi-scale. For example, in epidemic prevention and control, different regions can be classified into different levels of epidemic prevention based on geographical distance from severely affected areas, transportation, and socio-economic connections. These levels can include key prevention, moderate prevention, and mild prevention, with corresponding differences in prevention and control measures.

(4) **Structural scale**. In the process of modeling complex systems management problems, different divisions of structured, semi-structured, and unstructured problems emerge. Problems with stronger structures are relatively simple and easy to describe, analyze, and solve, while problems with less structure become increasingly complex and difficult. Correspondingly, the management methods and techniques for these problems also differ significantly. In order to reflect the objective situation of complex systems management, the concept of multi-scale structuring of problems can be proposed. Similarly, for a specific class of problems in complex systems management, attributes can be differentiated based on the objective uncertainty characteristics of the problems, such as distinguishing between general (simple) uncertainty, moderate (complex) uncertainty, and deep (serious) uncertainty, obtaining a multi-scale structural division based on uncertainty.

The above-mentioned several scale division approaches based on different dimensions illustrate that, on the one hand, scales and the concept of multi-scale are common in the management activities of complex systems. On the other hand, it is evident that scales, multi-scale concepts, and how to appropriately divide multiple scales are not only necessary in elaborated research on complex systems management but also enrich the complexity connotation in complex systems management activities.

7.4.2 Multi-scale from a Holistic Perspective

Above, we discussed the rationale and necessity of introducing the concept of multi-scale from the perspective of complexity in complex systems management activities. In this section, we will primarily discuss the significance of this concept from the holistic perspective of complex systems management.

The functionality and emergence of complex systems management activities are important aspects of the complex wholeness of management activities. The wholeness of management activities is constituted by elements, relationships, structures, processes, and particularly the complex interconnections and intricate structures within the activities. Therefore, in the context of the wholeness of management activities, there inevitably exist hierarchical and sequential multi-scale phenomena in the overall characteristics or attributes of the system. The irreversibility of reductionism in complex systems management illustrates that the overall functionality and emergence of the management activities are inseparable and irreducible in the multi-scale sense. In other words, once any part of the overall functionality is compromised, the overall functionality and emergence of complex systems management activities will be damaged or even lost. Therefore, whether one aims to fully unleash the presumed overall functionality of the system or optimize the system's functionality, it is necessary to engage in meticulous analysis and reconstruction of the information transmission and transformation between hierarchical structures within the system based on a multi-scale mindset, in order to achieve optimization of the overall functionality of complex systems management.

Next, we will use the multi-scale implications of the overall functionality of complex systems management as an example to illustrate this.

What kind of functionality does an artificially constructed complex system in complex systems management have that best reflects the intentions and objectives of human management? In practice, these artificially constructed complex systems often exhibit large temporal and spatial scales due to their long life cycles or large sizes, which also contribute to the high complexity of the "complex systems management-environmental composite system." Based on this, we can analyze the multi-scale characteristics of the functionality of this artificially constructed complex system (referred to as complex systems management functionality) using time scale and the complex degree of the "complex systems management-environmental composite system" as dimensions (as shown in the Fig. 7.2 below).

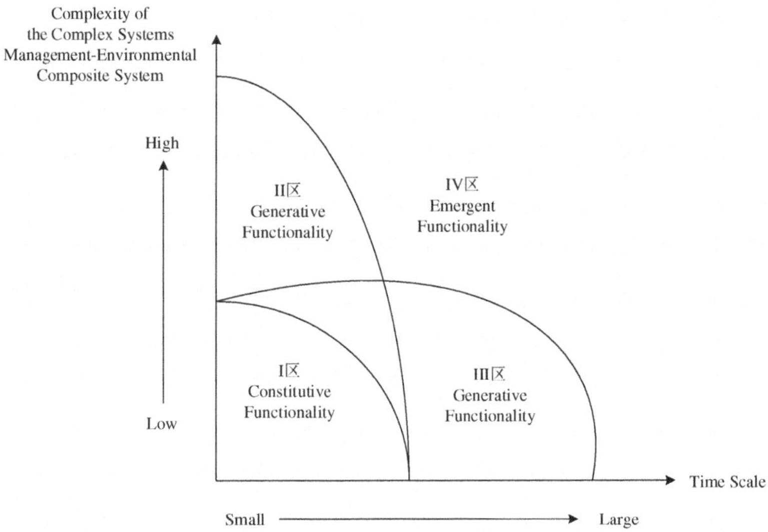

Fig. 7.2 Classification of complex systems management functionality

As shown in Fig. 7.2, we can classify the types of complex systems management functionality into different scale ranges. Region I represents the part of functionality in complex systems management that involves smaller time scales and lower complexity of the complex systems management-environmental composite system. In this region, the composite environment is relatively stable, and the disturbances and impacts caused by environmental changes on the normal release of management functionality are relatively small. The functionality in this region mainly consists of a basic and direct physical functionality formed by the system's hard resources, known as constitutive functionality. For example, the functionality of a completed large bridge project belongs to constitutive functionality.

Region II involves relatively small time scales in the context of complex systems management, but it encompasses the more complex part of the management-environment composite system. In this region, although the composite system experiences less variation over time and fewer disturbances and impacts on normal management functions, new scenario phenomena may emerge, directly or indirectly creating new composite system functions. These functions are not the physical functions initially assumed by the complex systems management subject but rather new functions that the complex systems management-environmental composite system self-organizes at a higher holistic level, generally referred to as generative functions. Generative functions typically do not belong to the most basic and direct preset functions of complex systems management but are indirect functions that radiate, derive, and expand from constitutive functions: For example, population clustering functions formed after the completion of major transportation projects or the economic boost from environmental management to real estate and tourism industries. Typically, in Region II, there are both constitutive and generative functions.

Region III represents the part of functionality in complex systems management that involves larger time scales but lower complexity of the complex systems management-environment system. In this region, the deep uncertainties associated with the large time scale may lead to significant changes and evolution in the composite environment, which may cause substantial disturbances and impacts on the functionality of the subject. It can also lead to new functional derivations and expansions in the complex systems management-environmental composite system, which are generative functionalities. For example, after the formation of new urbanization, long-term, gradual accumulations lead to regional population migration and cultural variations. Therefore, in Region III, both constitutive and generative functionalities exist.

Region IV represents the part of functionality in complex systems management that involves larger time scales and higher complexity of the complex systems management-environmental composite system. In this case, in addition to the independent influences of large time scales and complexity of the complex systems management-environmental composite system, there is also the interaction between the two, resulting in more complex scenario phenomena at different levels. For example, as complex systems management functionalities are continuously released and interact with new complex systems, the range of functional influence expands from local to global. This significantly increases the transmission distance of functional chains and may involve hierarchical leaps in functional causality, making the "causal" relationships between the functionalities at both ends of the chain more implicit. At this point, the "emergent," "peculiar," and "unnoticeable" functionalities generated by the new complex system will far exceed constitutive and generative functionalities. Correspondingly, their unpredictability and inexplicability will be greatly enhanced. They appear more as emergent or latent functionalities, collectively referred to as emergent functionalities. For example, a large-scale water conservancy dam project may lead to reduced water velocity upstream, causing changes in local ecological environments and promoting the growth of microorganisms and parasites. This results in the spread of infectious diseases and severe damage to human health and living environments, further triggering large-scale population migrations and the abandonment of previously prosperous population settlements. Additionally, the water diversion effect of the dam project, combined with reduced rainfall in the downstream catchment area, severely affects the "microclimate" and frequently causes drying up of lakes, leading to deserted towns and population migrations. Therefore, in Region IV, there are constitutive, generative, and emergent functionalities.

The overall form and characteristics of complex systems management functionalities mentioned above remind us that although management decision-makers always strive to design and construct comprehensive functionalities that are both extensive and good from their own perspectives, the inherent complex wholeness of the artificially constructed complex systems in complex systems management means that in addition to directly constructing basic functionalities, other types of functionalities may also emerge and manifest. Especially alongside the emergence of "good" functionalities, there is a significant possibility of "not-so-good" or "very bad" functionalities appearing. This tells us that while constructing complex systems management

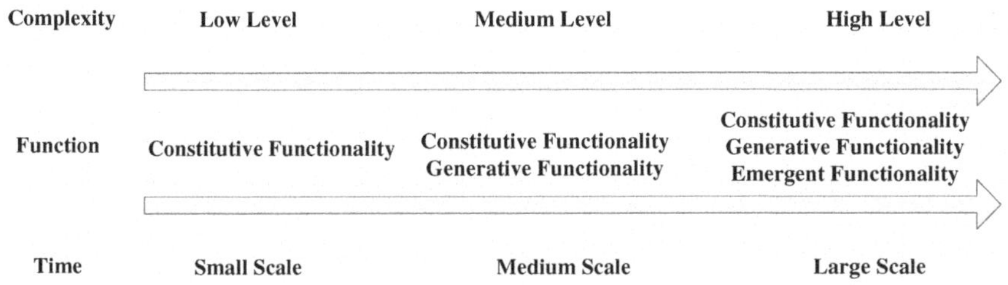

Fig. 7.3 The functional spectrum in complex systems management

explicit functionalities, we may unknowingly sow the "seeds" of latent functionalities, and whether these seeds will bear "bitter fruits" is often unknown. We need to pay special attention to and be vigilant about the possibility of generating emerging functionalities that have serious harmful consequences. Their appearance can significantly damage the initial functional expectations of complex systems management and lead to significant deviations and errors in management decisions.

From the distribution of complex systems management functionalities mentioned above, based on changes in time scales and the complexity of the complex systems management-environmental composite system, complex systems management functionalities exhibit a general trend of expansion from constitutive functionalities to generative functionalities and emergent functionalities. This can be seen as a universal law of the functional structure of complex systems management (as shown in the Fig. 7.3). We refer to the orderly arrangement of these multi-scale holistic functions in complex systems management as the spectrum of holistic functions.

The functional spectrum is an important attribute of managing the overall functionality of complex systems. It demonstrates that the holistic functionality of complex systems management can be decomposed and presented as a more refined and diverse complex microstructure in certain dimensions.

Initially, the assumed functions of complex systems management reflected human recognition of management goals. Therefore, the functional spectrum is also a concept closely associated with the management goals (purposefulness) of humans. Typically, as management evolves from general systems to complex systems, the subject's management goals also transition from single objectives or multiple objectives to a system of objectives. At this point, the dimensions or levels of the subject's goals continue to increase. Now, the concept of the functional spectrum not only enriches the meaning of complex systems management goals in terms of dimensions and levels but also expands and deepens the subtle structures of management goals on the same dimension.

In simple terms, the complex systems management objectives of the subject not only differ in hierarchy and dimension but also have a "spectrum" difference in multiple scales. The concept of functional spectrum tells us that even if people's designs for complex systems management objectives are "well-intentioned" and "thoughtful," there is a possibility of "going in the opposite direction" due to the generation and emergence of new functions. This not only enriches our understanding of the content of complex systems management objectives but also informs us that the design of complex systems management objectives carries a much higher risk compared to general systems management. In this sense, the functional spectrum in complex systems management can also be understood as the "objective spectrum," where some objectives are constitutive, some are generative, and some may even be emergent. In other words, complex systems management objectives include both the components of the subject itself and the multi-scale "self-organizing" components of human-made complex systems.

From the above analysis, it can be seen that there is a widespread existence of multi-scale phenomena in complex systems management and the significance of multi-scale as a complexity thinking and holistic behavioral criterion for complex systems management. The core idea is that in the context of complex systems management issues, there are certain order trends in the characteristics of management elements in a dimension, and this trend leads to different properties and features exhibited by these management elements. The differences in these properties and features need to be treated differently in practical activities and theoretical thinking of complex systems management, and they should not be treated as "one-size-fits-all." It is necessary to pay attention to their actual existence and the impact they cause.

In practical activities and theoretical research on complex systems management, practice has proven that establishing and using the concept of multi-scale can help us gain a more refined understanding, analyze, and address the complexity and wholeness of complex systems management issues. It reveals how to break the irreversibility of reductionism and discover deep-seated management principles, promoting the formation of behavioral criteria and enhancing the ability to manage complex wholeness.

7.5 Adaptability

The core content of this section is to discuss the close relationship between the concept of adaptability and the irreversibility of reductionism in complex systems management, as well as its important role in the management process.

7.5.1 Overview of Adaptability

The management subject and management problem are the most important and fundamental "poles" of complex systems management activities. Therefore, the irreversibility of reductionism in management is fully embodied and reflected through their behaviors and characteristics. For example, the subject must respond to changes in a complex environment and improve its ability to control the complexity of problems through a series of coherent, proactive actions. Meanwhile, management problems release various forms of complex wholeness due to the complex systems "*management-environmental composite systems*," fundamental uncertainty, multi-scale, and more. Management subjects and management problems exhibit the properties of complex systems management in such a mutually responsive relationship.

The following analysis shows that, from the perspectives of the overall complex systems management, the behavioral guidelines of the subject, capabilities, and the quality of solutions to management problems, "adaptability" plays a fundamental role in the complex systems management theoretical framework.

The term "adaptation" derives from the Latin word "adaptatus," which means adjustment or change. Although the specific meanings of adaptation differ across disciplines, their basic connotation remains the same. It refers to the proactive change in a subject's characteristics, behaviors, organizational patterns, and functions in response to changes in the external environment and conditions. This allows the subject to maintain harmony with the new environment, continue to survive, develop, play a role, and enhance its capabilities. This phenomenon and process are referred to as adaptation, and the magnitude of the adaptation ability is called adaptability.

Based on this basic concept, it is not difficult to understand that, in complex systems management activities, the subject enhances its cognitive, analytical, and control capabilities of complex wholeness through its adaptive behavior. Moreover, due to the variability and diversity of the complex systems management environment and activities, as well as the autonomy and adaptability of the subject's behavior, various types of complexity are created.

The first type of adaptability: Subject behavior adaptability

In complex systems management activities, the subject needs to actively adapt to the changes in the environment and problems, and grasp the ability to cognize, analyze, and control complex wholeness. However, any "individual" serving as a management subject has only partial and limited knowledge. Even a "group of individuals" involved in management cannot guarantee sufficient capabilities to address complex problems. This is a common situation in complex systems management activities. Therefore, the subject cannot rely solely on a simplistic and passive "stimulus–response" behavior principle. Instead, it needs to actively interact with various aspects, engage in continuous learning and experience accumulation processes, and develop and enhance its ability to address the complexity of management problems. For example, dealing with fundamental uncertainty requires solving uncertainty problems through uncertain pathways. In this process, the subject must learn and deepen its understanding of the essential attributes of the problem, progressing from ignorance to knowledge, from limited knowledge to more extensive knowledge, from partial knowledge to comprehensive knowledge, and from superficial knowledge to profound knowledge. In this way, the subject deepens its cognition of complex problems. Additionally, the subject will adjust and improve its management concepts, goals, organizational structures, and behavioral patterns appropriately and timely according to the needs of management activities, enhancing its analytical and problem-solving capabilities. It should be noted that this adaptability mainly stems from the self-learning and self-organizing abilities among the subjects within the complex systems management organization. It represents a type of proactive, adaptive learning behavior of the subject.

Indeed, the capabilities of an individual are never static or rigid. Practical capabilities that become truly evident in complex systems management activities emerge through adaptive learning and the accumulation of experience during the management process. This occurs under the combined influence of both theoretical and practical aspects of management thinking. This is mainly because new capabilities are often obtained through changes in the subject's behavior, organizational structures, and even organizational restructuring. Therefore, improving capabilities is both the purpose and the result of the subject's adaptive behavior.

In summary, the following conclusions can be drawn:

(1) Adaptability is the basic behavioral criterion for the management subject to cope with the complex wholeness of the environment and problems in complex systems management activities.
(2) The strength of the subject's adaptability is an important indicator of its capability and level of managing complexity.
(3) Adaptability is the result of the subject's self-learning and self-organizing capabilities. It represents the proactive and anticipatory response of the subject to the complexity of management problems.

The second type of adaptability: Adaptability of management solutions

The improvement of adaptability in the management subject's capabilities is not the ultimate goal; the ultimate goal is for the subject to be able to produce high-level and high-quality management solutions based on environmental changes.

The fundamental uncertainty and multi-scale nature of complex systems management activities require the functionality and effectiveness of important management and decision-making solutions to remain valid in the dynamic changes of the environment, including the composite system environment. This means that when the environment undergoes dynamic evolution, the functionality and utility of the solutions should remain robust without affecting the normal release of their functions. This quality of management solutions reveals another type of adaptability in complex systems management, called the adaptability of management solutions to environmental (including the composite system environment) changes, known as the adaptability of management solutions.

According to the principles of systems, any management solution for complex systems is an artificially designed complex system, and the utility of the solution represents the overall functionality of this system. In general, any solution has a range of environmental changes corresponding to its normal release of utility. If this range is large, it indicates that the solution has a strong ability to adapt to external environmental changes (resistance to interference) or, in other words, a strong adaptability to environmental changes. From the perspective of the empowering quality of management solutions, the larger the environmental changes, the lower the solution's ability to resist interference. This means that the adaptability of the solution weakens; just as a person's health deteriorates, their ability to resist changes in the external environment weakens. However, due to the objective nature of environmental changes in complex systems management, we cannot expect the changes to be limited to a small range that matches our subjective expectations. Thus, the ability of a complex system management scheme to adapt to environmental changes becomes an important measure of its quality. It also serves as a critical indicator for evaluating the competency of the decision-makers managing the system.

It is evident that the adaptability of management solutions is different from the adaptability of the behavioral subject. It is not an "active" response of the living subject to environmental changes but a reflection of the attribute of the adaptability capacity result of the behavioral subject, i.e., the reflection of the quality of the artificially designed complex system. Its essence is a holistic measure of the coupling degree between the functional adaptability of the management solution and the dynamic changes in the complex systems management environment.

Thus, we need to establish a set of methods to measure this adaptability. For example, research is needed on how to predict the complex overall changes in the complex systems management environment and how to measure the degree of coupling between solution functionality and environmental changes. Obviously, these techniques and methods are significantly different from the conventional prediction and evaluation methods used in general systemic management because these new problems are based on the overall behavioral and situational issues of the management subject under fundamental uncertainty, requiring methodological innovation to solve them.

Now, by integrating the above two types of adaptability, we form a logical chain based on the concept of adaptability: "Behavioral adaptability of the management subject-improvement of behavioral capacity-formation of behavioral results (formation of management solutions)-adaptability of management solutions."

It is not difficult to understand that the adaptability of the behavioral subject in management is manifested in various aspects of complex systems management activities, such as the adaptability of the management organization, the adaptability of the formation path of management solutions, and the adaptability of management objectives, among others. However, we believe that, for complex systems management activities, the adaptability of the behavioral subject is the most fundamental and dominant, while the adaptability of the management solutions to environmental changes is the quality representation of the behavioral subject's adaptability, and it is one of the most comprehensive attributes that reflects the quality of complex systems management. Therefore, we consider these two types of adaptability as the basic content of the concept of adaptability in the theory of complex systems management, while the other contents of adaptability are considered extensions and derivatives of these two basic types. This approach allows for a more concise representation of the basic concepts of the theory.

7.5.2 Significance of Behavioral Adaptability of the Subject

As mentioned earlier, the adaptability of the behavioral subject in management gradually improves through self-learning and continuous experience accumulation in the interactive processes between the subject and the environment, between subjects, and between the subject and the problem. Thus, the self-learning behavior of the subject becomes an important prerequisite and primary path for behavioral adaptability. Since the management subject in complex systems is generally composed of multiple subjects forming a "group," the self-learning of the subject should be understood as the self-learning behavior of the "group." An important indicator of this behavior is the formation of consensus within the subject group regarding complex problems and their solutions. In the following, we will provide a detailed analysis of this process.

Solving complex problems in complex systems management activities requires the formation of consensus within the subject group. Initially, there are often differences in cognition among different subjects, resulting in the phenomenon of "non-consensus" in the early stages of the cognitive process of the subject group regarding complex problems. At this stage, each individual's cognition generally contains both correct aspects of their knowledge and wisdom, as well as some degree of bias and errors. Therefore, it is necessary to gather and distill the scientific and correct aspects of different subjects' cognition while improving and correcting the biased and erroneous parts. The self-learning behavior of the subject group involves repeatedly revising, supplementing, and improving the cognition of "non-consensus," thereby continuously enhancing the scientific nature and focus of collective cognition within the subject group. Gradually, this leads to the formation of a consensus process within the subject group. The final consensus reached by the subject group is more scientific, comprehensive, and profound than any previous non-consensus, illustrating the significance of adaptive learning behavior for the subjects.

7.5.3 Interpretation of Self-organization Phenomena in Systems

From a systems theory perspective, "self-organization" refers to a process where a system, driven by its intrinsic mechanisms, autonomously evolves from simplicity to complexity, from coarseness to refinement, continuously enhancing its level of organization and order. Systems possessing self-organizing properties are termed self-organizing systems, and they are widely found in multiple domains, including physical systems, life systems, and social systems. The concept of self-organization originated in the late 1960s. The founder of Synergetics, Haken (1988), provided an example to explain it: "For example, a group of workers, without external commands but through a kind of mutual understanding, collaborate and fulfill their responsibilities in the production process. We call this process self-organization." The self-organization theory in the field of complexity science studies the laws of transition from disorder to order in systems.

Self-organization theory in the field of complexity science holds significant scientific value, particularly in revealing the unified mechanisms underlying complex self-organizing phenomena in various interdisciplinary fields of natural science. These frameworks have also been extended to the fields of social and economic sciences, playing a crucial role in transforming thinking and advancing theoretical development in these disciplines. Moreover, as the concept of self-organization closely resembles the concept of adaptability in complex systems management, it is necessary to provide a comparative explanation of these two concepts here:

1. The self-organization theory in complexity science primarily focuses on the phenomenon and problem of how material systems in natural science, such as physics, automatically transition from disorder to order, from lower level order to higher level order. The understanding of the problem and the basic principles heavily rely on specific theories in disciplines like physics and related concepts of complexity, and they are interpreted using additional complexity terms.
2. In terms of description and analysis methods, the self-organization theory in complexity science mainly utilizes dynamical equations as mathematical models. Since there are no specific external forces, the mathematical models of self-organizing systems can only be homogeneous equations, giving the object ontology a strong determinism. This methodological approach is not suitable for dealing with semi-structured or non-structured situations such as human psychology, behavior, culture, and values in social systems.
3. John Holland's famous assertion about "adaptive complexity" in 1995 largely summarizes the contribution of adaptive behavior of complex system subjects to the formation and evolution of system complexity [1]. The self-organization theory in complexity science mainly focuses on the overall ordered evolution of purely natural systems such as physics. If we inject purposefulness into this process with material elements, self-organization can also be seen as the system's adaptation to environmental changes. In the study of socio-economic management systems, human adaptation to the environment and a certain purposefulness are placed at a foundational position. In this sense, it is not only adaptive behavior that creates complexity, but we also need to experience and reveal the adaptive behavior that creates complex wholeness in the research of complex systems management.

Given the above points, it can be seen that for complex systems management in the field of socio-economic systems, the concept of adaptability is more essential and versatile than the self-organization concept of complexity science.

7.6 Scenarios

7.6.1 Overview of Scenarios

According to general intuitive perception, a scenario is a coherent "story" that unfolds with continuity in terms of the presence of objects, events, subjects, objects, connections, causality, changes, background, and plot. It is easy to understand that a scenario consists of numerous and diverse elements, with complex forms of element relationships, orderly structures, complete processes, and uncertain future development. Therefore, scenarios exhibit a distinct characteristic of complex wholeness in terms of their ontological attributes.

Based on complex systems thinking, the scenario of complex systems management activities refers to the state and orderly evolution of the management activity's elements and possible evolutionary paths under specific environmental contexts based on various mechanisms or related interactions.

This concept of scenarios emphasizes the following points:

(1) At any given point in complex systems management activities, the generation and evolution of scenarios exist in the "present," "past," and "future," forming a coherent, dynamic, and inseparable process.
(2) The generation and development of scenario forms are primarily the comprehensive results of the complex entirety, including humans, activities, and the environment, under the joint effects of organization and self-organization. It has extensive complexity, which is not possessed by scenarios in the general sense. The emergent complexity, the complex relationships between activity elements, the adaptive behavior of subjects, and various fundamental uncertainties are closely related.
(3) Various means and methods need to be applied to describe and analyze the complex wholeness of complex systems management scenarios. In today's highly developed information technology, we have more ways to "describe" scenarios, such as scenarios being a piece of text, some data, an image, a video, or an integration and fusion of these methods.

Now, let's provide an example to deepen the understanding of the concept of complex systems management scenarios.

Improving the navigation conditions and increasing the navigational capacity of the Yangtze River is one of the important functions of the Three Gorges Project in China. This requires predicting the future scenarios of water transportation in the surrounding areas of the project and designing the capacity of ship locks based on these predictions. In fact, since the impoundment of the Three Gorges Project in 2003, the annual growth rate of cargo throughput through the ship locks has reached 12.2%, and, in 2011, it exceeded 100 million tons. This actual scenario significantly exceeded the initial engineering design capacity by 19 years [2].

With the development of the socio-economy, the conflict between the actual demand for lock passage and the gap in actual passable capacity has become increasingly prominent, becoming a bottleneck problem restricting the development of the Yangtze River waterway. Based on this, experts have proposed the idea of constructing a new waterway for the Three Gorges Project, and this new project is complex in terms of site selection, technology, population resettlement, and ecological and environmental protection. It requires a long time to complete a series of preparatory work. Thus, it can be seen how important the accuracy of predicting future environmental scenarios is for the scientific decision-making and engineering robustness of major projects.

In addition, after the completion of the Three Gorges Project, a new composite system of major project management and the environment has been formed. This new composite system may give rise to new scenarios that have not occurred before due to the fundamental uncertainty of the environment. In fact, after the impoundment of the Three Gorges Project, the water downstream of the Yangtze River became cleaner with less sediment, but the water level of the Yangtze River decreased. As a result, the water intake pipes for residential and industrial water supply, which were originally stable, were unable to draw water due to the lowered water level. Additionally, riverbank collapses occurred along the Yangtze River, leading to instability in previously stable areas and the possibility of further potential disasters. This emergence of new scenarios in the "major project management-environment composite system" has a negative impact on the normal realization of the major project's functions and goals [3].

The actual scenarios of the Three Gorges Project reflect, in a concrete and vivid manner, the significant value of different types of scenario phenomena in the decision-making of major projects and risk prevention. These scenarios illustrate the generation and evolution of the project from real situations and highlight their importance.

7.6.2 Key Points of Scenario Thinking in the Context of Complexity and Wholeness

In the previous section, we primarily perceived the basic manifestations and characteristics of complex systems management scenarios from a vivid and intuitive perspective. Furthermore, it is necessary to have a scientific understanding of scenarios at the level of rational thinking, which is referred to as scenario cognition in the context of complex wholeness. Based on the aforementioned perception of scenarios, the key points of this cognitive process include the following:

(1) The scenarios in complex systems management activities form a coherent dynamic evolution process on the timeline, and it is not possible to isolate the "future" from the "present" and "past." However, the "future" is not entirely determined by the "present" and "past." Whether it's from past scenarios to

present scenarios or from past and present scenarios to future scenarios, both the final scenario state and the path of scenario evolution are uncertain. Therefore, it is not possible to entirely extrapolate future scenarios based on established rules and data, nor can one artificially constrain a single determined path as the real path to achieving future scenarios.

(2) Complex systems management scenarios encompass various real elements, such as physical, managerial, social, organizational, individual, behavioral, cultural, and psychological elements. Therefore, the scenario structure and mechanism consist of structural components reflecting objective natural and technological laws, semi-structural components reflecting social laws such as organization and behavior, and non-structural components reflecting cultural and value tendencies. Furthermore, there is a synthesis between these three aspects. In the dynamic changes of scenarios, there are both the influences of objective laws and the effects of human subjective will, as well as the self-organizing effects of the system. In short, scenarios in the context of complex integrative significance involve compositional components designed and defined by humans, newly associated and structured elements of the complex system, and emergent components resulting from the complexity of the whole. This requires us to use a cross-border, interdisciplinary, and comprehensive integration approach when describing and abstracting complex systems management scenarios.

(3) In general, scenarios are common and widespread phenomena, but scenarios in complex systems management are concepts with complex holistic implications. Specifically, scenario state spaces often have "ultra-high dimensions," scenario evolution paths are deeply uncertain, the subjects have incomplete knowledge, and scenarios exhibit characteristics such as emergence and evolution. All of these are closely related and mutually causal with attributes such as multi-scale, adaptability, and fundamental uncertainty in complex systems management activities. They are challenging for traditional scenario theories to explain and address. In summary, the transition from general scenarios to scenarios in complex systems management involves an expansion and deepening trend from systematicity to complexity.

(4) The fundamental uncertainty of complex systems management activities results in a "sufficiently large" space of possible future scenarios. This means that when we look forward on the timeline from the "today" point, we cannot find a definite specific scenario point corresponding to the scenario point of "today." What we can imagine is the existence of a sufficiently large future scenario subspace, where every point in this subspace could be the "tomorrow" of our scenario. Generally, the larger the time scale, the larger the future scenario space due to inherent uncertainty. Thus, our predictions of the future become more complex and difficult. Therefore, it is difficult for the subject to accurately predict future scenarios in a general sense. Instead, the focus should be on identifying possible future scenarios and understanding why they are possible.

(5) Due to the complex wholeness, we cannot solely rely on the subjective setting of different parameters, artificial limitation of uncertainty, or blindly expanding the future scenario space to predict and infer complex systems management

scenarios. Firstly, human cognition and imagination are limited, and we do not have the ability to possess complete knowledge about future scenarios. Secondly, the conversion and connection between the current and future scenarios are often lengthy, complex, and not fully apparent, and we do not have the ability to fully grasp all the mechanisms involved in scenario transformation and evolution. Thirdly, based on the fundamental uncertainty of future scenarios and the scenario realization path, the complex systems thinking paradigm suggests that our goal should not necessarily be a precise prediction of future scenarios but rather the recognition of the possibility of extremely abnormal scenarios occurring in the future and causing significant harm. In other words, the subject should focus more on establishing "bottom-line thinking" to prevent future scenario disasters.

7.7 Organizational Platform with Flexible Capability

Like all management activities, the core elements of complex systems management activities are the management subjects and the management organizations composed of these subjects.

7.7.1 Dominant Subject

The so-called dominant subject of complex systems management can be understood as a group of people who are responsible for specific management tasks (functions) at various stages of complex systems management. Therefore, management subjects are often referred to as subject groups. In reality, due to the heterogeneity and diversification of management subjects, it is necessary to have a subject within the subject group that plays a more leading and dominant role. This subject may have a higher administrative position, stronger resource advantages, or other reasons. In comparison, it has more authority and discretion in coordinating between subjects and has a greater authoritative and discretionary power. We refer to this subject as the "dominant subject" within the subject group of complex systems management. "Dominant" implies that it is better able to ensure and maintain orderly relationships among the subjects and the overall behavior of the system. The general subjects in the subject group of complex systems management may enter and exit the group based on changes in management functions and tasks, but the dominant subject is typically stable throughout the entire process of complex systems management.

The basic meanings of the concepts of complex system managing subject and principal subject can be summarized as follows:

(1) The complex system managing a subject is generally a concept of a "group" consisting of multiple relatively independent autonomous subjects. Each subject

carries out its management tasks within certain rules and processes, and within the subject group, there is typically a dominant subject that plays a leading role.

(2) Both the subject group and individual subjects can integrate and allocate management resources and determine and implement problem-solving approaches through compliant and appropriate means. When encountering difficult problems, the subjects negotiate in various ways, form and expand the scope of consensus, and demonstrate stronger management capabilities.

(3) The dominant subject adjusts relevant subjects based on the actual needs of the complex systems management environment and tasks, achieving optimization of the subject group. In other words, the composition and structure of the subject group in complex systems management are generally flexible and subject to changes throughout the entire management process.

(4) The capacity of the complex system to manage subjects refers to the overall performance of the subject group. It is neither the individual capacity of the dominant subject nor a simple aggregation of the capacities of individual subjects. Instead, it is the emergence of the overall capacity of the subject group system under the leadership of the dominant subject and institutional constraints.

7.7.2 Organizational Platform with Flexible Capability

As mentioned above, the complex systems management subject group is essentially an organizational system based on individuals (people), and the tasks of complex systems management are implemented and executed by such an organizational system.

Generally, in any management organizational system, the subject group forms an overall capability to handle problems through knowledge complementarity, experience complementarity, skill complementarity, wisdom complementarity, as well as effective communication and coordination mechanisms. The same applies to the complex systems management organizational system, where the overall organizational capability emerges from effective communication and coordination mechanisms within the management organization.

In relatively simple management activities, as long as the functional positions of the management organization are designed according to the needs of management tasks, with clear responsibilities for each position, the interrelationships between positions, and corresponding constraints and incentives, such a structurally rigid and behaviorally normative management organization possesses all the necessary management capabilities for conducting management activities. However, the activities and problems in complex systems management are much more complex. For example, major project management is a typical case of complex systems management. In the early stages of management, overall planning, overall demonstration, and overall design of the project are required, including the analysis of the relationship between the project and the environment, the design of project value and objectives, and the determination of project financing models and construction plans. During the construction phase, a comprehensive analysis of complex management problems is needed to establish construction goals, integrate resources, establish project management organizations, optimize technical routes, and ensure on-site execution capabilities, among other tasks. It is evident that these problems are not only highly complex in themselves but also closely interconnected. Therefore, to complete this series of management tasks comprehensively and effectively, a "rigid" management organization alone cannot directly and fully accomplish it. The management organization needs to demonstrate strong overall capabilities in the following aspects:

(1) Dynamically select and combine various unit entities within the management subject group and fully utilize organizational and self-organizational effects to enable the subject group to form the capability to handle the complex entirety of management problems based on different problem requirements. This is essentially the environmental (institutional) design of the complex systems management organization.
(2) Establish internal work rules and processes within the subject group to ensure the formation and operation of the aforementioned capability. This essentially involves the conditional (mechanism) design of the complex systems management organization.

In terms of overall and holistic significance, the main function of such a management organization is not to directly propose and determine specific solutions to complex problems but to provide the necessary environment and conditions for solving complexity management problems. Under the support of such an environment and conditions, the management organization can demonstrate its capability to handle complex problems as a whole. For example, the management organization can adaptively allocate and reorganize various management resources and utilize these resources to address complex problems. Therefore, such a management organization is essentially a "platform" that provides capabilities.

The term "platform" is borrowed from computer science, such as software development platforms and operating platforms, which primarily refer to an environment and conditions. With a "platform," it becomes possible to support, expand, and realize new system functionalities. Based on the analysis above, complex systems management organizations can be understood as a "platform" that generates and expands complex overall management capabilities because their essential function is not to directly or primarily provide specific methods and solutions to complex systems management problems but to provide the environment and conditions for the formation of capabilities and solutions.

So, what is the core capability and function of a complex systems management organizational platform? It should be flexible capability. As seen from the connotations of several core concepts presented earlier in this chapter, in order to effectively address the complex wholeness, an approach to respond is by continuously and passively supplementing its own resources and enhancing its own capabilities based on fundamentally unpredictable and uncertain scenarios. This approach would lead to unlimited expansion of the platform's scale and increasingly complex structure, which is not realistic. Another approach to respond is by using the platform's limited scale and resources but through the flexibility of its own structure and functions, obtaining adaptive mechanisms and capabilities to deal with such challenges. The generation of this adaptability does not originate from the design of an external organizer but rather from the self-organization of the organizational platform itself through its inherent adaptive capabilities., which is referred to as the flexible capability of the organizational platform. A management organization with flexible capability is called an organizational platform with flexible capability.

Chinese philosophy emphasizes "using softness to overcome hardness." Over 2,000 years ago, the Chinese philosopher Lao Tzu stated that the softest things in the world can conquer the hardest. For example, water is considered one of the weakest elements in nature, yet a drop of water can wear away stone, showing its ability to overcome significant challenges. However, to achieve the principle of using softness to overcome hardness, one must comprehensively analyze the complexity of a problem, identify its key aspects, integrate and allocate resources effectively, adapt and reorganize these resources as needed, and ensure that these capabilities remain adaptive and enduring as the environment and problems change. All of these capabilities collectively form the concept of "soft capability" within the framework of a complex systems management organization platform. The perspective on the soft capability of a complex systems management organization platform inspires us not to simply address the complexity of problems by building increasingly complex management structures. Instead, this implies the significance of developing a versatile capability platform characterized by adaptability, dynamism, and evolution and merging it with essential stringent rules and technologies to realize a balanced integration of flexibility and rigidity. Differentiating based on the "plasticity" of management platform capabilities, simple systems management platform capabilities are generally "rigid," while those of complex systems management organizations exhibit a significant degree of "flexibility."

7.8 Logical Correlations of Conceptual Framework

In this chapter, we propose the above-mentioned seven concepts as the fundamental core concepts of the theory of complex systems management. These concepts are a condensation and abstraction of the attributes of various key stages and important components of complex systems management practices. They are the basic units of people's understanding of complex systems management activities and can reflect the identity, universality, and regularity of complex systems management activities and problem attributes.

When presenting the above basic concepts, specific explanations and descriptions were provided for their practical management activity backgrounds, including phenomena, scenarios, and subject behaviors. This demonstrates that these seven core concepts all stem from the practical activities, phenomena, and problems of complex systems management. Therefore, they all have the "root" of the complex holistic attribute of complex systems management, meaning that all core concepts follow the principles of complex systems thinking:

(1) Complex systems management-environmental composite system. This concept abstracts the new composite system formed by integrating the artificially constructed complex system of complex systems management with the original environment. In this composite system, in addition to the original social, economic, engineering, and environmental ecology, new complex artificial systems are added. As a result, the "interactions" and "feedback" between the artificial complex system, the original environmental system, and the composite system will lead to new and profound essential attributes.
(2) Complex wholeness. This concept overall represents the essential attributes of complex systems management and the essential attributes of management environment, subject behavior, system relationships, and structure. It includes both the complex wholeness of the artificial systems in complex systems management and the complex wholeness of complex systems management-environmental composite systems. It encompasses the complexity of various stages and key elements of complex systems management activities, as well as their wholeness. Therefore, it provides guidance for accurately analyzing different types of complex wholeness issues and selecting different management approaches.
(3) Innate uncertainty. Innate uncertainty not only reveals the fundamental reasons for the emergence of complex systems management attributes but also provides a highly generalized description of the characteristics of subject behavior in complex systems management. Therefore, this concept deepens the condensation and description of the essential attributes of complex systems management.
(4) Multi-scale. This concept deepens and abstracts the phenomenon of distinguishable sequential changes in the form or characteristics of a class of management elements in complex systems management activities. It is of great significance for fine identification, analysis of complex systems management structures, and revealing the mechanisms of complex holistic integrity. In general, complex systems management activities and behaviors occur at different scales, so analysis and organization should be conducted in a multi-scale sense.
(5) Adaptability. This concept revolves around the complexity of subject behavior in complex systems management. It describes both the behavioral performance and characteristics of the subject and depicts the degree of overall coupling between the main results of subject behavior (proposed management plans) and environmental complexity. Therefore, this concept highlights the fundamental principles of subject behavior and the quality of behavioral outcomes, which are crucial in complex systems management activities.
(6) Scenario. This concept mainly refers to the macro states formed by the various elements of complex systems management activities under a certain environmental background, as well as the evolution of states and possible paths of evolution. In many cases, scenarios created by complex systems management activities themselves and human behavior give rise to the complex wholeness and the existence of phenomena that exhibit complex wholeness. The deepening

of the connotation of the scenario concept enhances our understanding of the complex wholeness of complex systems management activities.

(7) Flexible organizational platform. The management subject group in complex systems management is grounded in real-world management organizations engaging in actual management activities. Due to the complex wholeness of management problems, the core function of complex systems management organizations is no longer to directly solve various management problems but to provide an adaptive capability to handle the complex wholeness as a necessary environmental and conditional platform. This concept reveals the scientific connotation of complex systems management organizational models based on flexibility.

From the above analysis, it can be seen that the core concepts listed in this chapter are a condensation and abstraction of the complex holistic attributes of important elements and activity stages in complex systems management activities. Even though some scientific terms, such as multi-scale and scenario, already exist in other disciplines, they have been reconstructed and innovated here in terms of conceptual connotation. They not only introduce realism into complex systems management activities but also reflect the essential attributes of complex systems management activities.

Furthermore, the theory of complex systems management is the result of theoretical thinking about fundamental elements of management activities and issues, including the management environment, subjects, objects, goals, behaviors, and more. Therefore, it must adhere to the objective logic of management activities and issues in the real world and align with fundamental human thought processes. In other words, the core concepts of the theoretical framework should not be isolated or fragmented. They should not only provide comprehensive coverage of complex systems management activities and issues but also maintain close logical relationships among themselves. In the theoretical framework, core concepts should not only ensure that they originate from the practical aspects of complex systems management activities but also reflect strong physical, systemic, and logical correlations among these concepts. Without these connections, people would not be able to establish fundamental principles and scientific topics in the theoretical framework based on concepts and their combinations.

Now, let us examine whether the basic concepts proposed above have such multiform connections:

(1) Complex systems management-environmental composite system: the holistic abstraction of the complex systems management object (including the environment and complex holistic problem).
(2) Complex wholeness: the abstraction of the shared essential attributes of the complex systems management subject, object, and environment, directly reflecting the thinking principles of the theory of complex systems management.
(3) Innate uncertainty: the abstraction of the essential attributes of the complex systems management subject, behavior, and environmental characteristics, as well as the condensation of important causes of the essential attributes of complex systems management.

(4) Multi-scale: the abstraction of the criteria for complex systems management subject behavior and the quality of behavioral outcomes.
(5) Adaptability: the abstraction of the basic behavior and capability criteria of complex systems management subject and platform.
(6) Scenario: the abstraction of the phenomena carriers and attribute representation of complex systems management activities, environment, composite system form, and evolution.
(7) Flexible organizational platform: the abstraction of the functional attributes of complex systems management organizations.

It can be seen that these concepts cover the elements of complex systems management, such as the environment, subject, object, organization, goal, thinking principles, and behavioral criteria. They condense the core elements of complex systems management activities, particularly focusing on the core attribute of complex wholeness. Moreover, under certain logical correlations, they form the basic framework for complex systems management activities and complexity problems, clearly expressing their respective functions and roles in complex systems management activities.

The core concepts proposed in this chapter possess the required logical qualities of the theoretical system (see Fig. 7.4).

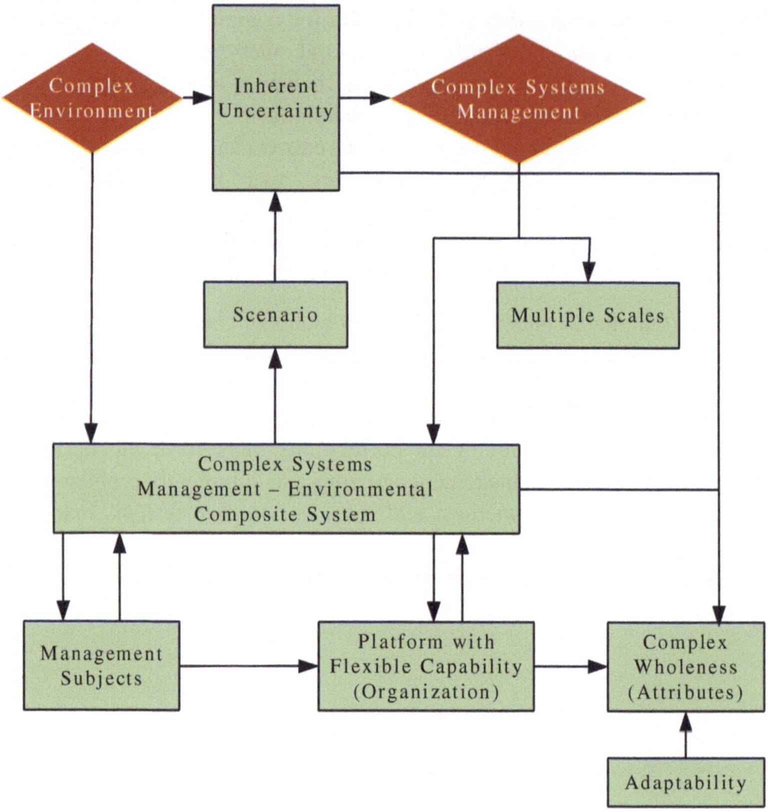

Fig. 7.4 The logical correlations diagram of core concepts

References

1. Holland JH. Hidden order: how adaptation builds complexity. Addison Wesley Longman Publishing Co., Inc.;1995.
2. Shuyan Guo SL, Tingzheng W. The three gorges water diversion project is an important development of the South-to-North water diversion project. Sci Technol Rev. 2003; 5:3–5.
3. Ding X. Research on decision-making methods for infrastructure engineering based on uncertainty;2019.

Chapter 8
Basic Principles of Complex Systems Management

The basic principles in the theory of complex systems management mainly involve the fundamental rules of management activities and subject behavior. They are the knowledge representation formed through the solidification and refinement of practical experience in complex systems management and the logical reasoning of core concepts. The formation of principles, especially the formation of a systematic framework, is an ongoing process that requires long-term exploration and continuous improvement.

8.1 Overview of Basic Principles

For an autonomous theoretical system of complex systems management, with the core concepts established, phenomena and problems in complex systems management practice can be analyzed based on these concepts or their combinations, combined with experience. The aim is to provide reasonable explanations and extract basic rules regarding physical relationships, logical relationships, causal relationships, and associative relationships in the phenomena and problems. This forms a declarative discourse known as principles. Most general phenomena and problems in complex systems management activities can be deduced, explained, and predicted using these principles. A principle is a relatively independent foundational knowledge unit, primarily addressing the behavioral principles, operational guidelines, and basic design for problem-solving paths concerning a specific class of complex systems management problems. Principles are typically expressed as affirmative statements.

The important role of principles in a theoretical system lies in the foundational and formalized rules they extract, which help people understand the essence of complex systems management activities accurately, in-depth, and not merely conceptually. They also establish behavioral guidelines and operational rules at the operational level.

Z. Sheng, *Outline of Complex Systems Management Theory—Based on Irreversibility of Reductionism Thinking*, International Series in Operations Research & Management Science 360, https://doi.org/10.1007/978-981-97-3361-3_8

Comparatively, some principles are more foundational, originating, and initial. They not only generalize the basic laws of a specific class of facts but also have guiding significance for various phenomena, problems, and even overarching theoretical thinking. Besides directly deducing theoretical conclusions, such principles can generate other principles. These principles are referred to as fundamental principles in the theory.

Basic principles are usually expressed as judgments and laws under certain conditions, and they concentrate on the logical deduction function of a theoretical system. The logical deduction function of basic principles is the "meta-function" of the theory. It can derive more specific judgmental-theoretical units, which can then form scientific issues and problems at a higher level.

In natural science theories, principles often manifest as causal laws in the physical world. However, there are not many such causal laws in the principles of complex systems management. This is because complex systems management activities and behaviors rarely possess clear and precise causal determinism, especially when they involve social aspects and human behavior. They are typically articulated as insights on addressing the key thoughts or behavioral principles of complex wholeness rather than as standardized processes and techniques at the technological and operational levels.

The theory of complex systems management is designed and developed at the intersection of natural science, engineering technology, social science, and humanities. Therefore, it possesses nature, sociality, and humanity simultaneously. Naturality primarily refers to the objectivity of natural laws and the normativity of technical operations in theory. Sociality refers to the holistic social characteristics in the theory and the social relationships of micro-level subjects. Humanity encompasses the psychological aspects, visions, and unique cultural values of individuals within the theory. Due to these reasons, the principles in the theory of complex systems management cannot be symbolized, formalized, and axiomatized like principles in natural science. Many times, they manifest as principles of management elements and subject behavior paradigms in a social context.

For example, the essential attribute of complex wholeness is the essence of complex systems management. Therefore, the "essence" of complex systems management theory lies in studying how the subject recognizes, analyzes, copes with, and harnesses this essential characteristic. It includes extracting and classifying various complex holistic phenomena and types and establishing principles and operational guidelines for targeted behaviors based on respecting the objective form and basic laws of complex wholeness.

Considering that the cognition of managing complex wholeness itself is "complex," it is necessary to propose corresponding basic principles from the aspects of the subject's basic cognitive thinking, behavioral principles, practical paradigms, and unique mechanisms.

Based on the composition of elements in complex systems management problems and constructing management activity paradigms, in order to improve the orderliness and effectiveness of complex systems management activities, people need to summarize and propose a series of behavioral-operational principles during the behavioral-operational process. Therefore, it is necessary to establish the "scenario-oriented" principle of complex systems management, which serves as the logical starting point for the study of complex systems management theory. Rounded in the fundamental characteristic of complex wholeness inherent in the management object, principles like "indivisibility relativity," "uniqueness contextualization," and "physical-system-management chain progression" are posited. Subsequently, the principle of "multi-scale management" is introduced to address phenomena occurring at multiple scales in management activities. Concerning guidelines for subject behavior, principles such as "adaptive selection" and "iterative generation" are advanced. Experience has demonstrated that these foundational principles are crucial for the theory of complex systems management.

8.2 Scenario-Oriented Principle

It is well known that problem orientation is a general principle in scientific research. However, for the study of complex systems management, we need to further reflect on this principle. When people initiate complex systems management activities, they first form an intuitive perception of a real problem through various senses and then abstract and refine the real problem into a scientific problem. Since the scenarios in complex systems management are the "substance" of real problems, any real problem exists within a scenario and is closely associated with it, forming an integrated whole. Therefore, both practical and theoretical problems are inherently linked to scenarios, possessing their "genes" and "lineage." Thus, when conducting theoretical research on real or scientific problems, it is necessary to separate them from specific scenarios.

From a theoretical perspective, this separation is reasonable and necessary because theoretical research involves abstract thinking, which abstracts the general principles of real problems. This requires abandoning the uniqueness and differences of individual specific details of the problem to reveal its essence and inherent laws. However, in this process, as we detach the imprints of scenarios attached to real problems, cutting the various connections between scenarios and scientific problems, it inevitably damages the authenticity of the attributes and characteristics of scientific problems. Thus, the scientific problems after detaching from scenarios are not entirely equivalent to the real problems in their original ecological scenarios. This results in a certain "loss" based on the essential or unique nature of the scenario; it is the cost paid to reveal the inherent laws of real problems [1]. This situation is particularly significant and serious for complex systems management, which possesses the attribute of complex wholeness. Therefore, we need to consider whether such operations are too "simplistic" and "rough" for the study of complex wholeness problems.

Now, let's switch viewpoints: the purpose of studying scientific problems is to solve real problems. However, if we directly apply the analytical conclusions and solutions of complex wholeness problems after detaching from scenarios to real problems, it may lead to serious inaccuracies or inappropriateness. To address this, one "compensatory" approach is to reintegrate the complex wholeness problems, along with the derived conclusions and solutions, "back into" the original scenarios. This allows the conclusions and solutions to be reintegrated, tested, and revised in the scenarios to the greatest extent possible. Only then can the inherent attributes and laws of the original real problems be restored through the "immersion" in scenarios. The true form and fundamental laws of a real complex wholeness problem are always the result of the combined effects and integration of abstract scientific problems and the scenario "substance." Therefore, in the process of theoretical research on complex systems management, if we strive to study scientific problems in the original scenarios or retain the role of the scenario matrix as much as possible, we can ensure the consistency between the "abstracted and refined" scientific problems and the essential attributes and fundamental laws of the "ecological" real problems. It can also minimize the damage to the attributes and fundamental laws caused by detaching from the scenario. Especially today, as the complex wholeness of complex systems management problems continues to increase, and the association between problem attributes and scenarios becomes increasingly close, it is even more necessary to study scientific problems in scenarios and reveal the attributes and fundamental laws of problems. Therefore, the study of scientific problems in complex systems management should further analyze and approach the problem within its corresponding scenario. As a result, the identified problem attributes, laws, and problem-solving solutions will be closest to the reality of the real problem. This clearly illustrates the guiding significancc of scenarios in theoretical research on complex systems management. Thus, being scenario-oriented becomes a fundamental-behavioral principle in complex systems management, a basic principle in the study of complex systems management.

So, how can we be scenario-oriented? First, from a logical perspective, the set of information for a problem should be a subset of the scenario information set. In this way, being scenario-oriented does not lead to a loss of information about scientific problems. Second, the scope and content of the scenario should be appropriately defined relative to the problem because excessively rich scenario information is not only unnecessary for problem research but also incurs high costs in scenario processing.

Therefore, the scenario-oriented principle does not mean an unrestricted expansion of the scenario space but seeks a balance between the cost of acquiring scenario elements and the precision requirements of problem research. Furthermore, since we are scenario-oriented, we must make the scenario not merely an abstract concept but observable, measurable, describable, and modelable. This involves the key technique of scenario modeling in complex systems management theory. Without this technique, if we only use traditional equations or statistical modeling techniques, the limitations of the technical capabilities will prevent the modeling of complex holistic scenarios.

In summary, in the theoretical research of complex systems management, when detaching real-world problems from their contextual existence to transform them into scientific problems, and during the process of analyzing the results and validating and refining the solutions to these scientific problems, we should not overlook the role and influence of the context on theoretical research outcomes. To address this, it is proposed to establish the following points:

(1) **Embeddedness of scenario reconstruction**. In the complexity of complex systems management theory, the scenarios corresponding to complexity problems are generally coherent in the temporal dimension. This means that the past, present, and future of scenarios evolve continuously. Therefore, in the study of complex systems management theory, researchers should not only focus on future scenarios but also consider past and present scenarios. In addition to scenario prediction, scenario reconstruction and reproduction should also be considered. In complex systems management activities, the formation and implementation of a management solution itself "embeds" a new complex system scenario into the original real-world scenario. This implies that we should not only recognize the uncertainty of future scenarios but also acknowledge that the past and present may have also been uncertain. Furthermore, when we connect the "future" with the "present" and the "past," it is easy to see that today includes the past, but today is not necessarily included in the past, and tomorrow is not completely contained within the past and today.
(2) **Inherent uncertainty of scenario prediction**. The future scenarios of complexity management problems are inherently complex and uncertain. Traditional scenario forecasting methods rely too heavily on decision-makers' subjective intuition, which can lead to human errors. Although managers can, to some extent, construct, predict, and imagine future scenarios based on experience, traditional knowledge, and deductive causality, it should not be assumed that humans can fully design and determine future scenarios based solely on their will and cognition. The future may bring unexpected and unimaginable scenarios that go beyond human imagination, posing significant potential risks to complex systems management.
(3) **Composite nature of scenario evolution**. The future scenarios in the complex systems management environment not only include the scenarios within the environment itself but also encompass new scenarios generated by the complex systems management-environmental composite system. In other words, these are new scenarios manifested by the physical system formed by complex systems management activities in conjunction with the original environmental system, emerging at the holistic level of the composite system, which did not exist in the original system. Therefore, management entities should not merely stand outside the management activities to observe the scenarios. Instead, they should recognize that, on the one hand, management activities are influenced by environmental scenarios, and, on the other hand, the management activities and behaviors of the management entities themselves can create new scenarios at the holistic level of the complex systems management-environmental composite system. It's like a person sitting on a boat, looking at the riverbank scenery.

While the shore is beautiful, the interaction between the person, the boat, and the scenery on the shore forms a new landscape altogether.

(4) **The path dependence of scenarios**. Any management activity has a retrospective effect. Therefore, under the dual influence of the original environmental scenario and the composite system scenario, the composition of the future scenario space and the evolutionary path to a specific scenario point in the future scenario space will be influenced by the dynamic evolution of these two types of scenarios, filled with inherent uncertainty. Therefore, managers need to take necessary precautions against scenario risks. They should not only focus on subjective scenarios that are perceived to have a greater likelihood of occurring (prospects), nor should they only focus on scenarios that they personally value and hope to occur (visions). Moreover, they should not subjectively exclude scenarios they do not wish to happen from the future scenario space.

Although the concept of scenarios or similar contextual concepts has appeared multiple times in other areas of management studies, the connotation expressed by the same academic language has an important relationship with its context. As mentioned above, scenarios appear in the context or scene of complex systems management, highlighting their complexity and integrality. From the perspective of complexity scenario thinking, we can see the potential impact of scenario detachment on real-world and scientific problems in complex systems management activities. According to academic logic, this detachment is inevitable in the methodological study of complex systems management. However, it requires us to either reintegrate abstract scientific problems back into their original scenarios for study to avoid the "deformation" of scientific problems due to scenario damage or, after studying abstract scientific problems, compensate for the impact of scenarios on research conclusions and solutions, and verify and improve them, thus "repairing" the damage caused by scenario detachment to the complexity and integrality of scientific problems. In recent years, relevant research has been conducted in academia, as reflected in Chap. 9 of this book on "robust management" and Chaps. 10–13 on "scenario modeling techniques," "scenario computability," and "data-driven scenario modeling," which all highlight the central role of scenario concepts and related principles in the study of complex systems management theory.

8.3 Principle of Indivisible Relativity

As mentioned earlier, the irreversibility of reductionism in complex systems management is also expressed as complex wholeness, non-additive wholeness, and indivisible wholeness. The term "indivisible" corresponds to "non-additive" because if it were additive and recoverable, the whole would be "divisible."

In practice, reductionism is generally divided into two parts: top-down decomposition and bottom-up integration. When we talk about the irreversibility of reductionism, we mainly refer to the inability to decompose the whole into multiple scattered parts and then integrate them back up to the original whole. Therefore, when considering how to manage the problems of complex wholeness, we should not stubbornly follow the path of bottom-up integration. Otherwise, if it works, wouldn't reductionism become reversible or divisible?

The above expresses a kind of abstract rational thinking used to understand the essential attributes of complex phenomena. However, any attribute must have a concrete embodiment, just like abstract points and lines have size and width. Similarly, "indivisibility" only exists in abstract speculation. The simplest explanation is that any whole with indivisible wholeness, as a form of a system, is composed of elements and has its own connections and structures. In this sense, no matter how complex the whole system is, from the perspective of elements, hierarchy, or structure, the whole system is divisible. Moreover, there is not only one path of separation, and the choice of which path depends mainly on our purpose of "separation." Therefore, the "indivisible" whole is also "divisible," and this is not contradictory because one is a theoretical abstraction and the other is practical concreteness; one is thinking, and the other is action.

Given this, it is not appropriate to absolutize the "indivisibility" of the complex wholeness in complex systems management activities, scenarios, and problems. In practice, anything with a structure is not a "rigid block." Therefore, based on certain requirements, the complex wholeness can be decomposed based on physical structure, system structure, or logical structure. This is the relativity of indivisibility in complex wholeness.

It is not difficult to see that the "indivisible relativity" in complex wholeness reflects the dialectical unity of the management subject's abstract thinking of "indivisibility" and the practical thinking of "relatively divisible" to enhance the ability to analyze and control complex wholeness.

In general, the reason why wholeness is "indivisible" is diverse. For example, the whole has a large number of constituent elements, complex relationships, strong interactions between internal hierarchies and scales, and close interactions with the external environment. All of these are what we commonly refer to as complexity. Therefore, when understanding complex wholeness from a macro perspective, the less divisible the wholeness, the more complex it is, and vice versa. This forms logical thinking: if we apply the principle of indivisible relativity of complex wholeness, allowing a certain degree of divisibility in wholeness, it will also reduce the complexity to a certain extent, naturally eliminating or alleviating the difficulties caused by the irreversibility of reductionism in complex systems management activities.

For example, a person's body exhibits typical complex wholeness. In general terms, this wholeness is "indivisible." However, if a person falls ill and a doctor needs to perform surgery to remove an organ (or a part of an organ), from the doctor's perspective, the indivisibility of the patient's body becomes relative. Through surgery, the complex wholeness of the patient's body is compromised to a certain extent, but, at the same time, the negative complexity (the illness) affecting the overall functionality of the body is reduced, ultimately enhancing the patient's overall well-being and safety.

Furthermore, when it's not possible to save a patient's life by removing the affected organ due to the principle of indivisible relativity, doctors can reconstruct the patient's physical integrity through organ transplant procedures. This serves to improve and strengthen the overall functionality of the patient's bodily organs, allowing the patient to regain a new level of overall well-being and safety. These seemingly contradictory approaches both achieve the same goal of optimizing the overall functionality of the complex biological system.

In the above analysis, we not only point out that complexity is often the main reason for the indivisibility of wholeness but also that complexity can be reduced through a certain degree of "relatively divisible" techniques. This inspires us: since complexity and wholeness are two different types of real forms and cognitive logic, neither completely equivalent nor entirely overlapping, they interact and work together in complex wholeness. Therefore, at the top level of complex wholeness, we can consider complexity and wholeness as the coarsest-grained components composing complex wholeness. By implementing "indivisibility," we divide the complex wholeness into complexity and wholeness, which are two separate elements. By further decomposing or partitioning complexity and wholeness, we can address complex wholeness scenarios where complexity and wholeness are closely intertwined. If such a relatively simple approach does not yield significant results, more sophisticated techniques can be employed to unravel complex wholeness. For example, by appropriately partitioning the temporal and spatial aspects of complex wholeness, we can temporarily "shield" wholeness and primarily focus on reducing complexity in a certain range and stage of complex systems management. This results in a phase of relatively simple complex wholeness, where complexity is reduced to some extent, thus alleviating the coupling and entanglement between complexity and wholeness. This "loosens" wholeness to some extent. At this point, when considering wholeness, it often leads to increased flexibility and options for feasible solutions related to wholeness. This process is then iterated, progressively achieving indivisible relativity based on complexity and wholeness. The essence of this approach is to transform the "resonance" of complexity and wholeness in the original complex wholeness into their indivisible relativity through iterative alternation.

Now, let's take a typical example: how to determine the bridge location for the Hong Kong-Zhuhai-Macao Bridge in China. This decision-making problem exhibits complex wholeness. Although the decision objective can be decomposed into dozens of sub-objectives, these objectives are closely interconnected and either strongly intertwined or in a state of "interconnectedness" [2]. The fixed landing points at both ends of the bridge render the northern direction impractical, while going south "immerses" the bridge in the vast ocean, making it technically infeasible and financially unsustainable. Therefore, the feasible area for the bridge location is limited to a specific sea area, which happens to be the habitat of the Chinese white dolphin in the South China Sea, a national first-class protected animal known as the "panda of the sea." As the white dolphin is strictly protected by relevant national laws, any decision to build a bridge in this area would conflict with the white dolphin conservation zone, posing a major dilemma caused by the complex wholeness in the field of large-scale engineering.

Since the bridge location inevitably traverses the white dolphin conservation zone, the decision-making for the bridge location transforms into a series of subsequent problems on how to resolve the crossing of the conservation zone. The white dolphin conservation issue and the bridge location issue belong to two different domains: ecological, environmental protection, and major national engineering construction, respectively. Each has its own rigidity but must be considered as a whole in the decision-making process, which is a typical representation of the complex wholeness decision-making problem in the Hong Kong-Zhuhai-Macao Bridge project.

Based on the basic analysis of this decision-making problem, we first propose the following degradation and integration strategies for the bridge location and white dolphin protection at the wholeness level: ① Degradation strategy: By fully utilizing the principle of decomposition, in the first stage of the bridge location decision-making process, we temporarily "shield" the white dolphin conservation issue in space, meaning we do not consider its impact. ② Integration strategy: Since the construction of the bridge must consider the impact on the Chinese white dolphins in the Pearl River Estuary, in the second stage of the bridge location decision-making process, if the bridge location crossing the white dolphin conservation zone is already determined, we then reconsider the white dolphin conservation issue and focus on how to address it, thus reflecting the complex wholeness of the decision-making problem.

From the entire process, in the decision-making process, the bridge location and Chinese white dolphin protection are integrated, and the complexity is appropriately reduced. However, a simple additive approach based on reductionism is not used to solve this complex wholeness decision-making problem. In the first stage of decision-making, the variability of cognitive virtual engineering is applied to address the decision dilemma of "spatiotemporal overlap" in reality. Through the thinking of "shielding" the white dolphin conservation area, the overall decision of the bridge location is initially decomposed into decision problems with reduced system complexity, such as diversified goals, multi-scale and multidimensional engineering target systems, and deep uncertainty in decision-making under various scenarios. Furthermore, in the second stage of decision-making, the heterogeneous decision goals of engineering value and environmental value in white dolphin conservation are "integrated" to form a time-ordered "spatiotemporal accompanying" decision sequence. By integrating the principle of exchanging economic resources for spatial resources, the decision problem is made holistic, ultimately resulting in a complex integrative decision solution for the real world.

From the above, it can be seen that the core idea of the complex integrative decision-making for the location of the Hong Kong-Zhuhai-Macao Bridge is to address the complex integrative decision-making goals arising from "spatiotemporal overlap." It first degrades the complexity by "shielding" the spatial goals that cause value conflicts, adopting "bottom-line thinking" in terms of engineering feasibility for the bridge location, that is, minimizing the impact on the Chinese white dolphins.

Then, it "reverts" the decision-making goal of protecting the white dolphins and integrates it into a temporally ordered "spatiotemporal coexistence" decision-making sequence. At the same time, it compensates for the loss of white dolphins' living space resources with other resources, such as providing economic and ecological compensation for the white dolphin conservation efforts, to ensure the transformation of spatial conflicts between engineering construction and white dolphin conservation toward overall coordination and balanced interests.

The above realization of the "indivisibility" of complex systems management in complex integrative decision-making is referred to as the principle of "relative indivisibility," where "indivisible relativity" serves as the main principle for complex systems management to deal with the irreversibility of reductionism.

Combined with the content before and after this book, the principle of indivisible relativity embodies the wisdom of the subject's behavior's flexibility and adaptive selection in complex systems management and fully reflects the role of unique details and context in complex systems management activities.

8.4 Principle of Uniqueness Contextualization

The earlier part of the book points out that complex wholeness problems possess both normativity and uniqueness. Comparatively speaking, there is less normativity and more uniqueness within these issues. Normativity mainly refers to natural and technological laws, which are characterized by physical forms that are observable and measurable. Uniqueness, on the other hand, primarily refers to the characteristics of social and cultural behaviors, which are characterized in engineering as understandable and perceptible. Although uniqueness is not equivalent to complex wholeness, it is often one of the causes of complexity. Therefore, understanding complexity cannot ignore uniqueness, and analyzing and solving complex wholeness problems cannot solely rely on normativity. It is necessary to fully consider the inducing role of uniqueness in complex wholeness. This is the main content of this section.

Uniqueness mainly induces complex wholeness through the following two channels:

The first channel: The complex wholeness of complex systems management activities and problems often stems from multi-scale phenomena, adaptivity, inherent uncertainty, emergent phenomena perceived as wholes, and self-organization. At the same time, all these phenomena lack data and information representation and indications and are devoid of complete knowledge support. They cannot be predicted based on historical data or cases. If future events could be predicted using historical data and existing rules, it would imply path dependence and the ability to extrapolate based on historical rules. However, uniqueness itself is non-traversable and non-computable.

In fact, the macroscopic structure of a system, the average information, or the generalization of historical data can only provide general and trivial characteristics of the system in an architectural and coarse sense. It is difficult to capture the uniqueness of complex wholeness scenarios and problems, including both certain common uniqueness and completely novel uniqueness.

So, how do we capture uniqueness? Let's start with an example: We can generalize the basic features of human facial features from a large sample of individuals from different ethnicities. We can also identify the fundamental differences in facial features between Eastern and Western populations by comparing them separately. However, if we want to pinpoint the uniqueness of John's or Mr. Zhang's facial features, we can only achieve this by conducting precise and detailed measurements and analyses of their facial details. In a certain sense, each person's facial structure possesses a uniqueness that results from the complex wholeness formed by their unique combination of family genetic traits and life experiences. Theoretically, every individual's facial features are one-of-a-kind in the world. To uncover the "uniqueness" of John's or Mr. Zhang's facial features, one would need to conduct thorough and precise measurements and analyses of the details of their faces among billions of people.

This inspires us to continuously align our research on the uniqueness of scenarios and problem complexity with real-world situations and pay close attention to scene details and their variations. Details, in this context, refer to the subtle elements or small components that emerge when breaking down the overall context of scenarios and problems. These details, due to their "small" nature, may be easily overlooked, but their significance lies in their ability to exert a profound influence on the overall state or state changes. In certain scenarios, details can impact the whole and even determine the broader picture, as exemplified by the "butterfly effect" discussed in Chap. 3 of this book, which is a typical case in complex systems where details determine the outcome at a larger scale.

Furthermore, details originate from reality, and reality requires specificity. It is not suitable to describe details using vague and coarse methods; instead, they should be interpreted through individual motivations, consciousness, and even irreproducibility. In this way, we will strive to use the real-world study (RWS) method to annotate details at micro-, local-, and time-scale levels as much as possible.

Taking humanization as an example, humans are the subjects in complex systems management. The adaptability of human psychology and behavior is the fundamental source of complex wholeness and the basic evolutionary level of complex systems management. Typically, people constantly adjust their actions based on changing environments to achieve their own goals. In most cases, apart from taking action due to inertia, people also tend to repeat actions that bring satisfactory benefits while abandoning actions that have previously produced adverse consequences. They try new actions based on existing knowledge in scenarios they have not encountered before, and they take emergency actions in response to unexpected events.

Behavior involves not only external connections but also internal processes required for self-behavior formation and evolution. Therefore, in the study of the complex wholeness of complex systems management, it is inappropriate to depersonalize and standardize human personality devoid of individuality. On the contrary, everyone has their own uniqueness, including how they learn, choose, regulate, and establish connections with the outside world. In the context of complex wholeness, it is necessary to respect and fully unleash the individuality and uniqueness of people, turning complex systems into complex adaptive systems embedded with uniqueness.

Therefore, it is not advisable to erase the uniqueness of individuals and simply classify them as economic men, rational men, etc. Instead, it is important to depict the inner world, psychology, behavior, and values of individuals in detail, including

(1) Memory elements: Data directly related to human psychology and behavior, such as memories and preferences, which can be stored in physical, chemical, or other forms. Memory elements are similar to human memory.
(2) Perception elements: Human activities of sensing, cognition, judgment, and other identification processes related to external inputs and internal memory elements. Perception elements correspond to the acquisition of external information by individuals, as well as cognitive activities based on their own attributes.
(3) Adaptation elements: Actions performed by individuals on memory elements under goal-driven circumstances, including copying and modifying. Adaptation elements correspond to human learning mechanisms.
(4) Action elements: Actions taken by individuals in response to inputs and decisions made based on memory, cognition, and learning. Action elements correspond to the practical actions taken by humans.
(5) Human mental model: It describes the psychological and intelligent factors that influence human decision-making, including physiology and instincts, psychology, pursuits, imagination, emotions, etc. The mental model corresponds to the decision-making mechanism of humans.

In the study of complex systems management, by revealing the unique micro-level details and the interrelated details of complexity among different individuals, we can explore what kind of uniqueness in micro-level details can make macro-level complexity "possible" rather than attempting to reveal the macro-level complexity as "inevitable" under certain microscopic dynamic mechanisms.

Therefore, the data sources supporting the description of uniqueness and details in complex systems management and complex wholeness must be extensive. They include observational research data conducted for specific purposes as well as non-research-related real data, such as various types of data related to problems collected through daily monitoring, recording, and storage. This includes both real-time large-sample data generated from actual management processes and small-sample data collected from general management processes. Based on data collection, various constituent elements of scenario details can be set in real-world settings to reflect the social, humanistic, and contingent aspects of complex wholeness. In methodological terms, this can be understood as achieving the complex wholeness of full scenarios and problems through sociality and humanization.

The second channel: since complex systems management is scenario-driven, uniqueness is a major factor in forming complexity, and this uniqueness is often embedded in the details of real-world scenarios or problems. Therefore, describing uniqueness is largely reflected in detail. Logically, there are two fundamental elements in the process of describing details: one is the describing entity (who describes), and the other is the described object (what the details are). It shouldn't be assumed that describing details is a straightforward task. First of all, details are the subtle elements of scenarios, and how to select, determine, and extract details, as well as the perspective from which they are defined, is closely related to the describing entity. In simple terms, details are both objective and influenced by the values of the entity, much like the immersive performance of the Beijing opera "Teahouse," where actors interact with the audience. Therefore, describing details is not about producing products that are standardized and identical to everyone but rather vary from person to person.

However, since details are the subtle elements of complex scenario wholes, ultimately, describing details is essentially describing scenarios. It involves the process of providing a holistic description, abstraction, and modeling of scenario elements, their relationships, the evolution of scenarios, specific time frames, small spaces, and small scenes. Due to the complexity of scenarios as wholes, this process typically results in a sophisticated structured model based on written language. This aligns with the inherent nature of describing details, as language itself is a behavioral pattern rather than a mere symbol of thought. Therefore, the meaning of any discourse cannot be divorced from the context in which the entity speaks or writes to convey the intended meaning, known as the context.

American economist Richard Bookstaber explained various sources of complexity when discussing complexity in his famous work 《*The End of Theory*》 [3]. He gave an example that can be considered to illustrate the important impact of the uniqueness of the context on the formation of complex wholeness. He said that in September 1998, Long-Term Capital Management (LTCM), due to financial constraints, issued an open letter to investors, trying to rationalize and tell the "respected investors" that there were excellent opportunities ahead for making money, but the company had recently suffered losses and "needed some funding." The content and wording of the letter were humble and sincere, but if we put this letter into the unique context of September 1998, the problem is not that simple.

As is well known, on July 2, 1997, the Asian financial crisis swept across Thailand, and, shortly afterward, it affected Malaysia, Singapore, Japan, South Korea, China, and other countries. On September 2, 1998, the Russian ruble devalued by 70%, causing a sharp decline in the Russian stock and foreign exchange markets, which in turn led to widespread volatility in the stock and foreign exchange markets of the United States and Europe. It can be imagined that the information expressed in the letter, "the company has suffered losses and needs some funding," when faced with such a harsh environment, would greatly frustrate the optimistic mindset and positive actions of investors who had some expectations of making money. The immediate reaction of many investors upon receiving the letter was to liquidate and short the company's funds in the main market, causing the fund's prices to fall and further exacerbating the company's losses. The main reason for this situation was not that the evaluation methods used by investors daily were wrong but rather that in the specific temporal and spatial context of the financial crisis, the thinking and behavior of countless individual investors would choose completely different courses of action from what the company anticipated based on traditional assumptions. This became a collective behavior that emerged at a macro level, giving rise to new complex, holistic, dynamic mechanisms. The role of this unique contextual environment not only rendered the company's predictive evaluation algorithms ineffective but also far exceeded the mere power of language in the letter. In short, investors made different rational actions based on different details of how the financial crisis would affect themselves and their families. However, the "sum" of all individual actions turned out to be a "disaster" for the company.

The main reason for this complexity is that although details may be subtle, they still have plots and scenes, vivid and specific characters, objects, cause-and-effect relationships, associations, contingencies, and developmental stages. To describe such rich and intricate content clearly, it cannot be achieved with just one or two words or symbols. It requires the construction of a clear and complete linguistic environment through basic language elements such as words and syntax. This includes not only carefully selecting appropriate words and phrases but also specifying and limiting corresponding characters, time and space, scenes, causes, and consequences. This is called the contextual environment, which plays a fundamental role in constructing the corresponding environment for describing scenarios (details). In practice, the context consists of various events, phenomena, and scenes closely related to the details, such as background knowledge, context, prefaces, and postscripts in the description. Furthermore, the contextual connotation can be expanded through deep exploration and understanding of implied meanings.

We can understand the significance of the contextual concept for exploring the uniqueness of complex wholeness from the following aspects:

1. It can provide heterogeneous, ambiguous mechanisms, unexplainable causality, and the complex wholeness that may arise from "irregular" or contingent possibilities.
2. It can provide the details of the plot and the actual events that immediately precede and follow it, helping to understand the causes of complex wholeness.
3. It can provide necessary but not sufficient factors for complex wholeness, and it can "fill in" the details of the formation of complex wholeness through the coincidental interaction and path dependence of other elements.
4. It can provide a basic rule for interpreting the evolutionary process of complex wholeness by presenting ordered sequences of scenario changes.
5. It can provide the underlying causes behind the complex wholeness, not just treating it as "accidental" but rather as "non-necessary" or "unexpected."

In summary, the uniqueness of complex wholeness often manifests in the micro-level, local, and detailed aspects of a scenario. The context is an important channel for understanding the inherent uncertainty, multi-scale nature, adaptivity, and other implied meanings within the details of a scenario. If we need to use new symbol systems and algorithms, including language and text, to describe complex wholeness, the context can help us unravel the order of unknown mechanisms, non-repetitive rules, and unverified functionalities of unstructured nature.

In conclusion, the core focus of this section is to explore the non-normative aspects of complex wholeness at the level of complex systems management activities and behaviors. It investigates how to paradigmatically handle the uniqueness within complex wholeness, establish principles for complex systems management activities and behaviors through detailed exploration and the contextual function of describing details in a given scenario, and find corresponding points of intervention. This part of the behavioral paradigm is referred to as the contextualization principle of uniqueness in complex systems management.

8.5 Principle of Physical-System-Management Chain Progression

So far, we have discussed at length how to analyze and deal with the behavioral norms of complex wholeness. However, for complex systems management, the core task and function are to provide solutions to complexity problems. Therefore, it is necessary to construct a complete paradigm for analyzing and proposing problem-solving processes, which encompasses the complete set of thinking and behavioral conventions for addressing complexity problems.

Since we are aiming for a "complete" quality requirement for the proposed conventions, we need to begin with the intuitive perception of the physical form of complexity problems, progress through abstraction and refinement at the system space level, and finally select and propose corresponding solutions within the manageable space. This sequential and orderly progression constitutes the fundamental principle of "Physical-System-Management Chain Progression" in complex systems management. Now, let's provide a detailed introduction to the relevant content.

The term "physical complex wholeness" primarily refers to the "vivid intuition" of the realistic forms of the scenarios, phenomena, and problems in complex systems management activities. Complex system scenarios, phenomena, and problems are composed of various material elements and combine to form a complex whole under certain natural laws, social relationships, and behavioral norms. The material structure, physical functionality, and dynamic mechanisms of this whole are referred to as physical complex wholeness and represent the first stage of physical morphology perception in complex systems management. Based on general cognitive regularities, it is essential to fully respect the reality and objectivity of complex holistic scenarios, phenomena, and problems at this stage.

According to the scenario-oriented principle, the term "physical" should not be solely understood as materiality. For example, the constituent elements of a scenario include not only material elements but also non-material, conceptual, and logical elements. Without these types of elements, a completely realistic scenario cannot be formed. We refer to this cognitive stage as the phase of physical complex wholeness.

To effectively manage complex systems, it is not sufficient to rely solely on an intuitive perception of scenarios and problems. It is necessary to further refine, explore, and understand the essential intrinsic connections and regularities of the attributes inherent in the scenarios and problems. Since attributes must be attached to specific subjects and cannot exist independently, this stage needs to be conducted in a system space that originates from and goes beyond the physical level.

To provide conditions for subsequent problem-solving, the refinement of attributes in this stage should not be conceptualized or generalized. It should focus on the complex holistic characteristics that cannot or should not be separated from the overall scenario and problem, such as the high openness of the environment, the diversity of subjects and inherent uncertainty; strong interconnections, multiple constraints, and scale among problem elements; the emergence, evolution, and adaptability of problem states or subject behaviors and functionalities, etc. In particular, the unique contextual factors within the context of complexity should not be ignored, and the language system of complex systems should be applied for expression. By accomplishing these aspects, not only does it "lock in" the scenario form and essential characteristics of the complexity problem at hand in a general and universal sense, but it also does so in terms of individual uniqueness. This is referred to as the progression from physical complex wholeness to system complex wholeness, demonstrating the entry into the stage of abstract condensation and logical deduction of complex systems management attributes. It also signifies the concretization of the paradigm shift to complex systems thinking, known as entering the stage of system complex wholeness. The significance of this stage includes, firstly, ensuring the revelation of the inherent qualities of the scenario and problem at the level of rational thinking; secondly, fully leveraging the aforementioned principles; and thirdly, using a common discourse system paradigm to reduce ambiguity in thinking and communication.

The ultimate goal of complex systems management is to formulate and propose management knowledge and solutions for complex wholeness, entering the stage of managing the complexity of solving problems. Without this stage, complex systems management activities will remain at the level of rational abstraction and systematic analysis, failing to achieve the ultimate purpose of complex systems management. To achieve this goal, this stage must adhere to the normativity of management thinking, research paradigms, and discourse systems, realizing a further progression from physical complexity to system complexity and then to management complexity. The key lies in, based on the previous two stages, synthesizing management knowledge for complex wholeness and forming management solutions for problem-solving in accordance with management thinking principles and fundamental principles (see Fig. 8.1 below).

The above content can be briefly understood as follows: The concept of physical complex wholeness refers to the complete "story" of real scenarios and problems. System complex wholeness refers to the synthesis and abstraction of the attribute and relational information of all elements in the "story." Management complexity refers to the orientation toward the complexity of the "story" and the search for problem-solving solutions based on the paradigm of complex systems thinking.

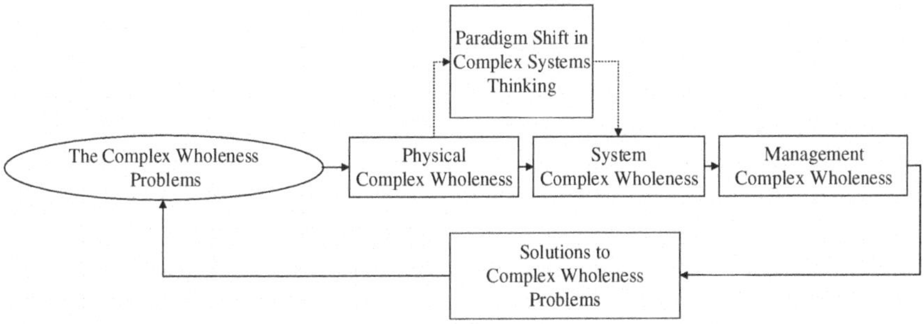

Fig. 8.1 Principle of physical-system-management chain progression

These solutions are immersed in the context of the scenario for validation and refinement.

In summary, these three stages constitute the physical-system-management chain progression model as one of the basic behavioral paradigms for analyzing and solving problems in complex systems management. It is also one of the fundamental principles of complex systems management and fully embodies Hsue-Shen Tsien's idea of "starting with the study of various specific complex systems, finding mechanistic solutions to the complexity of specific complex systems, and establishing new theoretical systems based on continuous accumulation."

Here, we provide an example of the "paradigm for decision-making in major projects." Major project management is a type of complex systems management within the field of engineering construction, also known as complex engineering systems management. The paradigm for decision-making in major projects refers to the standardized mode of implementing key decision-making activities in major projects according to certain behavioral conventions and processes. It should be emphasized that this paradigm is not set by individual decision-makers but is based on the transfer path of the complex systems thinking paradigm. It is important to emphasize that these patterns are not defined by individual decision-makers. Rather, they are established based on the paradigm shift towards complex systems thinking, leveraging a blend of theoretical and engineering thought processes. These patterns, while respecting general decision-making rules and embodying the unique intentions of the subjects, are designed and planned for decision-making strategies. This is done with the goal of clarifying and managing the complex and holistic attributes of decision-making problems. The paradigm can be divided into the following three stages, which fully embody the principle of physical-system-management chain progression:

Stage 1: Generally, in the process of decision-making in major projects, individuals first perceive the decision scenario and problem intuitively as material elements of the artificial complex system, the interrelationships and structures among these elements, and various forms of physical-level complex wholeness. Particularly, the vivid images of the decision scenario at the physical level are the primal substance of the decision problem. This represents the stage of physical complex wholeness in decision-making in major projects.

Stage 2: Because complex systems management is based on the paradigm of complex systems thinking, the analysis of decision problems in major projects cannot remain solely at the intuitive perception level. Instead, it should gradually abstract to the level of complex systems and employ complex systems thinking and discourse systems to articulate and refine the logical relationships and attribute characteristics of decision problems. This represents the transformation from the physical complexity of decision-making problems in major projects to the system complexity, as well as the abstraction and extraction of the core elements and structures of decision scenarios. It is evident that this transformation is primarily aimed at enabling decision analysis to take place at the level of rational thinking, facilitating the revelation of the inherent relationships and objective regularities of decision problems, enabling in-depth analysis and planning of decision solutions, and preparing for the formation of decision solutions. This stage essentially involves a significant paradigm shift in decision thinking toward complex systems.

Stage 3: Based on the refined system complexity of the decision problem and guided by management concepts and principles, the recognition of the system complexity and mechanisms of decision problems is once again transformed into management science concepts, principles, logic, and discourse systems. This process results in scientific management issues that encompass the essence of complex systems thinking while fully reflecting the essential attributes of complex wholeness decision-making. In this stage, it is particularly important to use multi-symbol system frameworks to create analyzable or computable decision problems and solution designs, incorporating unique contextual elements of the decision problem. This process involves reconstructing and predicting the scenarios and evolution of decision problems and solutions, deducing and analyzing the effectiveness and robustness of decision solutions within virtual decision scenarios. Through theoretical, empirical, simulation, and iterative approaches, the final solution to the decision problem is gradually obtained. This represents the transformation from system complexity in the process of decision-making in major projects to the complexity of management within the context of complex wholeness decision-making.

The transformation of these three stages forms the basic paradigm of decision-making in major projects based on complex systems management. For more detailed information, please refer to Chap. 14 of this book.

8.6 Multi-Scale Management Principle

8.6.1 Basic Connotation of Multi-Scale Management

As mentioned in Sect. 7.4, the concept of multi-scale in complex systems management is widely present and has significant practical significance. The core idea is that in complex systems management, the attributes of management elements in a certain dimension exhibit a certain order and differentiation trend. Moreover, this trend will result in distinct characteristics of the management element attribute, varying depending on the dimension. It is necessary for us to carefully distinguish and treat them in practical activities and theoretical research of complex systems management to improve the level and precision of management. For example, the coordination between decision-making and the environment in complex systems management is generally a spatiotemporal multi-scale problem. In the specific decision-making process, decision-makers need to consider the coordination issues across different spatial and temporal scales and integrate them for comprehensive evaluation. This requires coordinating different scales: should we reduce the "large scale" or enlarge the "small scale?" This means that multi-scale management in complex systems management refers to paying full attention to and distinguishing the feature differences caused by multi-scale attributes of the same-dimensional management elements. Based on these differences, corresponding management principles, processes, and methods are designed and constructed to enable management activities to better reflect and accommodate these differences.

Multi-scale management is based on the following idea: the differences caused by different scales of management elements have an important impact on the complex entirety and management effectiveness of complex systems management activities. Often, it is these differences that constitute management complexity. Therefore, these differences should not be simplified without distinction. For example, as mentioned in Sect. 7.4.2, multi-scale phenomena form different functional types in complex systems management, such as constitutive, generative, and emergent types, and are integrated into the concept of functional spectrum that deeply reflects the complex wholeness. If we fail to differentiate these multi-scale phenomena in practical activities or theoretical research of complex systems management, we may easily overlook the fine structure of the functional spectrum, leading to decision-making errors in complex systems management.

For instance, in natural geographic research, we often divide the Yellow River system in China into three scales: upper, middle, and lower reaches. However, when considering research related to the governance of the ecological-environmental water-sediment relationship of the Yellow River, such a scale division may appear coarse. Specifically, when it comes to the upper reaches, it would be more appropriate to further divide it into three finer scales: the headwater region, the canyon region, and the alluvial plain region.

8.6.2 Reductionist Division of Multi-Scale Management

How can we embody multi-scale management in complex systems management? We need to consider the following aspects based on the logic of the inherent connotation of multi-scale:

1. Is it necessary to divide into multiple scales?

Although multi-scale is a universal phenomenon to some extent, it is not necessary to divide the attributes of management elements into multiple scales at all times and for all problems. The need for division should be determined based on the following two points:

(1) It has a significant impact on analyzing and solving management problems. If the characteristics of the element's attributes at different scales have a significant influence on management problems, it is generally necessary to divide the element into scales. On the contrary, if the attributes at different scales have minimal impact, there is no need for scale division.
(2) It is based on the precision requirements of analyzing and solving management problems. Generally, analyzing complexity problems requires a careful examination of the impact of management elements on the problem, which emphasizes the role of multi-scale. However, if the analysis does not require such precision, there is no need for scale division.

2. How to divide into multiple scales?

How do we divide the scales of an element's attributes? Dividing different scales within the same dimension is like dividing intervals on a straight line. However, multi-scale division is much more complex than interval division. Firstly, scales generally do not have intuitive measurement features, such as interval lengths, and, in most cases, they are only expressed qualitatively or descriptively. For example, for the time dimension, scales can be categorized as near-term, medium-term, and long-term; for the spatial dimension, scales can be defined as small-scale, medium-scale, and large-scale; for uncertainty, scales can be shallow, moderate, and deep; for supplier location, scales can be centralized or distributed; for problem descriptions in science, scales can be structured, semi-structured, or unstructured, and so on. It can be seen that, behind the scales, there is generally some specific physical, systemic, or managerial connotation. Scale division is often fuzzy and flexible. It cannot be as clear in terms of quantity or geometric characteristics as mathematical interval division. Generally, scale division is understood as the definition of the boundary of inherent connotation features by the subject at the level of rational thinking.

3. How to extract scale features

This issue has actually been considered during the design of multiple scales. Initially, it was our discovery that management element attributes exhibit different characteristics at different scales within the same dimension, and the recognition that these characteristics in different scales have an undeniable impact on management problems, on the basis of which we introduced the concept of multiscale. Therefore, it can be seen that the influence of management element characteristics at different scales on management problems often serves as a guiding factor for conducting multi-scale divisions. This illustrates that the division of multiple scales is a highly purposeful and practical management behavior, and its important basis is to conduct a correlation analysis between management problems and different characteristics of elements under the premise of extracting different scale characteristics of management elements. Elements of the same type should generally belong to the same scale. For example, in a single-scale context, people often only focus on the direct physical functionality of major projects, which is the constitutive function. In the near to medium-term scale, people generally distinguish between constitutive functions and generative functions. If we add a long-term scale, we would also consider emergent functions. This comprehensively reveals the multi-scale representation of the functional spectrum of complex major engineering systems.

4. How to utilize multi-scale effects

The practical significance of extracting the features of management elements at different scales lies in fully revealing the influence of attribute features at different scales on management problems thereby conducting detailed analysis and solving complexity issues. This is the fundamental role and significance of multi-scale management. Specifically, based on the description of scale features, it is important to analyze and establish the correlation and mutual influence between various features and management problems. Without accomplishing and performing these tasks well, the practical significance of multi-scale management cannot be fully realized. Therefore, we need to extract the specific features of different scales and the complexity management connotations contained in these features based on theoretical thinking and consider the real relationship between different features and management problems from a practical perspective, clarifying the actual impact of features on management scenarios.

For example, multi-scale based on the time dimension is a common phenomenon, but it is not sufficient to simply divide the time dimension into near-term, medium-term, and long-term scales and believe that it reflects and accomplishes multi-scale management. The practical significance of this division lies in seeking the feature representation at different scales based on the scientific connotation of management elements. For example, the functional representation of complex systems management, as a management element, will exhibit different formation pathways at different time scales. It reveals the significant differences between these pathways. This requires us to "embed" the feature of different functional formation pathways into functional design management and carry out on-site management activities based on different formation pathways. Constitutive and generative functions can generally be addressed using traditional engineering planning design and comprehensive socio-economic analysis methods. However, these methods alone often fail to discover and predict emergent functions. In such cases, new methods are needed to discover and predict the large-scale temporal evolution phenomena of the environment of key projects and, based on this, to form emergent functions at a large time scale. It should be noted that even within the same dimension, such as dividing the time dimension into near-term, medium-term, and long-term scales, the extraction of features depends on the specific research problem, the relationship between the problem and the elements, and whether the relationship can differentiate more detailed features of elements, etc., and there is no consistent answer.

For example, in the context of environmental management, the influence of different time scales mainly manifests in the "severity" of environmental uncertainty. Therefore, it is necessary to divide uncertainty into three different levels: shallow, moderate, and deep, based on the near-term, medium-term, and long-term time scales. This is completely different from the concept of the functional spectrum based on different time scales. As for management elements in different dimensions, the extracted feature representations and descriptive methods in the context of multi-scale, as well as their impact on management problems, all present different management connotations, which need to be analyzed and clarified based on specific management problems one by one.

8.6.3 Overall Integration of Multi-Scale Management

The division of management elements on a single dimension into multiple scales can, in a sense, be seen as a means of reducing complexity in theoretical thinking by the entities. However, the practice of complex systems management requires us not only to have multi-scale divisions in dealing with abstract thinking but also to have multi-scale integration in dealing with concrete thinking within the same dimension. This demands that we synthesize management elements from multiple scales back to their original dimension and conduct a comprehensive study of management problems within the original dimension. This is an essential step in practical multi-scale management. Without this step, even if we refine the complexity of management elements within the same dimension using the multi-scale concept, we are merely dealing with "decomposed" complexity. Therefore, it is crucial to integrate it within the dimension to achieve a holistic understanding of complexity in that dimension.

In summary, multi-scale management consists of two main stages. The first stage involves the reductionist thinking of dividing the dimension into multiple scales, extracting different scale features, and analyzing the impact of different scale features of elements on management problems. The second stage involves holistic thinking, integrating multi-scale analyses at the dimension level, and forming a holistic understanding of the impact of management problems within that dimension. These two stages, which encompass both reductionist scale division and holistic dimension integration, constitute multi-scale management in the sense of addressing complexity as a whole.

Regarding the specific techniques and methods of multi-scale management, in the "reductionist division" stage, after the division into multiple scales, the management activities corresponding to a particular scale are relatively concentrated and straightforward in relation to the environmental background. The features, attributes, impacts, and effects of that scale are relatively clear. This means that at this stage, the focus can be placed on a specific part of the complexity within the management elements. From the perspective of the determinism and applicability of management-related techniques and methods, system analysis techniques and methods are generally used.

In the "holistic integration" stage, the focus is on the embodiment of holistic thinking principles. Therefore, methods such as comprehensive methods for management goals (functions) and various types of comprehensive evaluation techniques are generally adopted. The basic principle of comprehensive evaluation techniques is to establish a comprehensive utility function (indicator) that includes the subject's value orientation as well as the attributes, impacts, and effects at various scales. This function serves as a representation of the management elements in the holistic sense of the dimension. By combining specific methods, the subject's cognition based on the combination of objective attributes and subjective values can be obtained.

Commonly used comprehensive evaluation methods include the following:

(1) For a class of non-structured prominent problems, such as macro and strategic management issues, which are difficult to describe quantitatively and establish structured models, methods primarily based on expert qualitative evaluation are often used.
(2) Factor analysis, principal component analysis, and cluster analysis are used to conduct a comprehensive evaluation of multi-scale features and impacts.
(3) Evaluation methods such as scoring methods, correlation matrix methods, and analytic hierarchy process (AHP) are used for comprehensive evaluation within the system analysis.
(4) Fuzzy recognition, fuzzy comprehensive evaluation, and other methods are used for comprehensive evaluation.
(5) Process analysis and comprehensive evaluation are conducted through computer simulations.
(6) Interactive multi-objective comprehensive evaluation methods are formed by combining the objectivity and subjectivity of multi-scale features through human-machine dialogue.
(7) Integration and improvement of two or more comprehensive evaluation methods to obtain more comprehensive evaluation methods, such as integrating fuzzy evaluation methods with cluster evaluation methods to form fuzzy cluster methods or integrating with artificial intelligence methods to form fuzzy artificial neural network methods.
(8) Furthermore, there are comprehensive evaluation methods based on the method set combination and computer-based evaluation support systems.

Although the aforementioned methods are comprehensive evaluation methods that have been developed and successfully applied in various fields over the years, their basic ideas provide inspiration and applicability for addressing and handling the "integration" in multi-scale contexts. However, system synthesis and its evaluation have always been complex and challenging. It should not be assumed that, by directly applying comprehensive evaluation methods that have been successfully used in many other fields, we can achieve multi-scale integration in complex systems management. In addition to selecting and borrowing appropriate and mature comprehensive evaluation methods, it is more important for us to integrate theoretical thinking, practical thinking, and other types of thinking in the actual activities of complex systems management.

In conclusion, multi-scale management is based on the multidimensional analysis of management elements in the context of complex systems management. It involves dividing the management elements on the same dimension into multiple scales, analyzing the impact of differential element attributes and characteristics at different scales on management problems, and further integrating them into a holistic understanding at the dimensional level based on multi-scale analysis. This forms the basic principles of multi-scale management.

8.7 Adaptive Selection Principle

8.7.1 Connotation of Adaptive Selection

In Sect. 7.5, we highlighted that, in complex systems management activities, entities enhance their cognitive, analytical, and control capabilities of complex wholeness through their adaptive behaviors. Specifically, there are two fundamental adaptive behaviors of entities: entity actions and the management proposals presented by entities. Additionally, as in other domains of management, "choice" is the most common and fundamental-behavioral pattern in complex systems management. Therefore, entities' choices based on adaptive criteria become a type of "adaptive selection." This isn't just a practical aspect of entities' behavior in complex systems management activities; it also becomes one of the fundamental principles in these activities.

What is the scientific connotation of adaptive selection in complex systems management? Firstly, adaptiveness is the goal of the subject's selection behavior because the complex wholeness of the management environment and problems in complex systems require subjects to change their traditional optimization thinking of management goals in many scenarios and adopt new adaptive thinking, forming a new concept of adaptiveness as optimality. For example, the ability of subjects to exhibit highly adaptive behavior is considered strong, the management solution that can adapt to the changes in inherently uncertain scenarios is the optimal one, and the flexible organizational model that can adapt to the complex wholeness of different management problems is effective, and so on. Of course, aiming for adaptiveness will largely embed subjective value preferences of the subject, which can easily lead to biased choices. To prevent this, firstly, the subject needs to improve its own cognitive and analytical abilities through self-adaptive learning behavior, and, secondly, the selection process should be seen as an ongoing trial-and-error, iterative, and converging process rather than a one-time decision.

Furthermore, adaptive selection compensates for behaviors such as complexity reduction and indivisible relativity at the operational level because, in the process of complexity reduction and indivisible relativity, the assumption of "problem virtualization" is adopted, which inevitably damages the reality complex wholeness of the scenario. However, in real activities of complex systems management, the inherent complex wholeness of the situation or problem still exists. Therefore, when subjects face the remaining real complexity and need to develop problem-solving solutions, they need to restore the inherent complex wholeness of the situation or problem from the "virtual" level to the "actual" level. This requires the subject to make another selection through its adaptive behavior in order to restore and correct the "deviation" and "damage" caused by the inherent complex wholeness [4].

Moreover, the strength of behavioral adaptiveness is also an indicator of the subject's selection capability. In order to control the complexity and solve complex problems, subjects need to continuously accumulate experience and improve their capabilities through self-adaptive learning behavior. Therefore, the subject's own adaptive learning behavior is a prerequisite and premise for its adaptive selection and an important indicator of its capability. From the perspective of system science principles, any management solution for a complex system is an artificially created complex system by the subject, but it also creates new complex wholeness. This is the subject's proactive response to the complex wholeness of the problem. Therefore, the subject needs to compare and select solutions based on adaptive learning to ensure the feasibility and other qualities of the solution. Thus, whenever a solution is made, the subject should follow the path of adaptive learning, which is the fundamental connotation of self-learning behavior in the process of adaptive selection.

Therefore, adaptive selection becomes an important practical principle for subjects in the activities of complex systems management and one of the fundamental principles in complex systems management.

8.7.2 Path Strategy for Adaptive Selection

After clarifying the basic connotation of adaptive selection, the important question is how to design its selection path. Generally, adaptive selection mainly includes adaptive goal selection, organizational pattern selection, solution selection, and solution evaluation method selection under the adaptive principle. The following examples focus on goals, organizational mechanisms, and evaluation methods:

1. Adaptive Selection of Goals

In the practice of complex systems management, due to the complexity of management goals and the influence of entity preferences, a particular goal of the entity can only be a choice based on limited knowledge and experience at a certain stage. This may result in incomplete goals, unclear hierarchy and relationships, and poorly handled conflicting objectives. Therefore, there is a need for adaptive adjustments to the goals.

For example, based on the preliminary design of goals, a deeper screening and refinement of goals can be conducted to extract more essential core goals. A multi-objective system is not a simple superposition of multiple goals. For risks, it cannot be assumed that the sum of several small risks equals a large risk. Additionally, goals with different attributes often do not possess simple additivity. For instance, we cannot simply add management quality goals and safety goals. Instead, it is necessary to further reveal the complex correlations within the multi-objective system and comprehensively integrate the goals. The main challenge lies in coordinating conflicting goals.

2. Adaptive Mechanisms of Organizations

Based on the analysis above, as a flexible capability platform, a complex systems management organization needs to have appropriate operational mechanisms to provide adaptive selection mechanisms. The term "mechanism" originally refers to the structure and working principles of a machine. If a machine is understood as a system, its structure and working principles can be seen as the structure, processes, and flows of a system. Therefore, "mechanism" can be regarded as the constituent elements, interrelationships, and input/output procedures of a system or organization. So, how can a complex systems management organization platform form its basic adaptive mechanisms?

First, design the organizational platform's adaptive mechanisms based on the following four key elements:

(1) Scenarios (problems). Complex systems management encompasses a wide range of scenarios and problem types involving various domains and levels. Among them, there are significant "major scenarios" and "difficult problems." All adaptive selection behaviors of the subject must be guided by scenarios (problems). The inherent complex wholeness of the scenarios (problems) serves as the subject's adaptive direction and, even more so, the adaptive goal under the complex systems thinking.
(2) Subjects. Different scenarios (problems) require different capabilities from subjects. For example, "challenging problems" require subjects to have richer experience and knowledge. "Macro scenarios" require subjects to possess decision-making capabilities that match the scenario, such as the ability to make decisions and propose solutions according to relevant regulations and policies. This is often more important than having specific knowledge and experience because it directly affects the legitimacy and authority of complex systems management solutions. Moreover, there should be no overstepping or lacking the subject's capabilities.
(3) Resources. Here, the term "resources" mainly refers to the basic elements for building the organizational platform and the mechanisms that support the platform in unleashing its capabilities. The platform must establish an operational mechanism that integrates individual entities' capabilities into the collective capabilities of the entity group. This is the most important functional resource of the organizational platform.
(4) Environment. It refers to the synergy of interfaces and interactions within the organizational platform. For example, it involves selecting the appropriate mode of association and collaboration among subjects based on the characteristics of the scenarios (problems).

Second, design adaptive mechanisms based on the dynamic changes of the four elements mentioned above:

In practical activities of complex systems management, the four elements mentioned above undergo various changes. Therefore, the organizational platform should generate corresponding adaptive changes and functions. Among them, the changes in scenarios (problems) are fundamental and dominant, playing a guiding role. All changes in the organizational platform can be said to be caused by changes in scenarios (problems). For example:

(1) Changes in platform subjects. Changes in tasks inevitably impose different requirements on the authority, knowledge, and capabilities of the subjects. Therefore, the platform must have adaptive mechanisms for changing subjects and so on. Generally, there are more changes in the general subjects, while there are fewer changes in the principal subjects.
(2) Changes in platform structure. Changes in scenarios (problems) may lead to changes in the interrelationships and interactions among platform subjects. This directly requires changes or reorganization of the platform structure to emerge new necessary capabilities and functions, often manifested as changes in complex systems management patterns.
(3) Changes in platform mechanisms. They mainly refer to changes in internal workflow processes, principles, and execution procedures within the platform. For example, in different stages of complex systems management activities, the associated subjects and the modes of association will undergo adaptive adjustments to adapt to the changing scenarios (problems).

The various "flexible" qualities and capabilities exhibited by the complex systems management organization platform mentioned above fully reflect its basic principles of constructing its own capability to handle complex wholeness capabilities through adaptive mechanisms when facing complex scenarios (problems).

3. Adaptive Evaluation Methods for Management Solutions

In Sect. 7.5, we introduced the second type of adaptivity, which is the adaptivity of management solutions. Although it does not belong to the adaptivity of the subject's behavior process, the "quality" of the solution, as the ultimate result of the subject's adaptivity, becomes a direct representation of the "quality" of the subject's adaptivity behavior process. How to choose evaluation methods for solution quality based on the complex wholeness characteristics of scenarios and problems is a series of interrelated and sequentially ordered questions in the theory of complex systems management, which requires us to clarify the conclusions through the demonstration of basic concepts and principles under the theoretical thinking principles.

For example, what is the quality of management solutions? From the preliminary construction of concepts and principles, the quality of management solutions in complex systems management is mainly used to measure and evaluate whether the management solutions can continuously and normally deliver their predetermined functions (utility) throughout their life cycle. Specifically, it assesses whether the solutions can maintain their robustness in the face of deep changes in environmental scenarios.

This theoretical viewpoint reveals the connotation of solution quality in the complex holistic characteristics of complex systems management. It proposes using the overall matching degree between the intended solution functions and the depth of changes in environmental scenarios as the technical route for evaluating and measuring intended quality.

The changes in environmental scenarios here include two types of scenario changes:

(1) Scenario changes in the peripheral environment before the formation of the artificial subject system in complex systems management.
(2) Scenario changes emerging in the "complex systems management-environment composite system" after the formation of the artificial subject system.

This also involves how to discover and predict scenario changes in complex systems and how to measure the degree of matching between management solutions and scenario changes. The introduction to this aspect will be presented in Chap. 10 of this book.

In summary, adaptive selection is the basic behavioral principle and representation of the subject in complex systems management activities. In management practice, adaptive selection fully reflects the combination of theoretical thinking and practical thinking. Its main connotation is how the subject improves the adaptivity and robustness of management solutions through an adaptive selection of goals and adaptive mechanism design for the management organization platform.

The above content in this section represents the principle of adaptive selection in complex systems management.

8.8 Iterative Generation Principle

In the previous sections, we mentioned that adaptive selection is the most common and fundamental-behavioral principle of the subject in complex systems management activities. Furthermore, in this section, we will introduce the practical-operational procedures of subject adaptive selection and the principle of the management solution generation process.

8.8.1 Iterative Nature of Subject Behavior in the Selection Process

From the perspective of practical thinking, all selection behaviors of the subject in complex systems management activities or all the objectives of the selection are aimed at determining and proposing solutions to complex wholeness. For a relatively simple management problem with clear management goals and a high degree of problem structuring, one can generally use explicit goals, rigorous analysis, modeling, and optimization techniques to select the "optimal" solution from several feasible options to solve the problem. We refer to this type of solution generation process as the "optimization" generation process principle. However, for complexity problems in complex systems management, such an "optimization" generation process encounters obstacles due to the complex wholeness of the problem. This is mainly because

1. The goals of complex wholeness have characteristics such as multiple levels, multiple dimensions, multiple scales, strong correlations, and inherent uncertainties.
2. Complex wholeness is difficult to be fully expressed using structured models.
3. Even if a structured model is reluctantly constructed, it is often difficult to solve the model due to its complexity.

In complex systems management activities, although the subject still aims to solve problems, various subjective and objective reasons make it difficult for the subject to generate problem solutions using the "optimization" principle. At this point, the subject can only deal with the complex wholeness of the problem based on the adaptivity behavioral principle. Regarding the problem, the subject needs to decipher the complexity of the holistic problem, refine and synthesize management goals, and perform multi-scale attribute divisions on the problem properties during the selection process. So, how does the subject operate and implement all of these in practice? Let's analyze the practical process below.

1. Iteration at the "First Level" of subjects

The problem-solving ability of each management subject is based on its own learning, knowledge accumulation, and creativity. In general, it represents a broad self-learning capability of the subject itself. Learning is a complex process for subjects, especially when it comes to learning that enhances their understanding and mastery of complex wholeness. This type of learning involves not only understanding, comprehending, and accumulating knowledge but also original thinking, breakthroughs, and insights.

As the Gestalt theory of cognitive learning suggests, the ultimate goal is to form an organized whole in the cognition of the subject, a complete structure, rather than just forming associations between stimuli and responses. Regardless, the self-learning of the subject must gradually grasp the essential nature of the complex wholeness to be solved and immerse the learning outcomes into the context of the problems. This clearly represents a self-iterative process in the cognitive thinking of the subject. In this continuous iterative process, the information and knowledge mastered by the subject itself become increasingly enriched, leading to a more comprehensive, refined, and profound understanding of the problem and how to solve it. The length and effectiveness of this iterative process vary from person to person, depending on their level of expertise, the difficulty of the problem, and the trajectory of the learning process. However, no one can achieve mastery instantaneously, as this is determined by the laws of human learning and cognition. This type of iterative behavior occurring within individual subjects is referred to as "First Level" iteration. It represents the most fundamental type of iterative behavior in the selection process of complex systems management solutions.

2. Iteration at the "Second Level" of subjects

Due to the involvement of various fields such as politics, society, economy, technology, and culture, complex systems management problems cannot be addressed by an individual or a few subjects alone. Designing and determining complex systems management solutions requires not only a high level of cognitive, analytical, and synthesis abilities from individual management subjects but also the collaborative problem-solving efforts of a group composed of individuals from multiple domains. In practice, different domains have different jurisdictions and specialized knowledge required to address various problems. This implies that the composition of the complex systems management subject group cannot remain fixed. Instead, it needs to undergo continuous selection and appropriate restructuring of the subject group's structure, guided by the leading role of the dominant subject, in order to form a new organizational platform based on the nature of the problem.

From the perspective of the selection process, this represents an ongoing iteration of both the management subject group and the organizational platform in terms of reorganization or optimization. It is through this iteration that the organizational platform adaptively generates the jurisdictions and capabilities that align with the problems to be solved. This iterative behavior, occurring within the subject group during the selection process of complex systems management plans, is what we refer to as "second-level" iteration behavior. It refers to the iterative behavior at the organizational platform level or the management organizational level in the context of subject behavior selection in complex systems management. The iteration of the organizational platform is not an end in itself; however, it serves as an organizational guarantee for the high-quality completion of the management plan selection. Furthermore, it requires the subject group to effectively carry out the selection of management plans based on the dynamic iteration of this platform.

3. Iteration at the "Third Level" of Subjects

In general, for a management problem, the subject group initially forms one or more initial solutions based on the platform's operating mechanism. These solutions are then subjected to multidimensional analysis and evaluation. Based on the conclusions drawn from the analysis and evaluation, the subject group may modify and improve the original solutions, which represents the vertical iteration process of the solutions. Alternatively, multiple solutions may be compared, eliminated, and recombined, which represents the horizontal iteration process of the solutions. In practical management activities, both vertical and horizontal iterations are not limited to a single iteration; multiple iterations are usually required to reach the final solution. This is mainly due to the following reasons:

(1) Comparing complex systems management solutions involves a dynamic, comprehensive evaluation of the integrated functionality of the solutions. This

includes not only the transformation of single-item evaluations of solution functionalities into multiple-item evaluations but also comprehensive evaluations based on the overall significance of solution functionalities. Particularly during the evaluation process, revisions or even repeated revisions of the evaluations may occur as the subject's cognition deepens and values change.

(2) Relevant data, information, expert knowledge, and experience are crucial in the comparison of solutions. However, in practice, the relevant data and information can only become increasingly rich and complete through multiple comparisons, while expert knowledge and experience can gradually be unleashed.

(3) In general, the selection of solutions requires the formation of a consensus within the subject group. The "consensus" within the subject group signifies the gradual concentration and convergence of individual subjects' understanding of the complex wholeness of the problem. This process depends on the level of complexity inherent in the problem itself. The higher the complexity, the more difficult it is to achieve convergence, and the more iterations of comparisons may be required. Additionally, it depends on the evolution of the subjects' values. For example, if the subjects' environmental awareness is strengthened, they may reject a previously accepted low-quality environmental solution and propose a higher quality environmentally friendly solution. This indicates that the original "consensus" has turned into a "non-consensus." Such intricate and repetitive iterations are common in the practice of complex systems management.

In this regard, the selection process of complex systems management solutions, at the operational level, is primarily composed of vertical or horizontal iterations of the solutions. In this process, whether it is the comprehensive evaluation and cognitive improvement of the subjects or the formation of a consensus among the subject group regarding the solutions, it embodies a general pattern of "continuous comparison, gradual approximation, and ultimate determination." This is a type of iterative behavior that occurs in the formation of consensus among subject groups, which we refer to as the "third-level" iterative behavior, and it represents the highest level of iteration in the process of selecting complex systems management solutions.

To summarize, the behavior of selecting complex systems management solutions manifests as a comprehensive iterative pattern consisting of three levels and mutually reinforcing iterations: "individual entity self-learning iteration-entity group platform iteration-solution consensus formation iteration" at the operational level.

The specific procedure involves continuously comparing, adjusting, and revising a phase-specific plan vertically or horizontally by subjects. It may even involve the rejection of the original solution to redesign a new one, thus iteratively approaching the final solution through a sequence of solution iterations.

From a theoretical perspective, if there exists a logical "optimal solution" to a problem in complex systems management, the entities, in the process of practical solution selection, approach this "optimal solution" through a continuous process of comparison and revision. We refer to this as the "iterative generation" principle of complex systems management solutions and often summarize its overall operational procedure and process as "comparison, iteration, approximation."

It is evident that from the perspective of the technical approach of solution formation, the aforementioned "optimized" generation principle incorporates the complex wholeness of the problem into the selection behavior of the solution all at once. To accomplish this in practice, two conditions must be met: first, the problem cannot be too complex, and, second, the capabilities of the entities must be very strong. However, in the activities of complex systems management, these two points are often unattainable. However, the "iterative generation" principle decomposes the complex wholeness of the problem into multiple stages of solution generation. Not only does it enable the entities to encounter only a part of the complex wholeness at each stage, but it also employs a series of adaptive solution sequences to approximate the solution. This practical behavior reflects both the criteria for reducing the complexity of the entities and the criteria for adaptive selection.

In fact, the principle of iterative generation, which is manifested as the principle of comparing, iterating, and approximating the "optimal solution" to a complex wholeness problem, is also widely used in other disciplinary fields. For example, in mathematics, finding the circumference of a circle is a "complex wholeness" problem. Before the invention of calculus, when people's level of cognition was relatively low, they could first calculate the circumferences of the inscribed regular triangles, regular quadrilaterals, regular hexagons, regular octagons, etc., of the circle. By increasing the number of sides of the regular polygons, they formed an iterative sequence of the circumferences of the inscribed regular polygons to approximate the circumference of the circle. Of course, for any finite number of iterations, this method of approximating the circumference of a circle using the circumferences of inscribed regular polygons has errors, but the errors can be reduced by increasing the number of iterations. In the actual calculation process, the number of iterations depends on the desired level of accuracy.

8.8.2 *Iterative Nature of the Technical Route Selection Process*

As mentioned earlier, the iterative generation principle of complex systems management solutions is manifested at the operational level, primarily through the comparison behavior of the subjects. The key technical route used for comparison is comprehensive evaluation, which involves combining qualitative and quantitative approaches for evaluating objectives. While these two core key techniques are widely used in general management practices, they also exhibit distinct iterative characteristics in the actual operational process. Emphasizing this point has guiding significance for how we appropriately use technical routes. The specific explanations are as follows:

1. Iterativity of Comprehensive Objective Comparison

Whether it's a vertical or horizontal iteration of management solutions, the subjects should conduct equal-depth and identity comparisons of the solutions in the same environment. This requires the subjects to propose comprehensive evaluation techniques that involve comprehensive objectives. In addition, during the comprehensive evaluation process, the subjects need to filter objectives for each stage and make qualitative or quantitative judgments about the relationships between objectives. In the next stage, the subjects have to perform a similar iteration based on the evaluation of the previous stage. This reflects the "iterative" essence of the objective synthesis technique route adopted by the subjects [5].

2. Iterativity of Qualitative and Quantitative Integration

Due to the complex wholeness of complex systems management solutions, the subject initially needs to formulate the top-level design of the solutions. This stage primarily relies on existing scientific theories, empirical knowledge, and the integration of subject group intelligence to form intuitive judgments primarily expressed through language and written descriptions. This is known as the qualitative approach in solution generation, involving the overall analysis and design of solution objectives and functions, analysis of the relationship between problems and the environment, and the design of technical route for plan formation, as well as the design of comprehensive evaluation indicators and criteria for plan. Observation, description, case and document analysis, field investigations, and other qualitative methods are commonly used. The results of qualitative methods often include empirical hypotheses, planning, preliminary solution ideas, and technical route descriptions [6, 7].

However, to propose a solution for the complex wholeness of complex systems management, it is necessary to provide precise quantitative descriptions of the relationships between plan elements. General language or textual descriptions are often too coarse for complex wholeness plans. Hence, logical reasoning, modeling, data analysis, and simulation experiments are employed to precisely calculate and rigorously demonstrate certain parts of the problem and the plan. This is the quantitative method in plan generation. Quantitative methods generally build upon qualitative methods and their conclusions, employing rigorous logical reasoning and mathematical deduction for precise solutions and demonstrations. Detailed descriptions of different plans and parameter settings in the plans widely employ quantitative methods such as data collection and analysis, mathematical modeling and solution, and computer simulation. Therefore, in the selection of complex systems management plans, how to handle the relationship between qualitative and quantitative methods?

Firstly, when people initially comprehend and analyze the complex wholeness of a problem, they can only explain its external manifestations through perception. They primarily use language and written descriptions to express speculative content and establish conceptual frameworks based on the flexibility of experiential judgments and feelings. In this stage, people's induction, understanding, knowledge, and experience play a fundamental role, laying the foundation for further rigorous and detailed analysis of the problem using standard procedures and precise methods.

Secondly, quantitative and modeling techniques in complex systems management have a direct and significant impact on the complete and precise formulation of complex wholeness problem solutions, especially for highly complex wholeness with great significance. Due to their strong wholeness nature and close connection with inherently uncertain external environments, it is necessary to move beyond initial qualitative understanding and comprehension. This is achieved through data collection and analysis, rigorous logical reasoning, mathematical deduction, and various quantitative methods to precisely solve and argue the problem. It may also be necessary to revise the quantitative results themselves, all of which are accomplished through continuous iteration.

Finally, complex systems management of complex wholeness requires a combination of qualitative and quantitative approaches [8]. This is because such problems typically have interdisciplinary and cross-domain characteristics. Instead, they require the collaboration of experts from different fields, disciplines, and levels to form an expert system. This expert system relies on the integration of collective experience, knowledge, and wisdom to create a consensus within the group, which is the essence of qualitative integration. Moreover, when using quantitative methods, it is also necessary to integrate various models, calculation tools, and quantitative approaches based on the overall objectives [9]. This is quantitative integration. Both qualitative and quantitative integration methods primarily employ information technology and network-based techniques as tools and methods. This is achieved through the establishment of model libraries, databases, method libraries, rule libraries, and more. Additionally, decision support platforms are established to serve as intellectual support during the process of qualitative and quantitative integration. From this, it can be inferred that the selection of solutions for complex wholeness problems in complex systems management not only requires the use of methods that combine qualitative and quantitative approaches but also necessitates the application of comprehensive integration methods from qualitative to quantitative [10].

It is evident that, in the aforementioned process, both qualitative and quantitative stages require multiple iterations to be completed. Moreover, the iterations within the qualitative stage and the quantitative stage mutually reinforce each other, giving rise to new iterative demands and forming interactive iterative characteristics within the combined qualitative-quantitative framework.

Although various techniques and methods are employed in the process of selecting complex systems management plans, the combination of qualitative and quantitative approaches is the primary and commonly used technique and method. Whether it is within the internal qualitative synthesis integration, internal quantitative synthesis integration, or the transition from qualitative to quantitative synthesis integration, all of them exhibit a continuous iterative process [8]. It is precisely this multi-level iterative approach that not only leverages the strengths of qualitative and quantitative methods but also compensates for their respective shortcomings, resulting in enhanced selective capabilities. Furthermore, it reveals the fundamental principle of "iterative" generation manifested by the subject's operational behavior in the process of plan selection.

8.9 Logical Correlations of Basic Principles

Based on the previous seven basic principles, this section provides a logical correlations analysis as a whole.

According to the principles of theoretical thinking, if we abstractly analyze the phenomena and problems in complex systems management activities and extract the two most fundamental and universal elements, they should be the subject and the complex wholeness.

The subject refers to the individuals and groups who play a leading role in complex systems management activities and possess cognitive and practical abilities. Without the subject, there would be no complex systems management activities.

Complex wholeness is the most essential attribute in complex systems management. Without complex wholeness, it cannot be considered a complex systems management activity.

The principles in complex systems management theory should not only reflect the essence of complex systems management in terms of cognitive attributes in theoretical thinking but also embody the irreversibility of reductionism in terms of the value intention in practical thinking. Furthermore, the principles should achieve logical coherence through the combination of these two perspectives. Therefore, the principles must concentrate on the two most fundamental and universal elements: the subject and the complex wholeness, fully revealing the fundamental laws governing the behavior of the subject and the characteristics of the objects in complex systems management activities. The ability to achieve this is the primary criterion for assessing the academic quality of the basic principles in complex systems management theory. In the following discussion, we will elaborate on this issue.

In this chapter, we have proposed a total of seven basic principles: scenario orientation, indivisible relativity, contextual uniqueness, physical-system-management chain progression, multi-scale management, adaptive selection, and iterative generation.

Firstly, the scenario serves as the carrier of all information of complex wholeness in complex systems management activities and the substance in which problems are immersed. In terms of theory, complex systems management activities, scenarios, and problems have the same attributes. Therefore, scenario orientation ensures the authenticity and uniqueness of the complex wholeness in complex systems management.

In the context of complex systems management, the notion of complex wholeness encompasses not only general forms of complexity that may be relatively independent but also a distinctive characteristic of indivisibility. At the level of practical thinking, this primarily refers to the tight interconnectedness within such entire systems. It's important to note that, in practice, it doesn't necessarily mean that these entire systems are entirely rigid or inflexible. Instead, they can often be analyzed and addressed by making them relatively "separable" with respect to their structure, details, logic, and spatiotemporal scales without causing a fundamental qualitative change in their complexity. This is where the principle of indivisible relativity, as introduced in this book, comes into play, allowing for a more manageable approach to handling complex holistic systems.

Uniqueness significantly contributes to complex wholeness, manifesting as complex patterns that are highly non-normative, inherently uncertain, and challenging to verify. These distinctive patterns can serve as a starting point for refining corresponding uniqueness by delving into the details of the scenario and examining the contextual functions that describe these details. In this regard, contextual uniqueness becomes a new tool to cope with the complex wholeness based on uniqueness.

The above points mainly provide us with methodological principles for complex systems management in a macroscopic global sense, enabling us to delve into the management at the middle and micro levels. Firstly, complex systems management is a type of complexity management, and complexity management is the refinement of management mode attributes through the paradigm of complex systems thinking. Therefore, complex systems management inevitably manifests as a behavioral law that progresses in an orderly manner from physical reality and system connotation to management activities. This law abstracts the physical perception of complexity issues under the paradigm of complex systems thinking and realizes its "implementation" in practical management activities, thereby achieving an integrated chain progression to solve complexity issues.

The adaptability of the subject's behavior not only creates complexity but also becomes a means to "deal with" complex wholeness. Therefore, entities can incorporate the principle of adaptability into processes such as the selection of solutions for complex wholeness problems and the determination of organizational platform mechanisms. This fosters a type of behavior at the management operation level that uses "adaptability" as a guiding principle. This approach serves as a compensatory mechanism for the inseparability, uniqueness, and contextual nature of these processes, offering greater operational and implementable value than other principles.

Furthermore, the subject can make necessary scale divisions of the commonly existing multi-scale phenomena in complex systems management activities, analyze the influence of different scale characteristics on management issues, refine the complexity of management elements and integrate them into a multidimensional sense, and carry out "multi-scale management" activities.

In this way, with the joint influence of the subject's adaptive selection and multi-scale management as two behavioral criteria, the subject's management behavior has a basic "handle" to follow, enhancing the subject's practical ability to handle complexity.

Furthermore, all the behaviors of the complex systems management subject and the purposes of these behaviors are aimed at designing and proposing solutions to complex wholeness. In real management activities, due to the limitations of subject capabilities, the subject can only decompose the overall complexity of the problem into various stages of solution generation. This allows the subject to face only a portion of the overall complexity at each stage, thereby dealing with subproblems and their corresponding solutions that are relatively less difficult at certain stages. These sub-solutions are then organized into a coherent sequence to approach the final

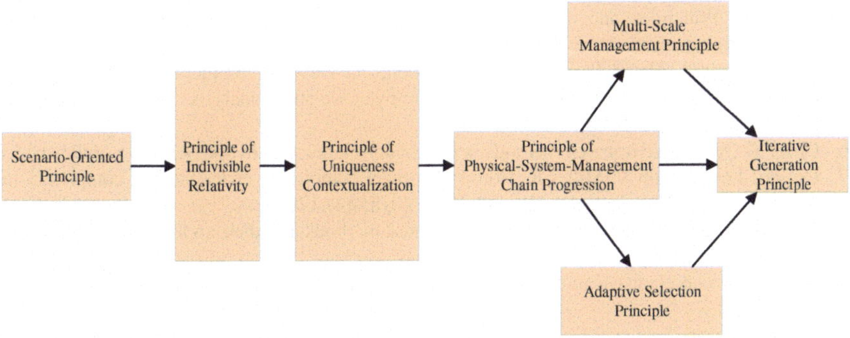

Fig. 8.2 Logical correlations of basic principles

solution to the overall problem. This iterative generation of solutions by the subject in management activity practice fully embodies the criterion of adaptive selection and represents a common and realistic operational principle in actual complex systems management activities.

Specifically, in complex systems management activities, the subject first adopts a scenario-oriented approach guided by the complex holistic scenario. Through the "variability" of cognitive understanding of the scenario, the subject establishes behavioral principles such as indivisible relativity, unique contextual principles at the macroscopic global level, and the physics-systems-management chain progressive paradigm. Building upon these principles, the principles of multi-scale management and adaptive selection form more actionable behavioral criteria. The principle of iterative generation then integrates all of these principles together, forming general rules for subject behavior and operational processes.

Seven basic principles proposed in this chapter are derived from the practice of complex systems management and closely focus on the fundamental management elements of the subject and complex wholeness. They fully reveal the logical relationships, causal relationships, relevant relationships, and subject behavioral criteria in complex systems management phenomena. There is a close logical correlation among them, forming the basic principles system of the theoretical framework of complex systems management (see Fig. 8.2). Based on this, it is possible to further derive complex systems management scientific issues with scholarly quality and theoretical value using the core concepts and basic principles presented earlier.

References

1. Sheng Z. Problem-oriented: the driving force behind the development of management theory. J Manag Sci. 2019;22(05):1–11.
2. Shuai Jin ZS, Ding X. Evolution and enlightenment of the coordination decision-making system of the Hong Kong-Zhuhai-Macao Bridge project. Constr Econ. 2013;(12):27–31.
3. Bookstaber R. The end of theory. Princeton University Press; 2017.

4. Cheng S. Principles and strategies of adaptive selection in the management of major infrastructure projects. Oper Res Manag. 2017;26(02):153–7.
5. Owens J, Ahn, J., Shane, JS, Strong KC, Gransberg DD. Defining complex project management of large U.S. transportation projects: a comparative case study analysis. Public Work Manag Policy. 2012;17(2):170–88.
6. Chen X. Qualitative research methods in the social sciences. Chin Soc Sci. 1996;06:93–102.
7. Lessard D, Sakhrani V, Miller R. House of project complexity understanding complexity in large construction projects. Eng Proj Organ J. 2014;4(4):170–92.
8. Yu J. From qualitative to quantitative, integrated methods and their applications. China Soft Sci. 1993;05:31–5.
9. Sheng Z. Comprehensive and integrated management of large-scale projects. Beijing: Science Press; 2009.
10. Sheng Z. Sutong bridge engineering system analysis and management system. Beijing: Science Press; 2009.

Chapter 9
Foundational Scientific Issues in Complex Systems Management

9.1 Overview of Scientific Issues

9.1.1 Definition of the Issue

In academic research, it is common to refer to "scientific problems" rather than "scientific issues." Why is it not called "scientific problems in complex systems management" but "scientific issues of complex systems management" here? Firstly, this is because "problem" and "issue" are not the same concept. Secondly, it is related to the early stage of development in complex systems management.

Typically, a "scientific problem" refers to the unresolved issues in scientific knowledge and practice that scholars propose within a specific knowledge background. It includes certain objectives and a response domain for solutions but lacks definitive answers. Therefore, maximal efforts are needed to seek and explore solutions.

The formulation of scientific problems requires clarity and focus, avoiding being too generic. In the theoretical system of complex systems management, there are indeed numerous scientific problems. Researching and solving such problems are crucial tasks, usually expressed by scholars through academic papers or highly specialized academic works. As an introductory book on foundational knowledge in complex systems management, it is not suitable to dedicate an entire chapter to such a heavy task.

On the other hand, "scientific issues" refer to questions or topics that scholars propose with a certain scientific basis and relevance to objective reality. They are often broader in meaning than problems. Sometimes, an issue can encompass problems with similar or closely related connotations and characteristics. The purpose of raising scientific issues is not necessarily to immediately obtain answers but rather to attract attention and research interest from the academic community.

For a new field such as complex systems management, which has been constructed by a paradigm shift of complex systems thinking, this is a reflection of the "crossing

Z. Sheng, *Outline of Complex Systems Management Theory—Based on Irreversibility of Reductionism Thinking*, International Series in Operations Research & Management Science 360, https://doi.org/10.1007/978-981-97-3361-3_9

the river by groping the stones" mode of exploration. It involves thinking and groping from a broad perspective while not yet having enough ability to "step on stones" precisely.

The formulation of scientific issues requires a higher level of thinking than proposing specific scientific problems. For example, the establishment of the discourse system for the knowledge morphology of complex systems management, as presented in this chapter, or combining thinking paradigms with specific scientific problems to form scientific problem categories with methodological significance, such as robustness management, all fall under scientific issues. Foundational scientific issues are closely related to the whole properties of complex systems. As the theory of complex systems management continues to develop, these foundational issues will deepen and expand. This progression is an important sign of the maturation of the complex systems management theory.

Based on the above ideas, this chapter does not select specific complex systems management scientific problems such as decision-making, technological innovation, and supply chain to introduce. Firstly, the number of such management problems is vast and difficult to enumerate comprehensively. Secondly, specific problems require consideration of unique contexts and tailored solutions, making them less universally applicable. Therefore, this chapter takes "issues" in its original sense and proposes broad and meaningful topics such as "management quality," "essential management," "management organization," and "knowledge morphology of management" as the objects of exploration.

9.1.2 Complex Systems Management in Its Infancy

Realistic and theoretical analysis indicates that although complex systems management has tremendous potential for development, it is still in its early stages and has only formed the rudimentary foundation of the field. Therefore, special attention should be paid to the theoretical norms and applicability of paradigms in the development process of complex systems management.

Firstly, the high openness of academic thinking is conducive to encouraging complex systems management to learn from various academic fields in many aspects and serves as the fundamental cognitive paradigm and logical starting point for complex systems management. It should promote interdisciplinary cross-fertilization and integration under unified thinking principles. However, this guideline should not, cannot, and must not use the academic framework of another field as a benchmark, "projecting" complex systems management academically onto that framework and then constructing complex systems management using the academic thoughts, concepts, and discourse of that field. Such a "parasitic" approach to academic development cannot become a viable path for the standardization of complex systems management theory. Moreover, without a unified thinking principle, it is insufficient to merely rely on ready-made conclusions and methods from one or more other fields and "assemble" them into an academic system of complex systems management.

The above paths are inappropriate because they fail to deeply reflect the essential attributes of complex systems management activities and problems at the practical and cognitive levels. Therefore, they cannot ensure the true formation of complex systems management knowledge and its systematic and logical development. As for using new concepts and terms to interpret and metaphorize complex systems management phenomena and knowledge, it may have limited or isolated significance, but it also leads to the internalization of scientific research due to the lack of philosophical thinking and the substantive connotation of problem attributes. All of these factors remind us that conducting complex systems management research must adhere to the basic laws and paradigms of academic development.

Regarding the understanding of the complex systems management theory development, it should be recognized that in any theoretical system, the truths it discovers and the regularities it reveals are relative. That is, they are relatively correct, relatively profound, and relatively comprehensive. Therefore, theories are only relatively true, and it cannot be expected that relying on a theoretical system alone can solve all practical problems in complex systems management. Especially in the practice of complex systems management activities, in addition to theoretical thinking, there are practical thinking, innovative thinking, and various other types of thinking required. Therefore, in the analysis and resolution of complex systems management problems in practical activities, theories are indispensable, but the idea that "theories are omnipotent" should not exist. In this regard, human cognition, experience, and insight into the complexity of reality often play a significant "clinical" role.

Furthermore, the complex systems management model should not be understood as the only one. Similar to other management paradigms in the field of management, the process of constructing and developing complex systems management theory should be open and inclusive. It should foster a situation of "letting a hundred flowers bloom and a hundred schools of thought contend" to facilitate the development and progress of complex systems management academically.

Constructing complex systems management theory is not an easy task, and its revision, improvement, expansion, and elevation require long-term and arduous exploration. Considering that the management of complex systems is a broad and profound domain-specific concept, it intertwines closely with aspects such as time, region, practice, environment, culture, institutions, history, and policy. Moreover, it is also influenced by the perspective from which the subject considers the issue, as well as their approach and depth of thought regarding the problem. Thus, the academic development of complex systems management is inevitably a task that requires long-term effort and may never be completely finished.

In summary, the current reality of complex systems management presents an increasingly strong demand for the construction of management theory. At the same time, the growing complex systems management practices, the expanding research teams, and the accumulating research results have provided many foundational conditions for constructing the academic system of complex systems management. In this sense, foundational scientific issues are advancing on the path of complex systems management practical and theoretical development.

Next, we will introduce seven foundational scientific issues in complex systems management.

9.2 Issue 1: Complex Wholeness Cracking

9.2.1 Overview of Complex Wholeness Cracking

In Sect. 8.3, we contemplated how to address the complex wholeness in complex systems management using the concept of "indivisible relativity." The key points of this thinking include the following: (1) it is a theoretical thinking that expresses thinking principles, and (2) it is a methodological approach at the macro level of management. However, as a management activity, complex systems management also requires "operation" at the thinking practical level and "scenario-oriented" in theoretical research. Therefore, it is necessary to transform the "indivisible relativity" of management into practical issues and operational methods. This is the scientific issue of "complex wholeness cracking" proposed in this section.

The guiding principle of "complex wholeness cracking" is the principle of "indivisible relativity." In this regard, the following key points can be identified:

1. Complex wholeness is an essential attribute of activities, situations, and problems of complex systems management.
2. Since complex systems management involves system practices and forms composed of different elements with their own structures and hierarchies, it implies that it is "divisible." The term "indivisible" is not absolute but relative.
3. By employing "indivisible" abstract thinking of complex wholeness and the dialectical unity of "relative indivisible" practical thinking, one can design behavioral guidelines to manage complex wholeness effectively. This helps alleviate the barriers caused by the irreversibility of reductionism in complex systems management activities.

Thus, at the practical management level, complex wholeness can be relatively divided into multiple levels, dimensions, stages, and scales or divided into relatively independent complexity, relatively independent wholeness, complexity based on wholeness, wholeness based on complexity, and complex wholeness in which complexity and wholeness are tightly "coupled." If we can achieve these divisions, it will naturally reduce the challenges posed by complex wholeness. Furthermore, if we can ensure that the above decomposition process does not compromise the essential attributes of complex systems management through certain "compensatory" measures, it indicates that we have embarked on a path of cracking complex wholeness with quality assurance.

Therefore, at the level of management practice and subject behavior, it is necessary to moderately reduce the relative independent complexity as part of cracking complex wholeness. This is known as complexity degradation in the complex wholeness cracking. In addition, the relatively independent wholeness needs to be moderately "dissected," which is referred to as wholeness dissection in the complex wholeness cracking. For the complex whole parts that complexity and wholeness are tightly "coupled," comprehensive treatment is required. For example, complex systems management comprehensive objectives clearly demonstrate the tightly coupled form of complexity and wholeness. In such cases, we can apply complexity degradation, wholeness "dissection," and sequential iteration in a sequential and spatially alternating manner to the target set, approaching the cracking of complex wholeness.

In summary, the above content can be divided into two parts: the first part pertains to the degradation of relatively independent complexity, while the second part focuses on the partitioning of wholeness (including wholeness based on relative independence and wholeness tightly "coupled" with complexity). Whether it is "degradation" or "partitioning," it utilizes the "variability" of the system's virtual agent to appropriately and reasonably "reduce" or "partition" complex wholeness. This helps alleviate the difficulties and limitations in cognizing and handling processes related to complex wholeness. Therefore, these aspects constitute the fundamental behavior of the subject in complex systems management activities and serve as foundational scientific issues for academic research in complex systems management.

9.2.2 Paradigm for Complexity Degradation

In general, complex systems management activities and problems are composed primarily of various material resources and represent artificial complex system subjects with unique specifications. As a result, they exhibit physical complex wholeness composed of hard resource subjects. A concrete and determinate complex system subject possesses a specific and tangible physical complex wholeness. At the same time, any specific complex system subject evolves from concepts to subjects and from abstract to concrete through human cognition of the attributes and relationships of system physical elements.

Therefore, at the abstract stage of conception and cognition, people mainly engage in activities at the level of thinking. In this stage, people use theoretical thinking to abstract the attributes of the hard system and systematize the relationships between these attributes, forming a logical system of complex system attributes. To distinguish it from system subjects, we refer to this system as a system's virtual agent. Clearly, unlike the physical complex wholeness of complex system subjects, the system's virtual agent primarily represents the complex wholeness of system element attributes.

Thus, through the logicalization and systematization of system element attributes, the system's virtual agent establishes and enriches the subject's understanding of the complex system whole attributes and functions of complex systems. It also supports the subject in decision-making and planning for constructing system subjects. Additionally, by inferring various relationships between system element attributes, the system's virtual agent guides complex systems management activities.

Additionally, since the system's virtual agent is a logical framework of element attributes and relationships established based on theoretical thinking, individuals' personalized value orientations and cognitive levels will, to a large extent, influence and determine how system element attributes are abstracted and how attribute relationships are defined. In other words, even for the same physical complex systemic form, different virtual entities may be formed due to differences in subjective thinking and values. We can leverage this "variability" in the process of forming a virtual agent to "reduce" and "decompose" the complexity of a system without altering its inherent physical complexity. Particularly, many times, this "reduction" and "decomposition" occur at the level of abstracted attributes and relationships of system elements, manifesting only as a conceptual and logical form, similar to ancient Chinese "armchair strategists" and modern "war simulations," without truly disrupting the system's objective, inherent, and actual complex form. However, it can help us clearly and simplistically understand and analyze the complexity of the system to some extent in a logical sense.

Of course, in real complex systems management activities, especially in the concrete implementation phase of system entities, the management subject must fully confront the material aspects of complex system entity elements and their physical interconnections. Hence, any practical complex systems management activity should avoid relying solely on abstract concepts, assumptions, and theoretical discussions akin to "armchair strategists." Similarly, it should refrain from solely embracing the notion of "reducing" and "decomposing" complexity. Instead, it should genuinely be based on entity thinking, integrating various attributes, relationships, and the inherent objective complexity of complex systems into a complete physical system entity.

This means that when the complex systems management subject understands and analyzes the attributes and correlations of complex system elements, it can fully utilize the "variability" of abstract logical systems to appropriately and reasonably "reduce" or "decompose" system complexity (referred to as **complexity degradation**) through various feasible paths. This helps us understand the laws and principles behind complexity. However, this is a thinking assumption and idealization adopted to enhance the cognitive ability of the management subject within the system's virtual agent logical system. It is a means of supporting the management subject's understanding of complexity, alleviating the difficulties and limitations of cognitive complexity in management activities, and assisting the management subject in discovering the laws of management complexity as an intermediate process.

Furthermore, in practical complex systems management activities, the management subject needs to use practical thinking to face the concrete physical subjects of the system, "reducing" their real complexity, and must not let counterfactual thinking cause any authenticity "damage" or "harm" to the objective complexity of the complex system subject.

From this, it is evident that the complete process of complexity reduction needs to fully embody and ensure the principles of combining counterfactual thinking with entity thinking and merging the individual characteristics of complex systems with general attribute patterns. This forms the whole thinking framework for complex systems management activities. In summary, during the early stages of complex systems management, the subject can utilize the principle of "variability" of virtual agents to assist and support cognitive processes and analytical approaches by employing hypothetical and idealized "decomposition" actions. However, in the later stages of complex systems management, it becomes essential to use entity thinking to "restore" the inherent physical complexity of the system, ensuring the authenticity and integrity of the creation of complex system entities. This constitutes the fundamental paradigm of complexity degradation.

It can be considered that complexity degradation is the primary and guiding behavioral principle for the management subject when facing complex wholeness problems in complex systems management activities. Specifically, complexity degradation behavior can be understood from the following three aspects:

(1) The reduction behavior included in "degradation" primarily refers to the improvement of the subject's capability, such as enhancing the ability to analyze and manage complexity through self-learning. In this case, the inherent physical complexity form of the system can be fully maintained. Therefore, to some extent, reduction reflects a "wholeness" thinking.
(2) The decomposition involved in the "degradation" primarily entails dividing the inherent complex wholeness of the system at the level of counterfactual thinking to reduce its original complexity. Therefore, in a certain sense, decomposition predominantly reflects a "reductionist" mode of thinking [1].
(3) In either case, there is no actual damage to the inherent physical complexity of the system subject, but both can help the management subject improve their ability to analyze and manage complexity. It is important to note that regardless of the type of degradation behavior adopted by the management subject, the inherent complexity of the system must not be damaged, nor should there be any qualitative changes in the characteristics of inherent complexity [2].

9.2.3 Pathways of Complexity Degradation

The complexity of managing complex systems is a comprehensive cognitive construct. It integrates the complexities across multiple fields, levels, dimensions, and scales that are related to the environment, organization, entities, and problems of complex system management activities. Moreover, it incorporates the physical complexity, system complexity, and management complexity of the situations and problems within complex systems. Therefore, it is crucial to design specific degradation paths based on the sources and causes of complexity and to find the right balance in the degradation process, striving for an approach that is "just right." Here are several fundamental degradation paths.

1. Degradation path to enhance the cognitive ability of the management subject.

The complexity of complex systems management problems mainly stems from two aspects: ontological viewpoints, which are the complexities arising from the management environment and the problems themselves, and subjective viewpoints, which are the complexity arising from cognitive deficiencies of the management subject, including insufficient knowledge, experience, and abilities. If the cognitive ability of the management subject improves, the complexity of the problem will relatively decrease. It has been proven that self-learning of the management subject is an important way to enhance cognitive abilities and reduce problem complexity. In addition, obtaining abundant information resources through modern information technology to reduce problem complexity caused by information deficiencies and insufficient information processing capabilities is also crucial. Rich management information resources and the subject's resource management capabilities can effectively reduce cognitive errors caused by psychological or physiological factors and reduce cognitive limitations caused by the subject's background, personality, emotions, thinking patterns, and values. Therefore, the use of information technology can effectively assist the management subject in degrading complexity [3].

2. Degradation path of improved management methods

In order to reduce complexity, degradation can also be achieved through improved management methods, such as

(1) Refining and coordinating management objectives

Refining management objectives means selecting, extracting, and refining constitutive objectives, generative objectives, and emergent objectives of complex systems management based on goal design, highlighting and ensuring the status of strategic and foundational objectives.

Coordinating management objectives builds on the refinement of complex systems management objectives, aiming to achieve an overall balance between objectives by eliminating, restricting, or compensating for some objectives. It strives to balance the direct and indirect, immediate and long-term, and functional, social, and strategic objectives.

(2) "Compression" of future scenarios

The concept of scenarios tells us that the future scenario space of complex systems management is "sufficiently large" and fundamentally uncertain. It not only includes scenarios that management subjects can imagine but may also encompass scenarios that are difficult or impossible for them to imagine. Moreover, the possibility of any future scenario occurring is difficult to ascertain, and it is challenging to predict the paths through which the current reality will evolve into future scenarios.

We cannot eliminate this objective condition because it arises from the inherent uncertainty of complex systems management and the complexity of the complex systems management-environmental composite system. Humans cannot fully ascertain all scenarios and possible paths from the present reality to the future.

In this case, we can use the following methods:

a. Based on the experience and knowledge of the management subject group, identify specific and meaningful scenarios within the future scenario space.
b. Regard future scenarios as predictions of environmental states and generate specific future scenarios by setting conditions and parameters.
c. Define a subspace within the future scenario space and assume that the possibility of future scenarios in this subspace is much greater.
d. Set certain meaningful scenarios as thresholds for future scenarios, specifying that any scenario within a small range near the threshold is acceptable or unacceptable, and so on.

It is evident that the ideas behind these methods aim to reduce management complexity by reducing the inherent uncertainty of future space in complex systems management.

3. Decomposition path of associative segmentation

The management elements of a general system are often hierarchical, with relatively few lateral associations. The goals are mostly predictable and directly achievable, and causal relationships are generally explicit. Therefore, management activities can be decomposed into several relatively independent parts horizontally. In this way, the complexity of the system is "limited" within the relatively independent parts and functional departments. This situation allows us to address localized issues based on reductionism and solve problems within each functional department. Ultimately, the overall complexity problem can be resolved through an "integrated" approach, where solutions from different parts are combined.

However, complex systems management increases the integrality between the parts, with closer lateral interactions than the interactions between vertical levels of the complex system. In terms of influence, the interactions between some elements are no longer local but may become global, even gradually expanding to dominate the overall behavior. At this point, the functional design of the complex system is no longer completely predictable, and causal relationships become less direct and obvious.

For example, due to the high integration between elements of a complex system, minor changes in individual elements can be amplified, leading to systemic consequences and the development of failures into accidents. This is why the causal relationships in complex systems management become ambiguous, risks become difficult to foresee, and unforeseen problems occur frequently. The complexity caused by the strong correlations and high integration of management elements in complex systems is referred to as **strong correlation complexity** in complex systems management. Clearly, strong correlations enhance the "degree of complexity." Therefore, it is possible to consider appropriately segmenting the "local" and "global" correlations of complex systems management to reduce complexity.

How can we segment the correlations of management elements in complex systems? This is a problem that requires both theory and experience. The basic ideas are as follows: First, it is important to recognize that the so-called segmentation of correlations is only a physical or logical structural division of the affiliations, inclusiveness, parallelism, causality, and relevance among attributes within a virtual agent's problems, which is a cognitive assumption and idealization rather than a dissection of any inherent physical complexity of the complex system subject.

Before segmenting the correlations of management elements in complex systems, an analysis of the correlation modes should be conducted, following the principles below.

(1) For cases where the correlations in a particular network are relatively weak, segmentation can be applied from the relatively weak points. After segmentation, the complexity of each part will be reduced, and then they can be "reassembled" to restore the original system.
(2) For cases where the above-mentioned relatively weak correlations do not exist, the correlation patterns and strengths can be simplified for "hidden" segmentation, followed by complexity reduction and integration.

Both of these two principles essentially involve the process of "correlation-segmentation-relation." Initially, the management subject attempts to decompose the complexity of complex systems management as a whole (equivalent to a complex machine as a whole) into multiple parts with relatively lower complexity (equivalent to different components of a machine). Based on the analysis and study of each part's complexity, they are then "reassembled" to restore the original machine (equivalent to the restoration of the wholeness of the machine).

It should be emphasized that regardless of the segmentation method used, it is never appropriate to truly disintegrate the physical relationships among the actual management elements in complex systems management activities. Doing so would damage the inherent physical complexity of the complex system.

9.2.4 Entire Decomposition Mode

This section mainly discusses how to decompose wholeness at the macro and micro levels, especially from the perspective of reality and the source, to understand it.

Wholeness from the perspective of reality often manifests as strong correlations, structural organization, or unclear logical mechanisms, giving the impression of "seeing flowers in the fog" or even "chaos." We should initially investigate the root causes that give rise to or trigger wholeness and then focus on unraveling or eliminating these causes to dissect the wholeness.

For example, fundamental uncertainty and strong correlations are the main causes of complex wholeness forms such as non-traversability, emergence, and self-organization in future scenarios of complex systems management. Therefore, the "future" is no longer an extrapolation of the "past" and the "present." Even if all the known rules are valid for the "past" and the "present," they cannot guarantee an accurate calculation or clear explanation of the "future" due to the inherent uncertainty in the longitudinal whole between the past and the future. In this case, to study the "future," which is a complex wholeness, we cannot solve uncertainty problems with deterministic paths but must use uncertain paths to solve uncertainty problems. Even the so-called uncertain paths are not traditional random paths with a known probability distribution but a path of trial and error, feedback, correction, and iteration that gradually converges, resembling "crossing the river by feeling the stones" to solve problems.

Similarly, the amount of information possessed by a subject determines what kind of uncertainty it can overcome and control. The more information one has, the less uncertainty in understanding the wholeness. However, since the uncertainty of the problem is "inherent," the subject can never grasp all the information about the problem. Even if the subject knows what it knows and what it doesn't know, it cannot avoid not knowing what it doesn't know. This leads to the complete understanding of future wholeness only becoming possible when it becomes a reality. In other words, we can only understand a higher dimensional whole in a lower dimensional space of information. Apart from pursuing how to know the scenarios in higher dimensional space, it is more practical to acknowledge reality and minimize cognitive biases.

These examples and others provide us with insights into the "origin and development" of wholeness in complex systems management and outline the behavioral framework for dissecting wholeness.

Finally, we briefly explain why "decomposition" is used to "decipher" complex wholeness. The basic meaning of "decomposition" is to separate a material whole with a knife. Therefore, "decomposition" of wholeness mainly emphasizes the segmentation of physical wholeness in complex systems, while "deciphering" involves "changing," "unraveling," and "removing" abstract attributes of not only physical wholeness but also systemic wholeness and logical wholeness. This corresponds well to the characteristics of complex systems management, which include both theoretical thinking and practical thinking. Hence, our path and objective involve transitioning from intuitive physical perception to a rational process of "deciphering" wholeness.

Next, to facilitate the reader's intuitive comprehension, we will utilize a specific problem in complex systems management as an example to elucidate the central concepts of this section.

It is well known that any management activity has clear management goals, which serve as the "leader" of management activities, determining the direction, understanding of management problems, and the management methods to be adopted. This also applies to the field of complex systems management. Generally, goals are the concretization of objectives and subjective expectations formed in the minds of individuals regarding the desired outcomes of management activities. Regardless of the type of management activity, management goals have a guiding, directional, and regulatory significance in a global sense.

In complex systems management activities, management goals not only possess general properties such as systemicity, feasibility, and achievability but also present a series of complex wholeness challenges. These challenges stem from inherent uncertainties, extensive temporal and spatial scales, adaptability, non-traversability, the coexistence of material elements (e.g., physical functions) and non-material aspects (e.g., social responsibilities) within goals, as well as the presence of multi-level, multi-dimensional, and multi-scale objectives. Furthermore, the interplay and entanglement among different goals make it impractical to decompose the goal system in isolation.

As mentioned in the previous section, the wholeness of complex systems management activities and problems refers to their correlation, functionality, and logical integration. Therefore, based on the premise of holistic thinking, we can consider the separability of internal local correlations, structures, functionalities, and dynamic changes within wholeness. This implies that the seemingly "indivisible" wholeness in abstract thinking is relatively "divisible" in a certain sense, and this does not undermine the wholeness of reality. If, in the later stages of management activities, the original wholeness is maintained through certain integration methods, the wholeness of complex systems management will be dissected to some extent.

Taking the example of the complex systems management activity of water environment governance in lake basins, the overall governance goal is to achieve a "community of shared destiny between humans and the water environment in lake basins." It is easy to understand that this is a strategic vision goal for governance or the ultimate goal of governance. Clearly, investing in the current state of lakes and ultimately achieving the ultimate goal will unavoidably entail a prolonged and continuous process. Along with this process, the water environment governance of lakes will go through multiple stages, each with its own characteristics and corresponding governance goals. After completing the tasks of each stage, dynamic integration and adjustments over multiple time periods will be carried out to gradually approach the final goal. Hence, the comprehensiveness and progress of lake water environment management should be delineated in specific temporal and spatial phases and within relevant contexts to ensure appropriate and transparent meanings. Based on this basic understanding, it is necessary to have a clear division of stages in the overall process of lake water environment governance and to clarify the goals of each stage, especially the "current" goals of lake governance, which have practical significance.

Therefore, based on the paradigm shift in complex systems thinking, it is necessary to clarify the deeper goals of "current" lake governance. Since it is water environment governance in lakes, in a direct sense, controlling and reducing the total amount or concentration of pollutants in lake water (such as biochemical indicators) is naturally a common goal in all stages of governance. The difference lies in the control and reduction standards proposed in different stages. We may refer to this "control and reduction of the total amount or concentration of pollutants in lake water" as the most direct, fundamental, and explicit governance goal of lake governance. While such a goal statement may potentially encourage short-term behavior, the pursuit of immediate benefits, and an action-oriented governance approach by the governing body, it remains the most practical, normative, and actionable "catalyst" for lake water environment management.

Furthermore, considering the inherent uncertainty, large spatiotemporal scales, adaptability, and non-traversability that incur complex wholeness, and deep-level characteristics closely related to regional socio-economic factors in the process of lake governance, it can be found that this goal is difficult to reveal the dynamic principles of lake water environment, the formation and operation mechanism of pollutant supply chains, and the multi-scale evolutionary laws. Therefore, based on the above normative goals, it is necessary to propose goals that reflect the essential attributes of complex wholeness and evolutionary emergence in different stages of lake governance under the guidance of complex systems management thinking, forming a deeper goal system and goal sequence for lake governance in each stage.

So, what are the deeper goals of lake governance in the "current" stage? This requires us to be guided by the main characteristics of the complex wholeness of the water environment in lakes at the "present" stage.

Generally, lakes at their current stage often exist in a long-term state of overload within the context of the water environment and regional socio-economic complex system. The environmental capacity has reached its limit, and both stability and resilience are close to the bottom line. This is the fundamental reason why problems such as fragility, recurrence, instability, and imbalance persist in the lake water environment that we have observed over the long term. Therefore, in the thinking of governance decision-making for such highly sensitive, fragile, and overloaded water environments, in addition to the above normative governance goals, it is necessary to enhance the goal of environmental-economic compensation mechanisms that reflect the resilience and robustness of the water environment, especially under extreme scenarios, based on the complex wholeness scenario. Efforts should be made to transform the focus from "pollution governance" to "water governance" at a deeper level, which means addressing the governance of the "water environment" at a deeper level.

Of course, this viewpoint does not negate the importance of controlling and reducing the total amount or concentration of pollutants in lake water as an important category of governance goals. The formulation of the aforementioned new objectives primarily reflects a comprehensive understanding of the complexity of lake water environment management. It embodies the concept of "generalized water environment governance," emphasizing complexity, comprehensiveness, resilience, and a large-scale perspective.

The "generalization" here fully reflects a paradigm shift in complex systems thinking management in the decision-making of lake water environment governance. The first level of the new thinking is the safety of the water environment as support and guarantee for the development of the regional socio-economic system, including the robustness and sustainability of such support and guarantee. The second level pertains to the self-organization, self-development, and maintenance of a favorable ecological evolution within the water environment system itself. The third level is the positive response of the water environment to protection and restoration through compensation mechanisms designed by the governing body. Therefore, the goal of "generalized water environment governance" is comprehensive governance that embodies the self-organization, self-development, and induced and controllable governance of the Taihu Lake water environment. It has far exceeded the narrow definition of traditional governance based on biochemical indicators of water bodies.

Based on the principle of indivisible relativity, we have explored the paths to degrading complex wholeness from two perspectives: complexity and wholeness. By synthesizing these paths, we have formed the fundamental issue of unraveling complex wholeness. However, complex wholeness in complex systems is always intertwined. It is impossible to completely separate them into "complexity without wholeness" or "wholeness without complexity." Logically speaking, if such separation were possible, complex systems and complex systems management would "degenerate" into simple systems and simple systems management. This should not be interpreted as theoretical "laxity" but rather as a reflection of the necessity for a paradigm shift in complex systems thinking when dealing with complex wholeness. It requires us to simultaneously consider and integrate theoretical thinking, practical thinking, and other types of thinking. Relying solely on an "absolute" and singular mode of thinking is unsuitable and biased for both the theoretical research and practical activities of complex systems management.

9.3 Issue 2: Designing a New Paradigm for Qualitative Transitions

9.3.1 Definition

In reality, when people face a complex systems management problem, they first perceive the problem intuitively as "complex" or realize that using reductionist methods to solve the problem yields little effect. The reason is that the problem not only exhibits general systematicness but also possesses a new essential attribute known as complex wholeness. We refer to the transformation of management activities and problem attributes from general systemic to complex wholeness as a qualitative transition based on complex wholeness.

"Qualitative," or simply "quality," fundamentally means essence or intrinsic nature, which refers to essential attributes. The qualitative transformation of management issues based on complex wholeness reflects the objective laws of the evolution of issue attributes. It requires the management subject to adaptively propose new behavioral guidelines in response to the new quality. In management practice or theoretical research, such adaptive behavioral guidelines have surpassed the perception and experience of individual subjects and solidified into a new paradigm design for qualitative transitions based on complex wholeness. Theoretically, it represents a fundamental scientific issue in effectively conducting complex systems management.

9.3.2 Basic Connotations

The fundamental reason for proposing this scientific issue is that once management activities and problems undergo a qualitative transition based on complex wholeness, the previously effective traditional systemic paradigm will reveal fundamental thinking flaws and capability shortcomings in terms of problem epistemology and problem-solving methodology. Consequently, it becomes difficult to address the new challenges arising after the qualitative transition.

In general, the design of a new paradigm based on qualitative transitions requires a comprehensive approach at three levels: ontology, epistemology, and methodology. It involves how the subject perceives, comprehends, and acts upon the qualitative transition. The new paradigm design should utilize complex systems thinking, knowledge, and discourse systems to reveal the new characteristics, basic laws, and principles arising from complex wholeness that affect management activities and problem formation.

During this stage, it is important to focus on two aspects as much as possible:

1. Clarify the purpose of a paradigm shift, especially the "landing point" after the shift. For example, addressing the mechanism rupture between levels and the ambiguity of causation in holistic phenomena caused by the irreversibility of reductionism. To achieve this, it is necessary to provide the necessary new knowledge and methods.
2. Supplement new details based on the unique scenarios and contexts formed by complex wholeness, enriching and solidifying the practical uniqueness of complex wholeness.

Based on the preliminary work mentioned above, the design of a new paradigm based on qualitative transitions enters the stage of multi-principle integration. This stage mainly includes the following:

1. At the macro level of complex systems management activities, first, guided by the complex wholeness scenario, establish the behavioral principles of the physical-system-management chain progression paradigm through principles such as the "variability" of virtual cognition, indivisible relativity, and unique contextualization.
2. At the micro level, "embed" adaptive principles and multi-scale division into behavioral norms, such as the selection of solutions for complex problems. Through the combined effect of adaptive selection and multi-scale management, decompose the overall complexity of the problem into different stages of the solution generation process. Then, iterate and converge the sub-solutions from each stage toward a final solution to address the overall problem.

By integrating the above points, the connotation of designing a new paradigm based on qualitative transitions rooted in complex wholeness is formed.

9.3.3 Seizing the "Window of Transition"

From a theoretical perspective, qualitative transition based on complex wholeness represents a fundamental change in the attributes of complex system activities and problems, transitioning from general systemic to complex wholeness changes. This abstraction signifies the transition of people's understanding of the essential characteristics of things from intuitive perception to rational thinking, constituting a change in the fundamental nature of phenomena.

The occurrence of such significant attribute transitions in a system undoubtedly has profound underlying mechanisms, and these driving forces and paths should be observed and sought from the basic evolution of the system. As mentioned earlier, if major new situations arise in a scientific field, they can trigger significant changes in the attributes of the research objects and lead to qualitative shifts in the objects. In this sense, whenever a crisis or dilemma arises in the previous form of scientific development, a revolutionary transition in scientific development often emerges, requiring new paradigms to serve scientific revolutions.

Therefore, we cannot detect and measure the quantitative changes in the complex wholeness using observation instruments, as done in experimental sciences in natural science, and determine whether the threshold for qualitative change has been reached based on numerical changes in quantity. The practical approach focuses on the macro level to determine whether unconventional scientific events have occurred, anomalous scientific phenomena have been identified, or even if crises in scientific development have been recognized, any of which could subsequently cause the original paradigm to become unsuitable or not adaptable. If such phenomena begin to occur frequently and gradually in reality, it is highly likely that the "time window" for qualitative transitions is forming.

Richard Bookstaber titled the first chapter of his book *The End of Theory* "Crises and Sunspots," leaving aside many headache-inducing details. Just the title of this chapter and the summary of its important content are enough to provide us with evidence supporting the viewpoint of this subsection.

At the beginning of this chapter, Bookstaber straightforwardly discusses the global financial crisis that ravaged the world in 2008. Queen Elizabeth of the United Kingdom visited the London School of Economics and asked a question: Why did no one foresee the financial crisis? Robert Lucas, an economist at the University of Chicago, candidly replied, "Economics cannot provide useful services for the 2008 economic crisis because the establishment of economic theory makes it unable to predict such a crisis." This question and answer alone carry profound implications. It indicates that not only did a major crisis occur in the real economy, but economics itself also faced a major crisis. It is worth noting that economics, as a scientific field with a long history, renowned scholars, diverse schools of thought, and abundant theories, has a significant influence on global economic development. Yet it "cannot predict a crisis." One can imagine how disappointed economists would be. No wonder Bookstaber named his book exploring this financial crisis *The End of Theory*.

Contrasting with the previous paragraph in this subsection: "If major new situations arise in a scientific field, they can trigger significant changes in the attributes of the research objects and lead to qualitative shifts in the objects" and "If unconventional scientific events, discover anomalous scientific phenomena or encounter a crisis in scientific development at the macroscopic level, causing the failure of the original paradigm, if such phenomena begin to occur frequently and gradually in reality, it is highly likely that the'time window' for qualitative transitions is forming." Bookstaber is precisely following this paradigm shift and demonstrates it in his subtitle "Financial Crises, the Failure of Economics, and the Triumph of Interdisciplinary Interaction." Although traditional economic theory has "ended," a new "window" of scientific paradigm has opened. *The End of Theory* is, in fact, Bookstaber's design of a new paradigm for qualitative transitions after opening this "window." Surprisingly, this renowned economist, at this crucial turning point in major theoretical development, did not reinforce the theoretical dam to withstand the impact of the next crisis using traditional theoretical paradigms. Instead, he approached the essence of economic theory from a completely new height and perspective of complex systems thinking, even proposing the hope of predicting and avoiding economic crises and helping us recover from them through a subject-based modeling methodology [4]. This approach does not assume that the world is governed by mathematically defined automata; on the contrary, it stitches together the achievements of recent scientific research on real-world complex systems, particularly four concepts that carry both technological allure and significant inspirational significance: emergent phenomena, ergodicity, radical uncertainty, and computational irreducibility.

By opening this new "window," Bookstaber not only provides an example of the new paradigm design for qualitative transitions but also directly and precisely chooses the methodology of complex systems thinking and complex systems management. This undoubtedly enhances the "popularity" of this book.

9.4 Issue 3: Complexity Spectrum Analysis

Although "complex wholeness" is an integrated concept formed by the combination of the relatively independent concepts of "complexity" and "wholeness," the term "complexity" here does not refer to the usual meaning of being numerous or mixed but specifically refers to the "indivisible" wholeness attribute. However, as discussed earlier in this book, many of the causes of indivisibility are driven and induced by various complex mechanisms and correlations in complex systems management activities, scenarios, and problems. Conversely, the feature of "indivisibility" also gives rise to and strengthens the complex forms, patterns, and degrees of complex systems management. This is what we discussed in Chap. 8—the independent complexity that is caused and strengthened by complex wholeness or further triggers the entanglement and emergence of complexity between complexity and wholeness.

This perspective tells us that although complex wholeness is an essential characteristic of complex systems management, the impact and role of relatively independent complexity on complex systems management activities and problems are comprehensive, profound, and throughout the entire process. In particular, since complexity often provides a more vivid and intuitive perception than wholeness, and because the complex forms, patterns, and degrees of complex systems management are "complex" in themselves, it is necessary to carefully analyze "complexity" at multiple scales and examine the impact of different scales of complexity on complex systems management and problems. This provides more precise and effective support for complex systems management and gives rise to the fundamental scientific issue of complexity spectrum analysis.

9.4.1 Overview of Complexity Spectrum

In Sect. 7.4.2 of this book, which discusses the "multi-scale perspective of wholeness," it is proposed that the functional structure of complex systems management has an expanding trend from constitutive functions to generative functions and emergent functions. This evolutionary law is referred to as the complexity spectrum of complex systems management.

Similarly, based on certain criteria, we can systematically discriminate and measure the complexity patterns, causes of formation, and their impact on management in complex systems management. This is the basic idea of the complexity spectrum in complex systems management theory. Essentially, this idea subdivides comprehensive complexity into multiple scales to reveal the different roles of complexity with different forms and characteristics in complex systems management activities. It promotes the enhancement of the subject's cognition and control capability of overall complexity.

According to the above idea, the introduction of the concept of complexity spectrum aims to enhance the identification of fine structures and subtle differences in complexity. Therefore, it primarily addresses two questions: what criteria to use and how to construct such criteria to distinguish fine structures and subtle differences in complexity. Since complexity in complex systems management originates from the irreversibility of reductionism, it is not possible to define a precise quantitative standard and design an algorithm to measure the degree of complexity. The practical approach is to analyze the causes of complexity formation in a given context. If there are more causes, the complexity tends to be more severe and essential. This is equivalent to constructing a multi-factor single-value mapping, where the various causes of complexity correspond to different factors, and the mapped value represents the degree of complexity, forming the complexity spectrum.

9.4.2 Complexity Spectrum Analysis

The identification of reasons that are sensitive to the degree of complexity and the visualization of fine structures of complexity constitute the task of complexity spectrum analysis. It is analogous to constructing the aforementioned "multi-factor single-value mapping." The analysis involves the following aspects:

1. Complexity spectrum quality of system element characteristics. Generally, if the number of constituent elements in a system increases, if the elements evolve from homogeneity to heterogeneity, or if the heterogeneity within the elements becomes stronger, such as in a city system where the scale and variety of constituent elements, as well as the diversity of interests, goals, and value demands among urban subjects, increase, the complexity of the urban system will be higher.
2. Complexity spectrum quality of system element correlations. If the correlation among elements within a system shifts from low to high, from a single mode to multiple modes, and the content of the transmission and transformation between system elements becomes more diverse, not only in terms of material, but also informational, energy, and logical transmission, and the transmission speed accelerates from slow to fast, the complexity of the system will increase.
3. Complexity spectrum quality of system hierarchical structure. Generally, if a system exhibits a multi-level structure at the macro level, it can be decomposed into micro, meso, and macro levels. Within different levels, certain elements and their correlations can form relatively independent systems. If these elements are highly heterogeneous and closely correlated, "strong correlations" will give rise to distinctive horizontal hierarchical complex forms. Furthermore, between different levels, particularly through some dynamic mechanisms from the micro to the macro level, complex wholeness emergence phenomena may arise vertically. In both cases, whether it is the horizontal or vertical scenario, the overall complexity of multilevel systems tends to increase.
4. Complexity spectrum quality of system multi-scale content. The complexity spectrum fundamentally falls within the scope of multi-scale analysis. Thus, regardless of the specific characteristic dimension, as long as it can be divided into different scales, it can be represented by a complexity spectrum. For example, at different time scales, artificial systems exhibit different types of functional phenomena, such as constitutive, generative, and emergent functionalities. This can be regarded as a representation of a functional complexity spectrum. Another example is in Sect. 9.2.3, where considering multiple time scales in the design of lake water environmental management goals, it progresses from a single "pollution governance" goal to a "water governance" goal based on complex wholeness scenarios, especially considering the robustness and resilience of water environments under extreme scenarios and the water environment economic compensation mechanism. Finally, it evolves into a "generalized water environmental governance" goal that embodies the self-organization, self-development, and induced and controllable organization of the comprehensive lake water environment. The complexity level increases as a result.
5. Complexity spectrum quality of system-environment uncertainty involves the connotation of uncertainty. Uncertainty can be divided into intrinsic uncertainty of the system, such as uncertainty exhibited by stochastic systems or uncertainty exhibited by deterministic systems with nonlinearity, the uncertainty of environmental scenarios, and uncertainty arising from the complex

systems management-environmental composite system, resulting in the emergence of new scenarios. As long as these forms of uncertainty exist, complexity is inevitable. The problem lies in the fact that, among various types of uncertainty, the more irregular and difficult to predict the uncertainty is, the more "severe" the complexity becomes. Moreover, in these scenarios exhibiting uncertainty, if there are also strong correlations, multiple scales, adaptability, or even essential uncertainty and unpredictable phenomena arising from the complex systems management-environmental composite system, then such complexity will inevitably manifest as extreme complexity.

6. Complexity spectrum quality of system subject adaptability. Section 7.5 of this book points out that the so-called subject adaptability refers to the ability to actively change their characteristics, behaviors, organizational patterns, and functions in response to changes in the external environment and conditions in order to maintain coordination with the new environment and continue their ability to survive and develop. Based on this understanding, the adaptive behavior of subjects inevitably gives rise to new phenomena, scenarios, and problems. Moreover, these phenomena, scenarios, and problems often exhibit characteristics such as suddenness, lack of precursors, and emergence. Therefore, adaptability is not only a form of uncertainty and even inherent uncertainty but also a representation of complexity.

   According to this logic, the capacity and intensity of the subject's adaptability will become the driving force behind the degree of complexity. For example, if the subject's association with the external environment is primarily manifested as a "stimulus–response" form, then by focusing on the complexity of environmental changes, we can understand the complexity of the subject's behavior. However, if the subject has a certain degree of adaptability to the environment, and the adaptive responses are relatively passive, simple, weak, and slow, it can be considered that the resulting complexity is generally predictable and manageable to a large extent. However, if all the aforementioned capacities and energies of responses are significantly enhanced, and the subject, in order to gain the initiative in survival and development, changes from a passive state relative to environmental changes—for example, actively creating changes even when the environment has not changed, or creating significant changes from minor ones, or even causing disruptions and extreme scenario changes—and wins through its dominant adaptive strategies, then if this new complexity spectrum dynamic strategy is applied to the complex systems management, it will result in multiple complexities caused by multiple factors such as multiple scales and essential uncertainty, and the consequences will be more severe. This is often adopted in major complex game situations and is worthy of our attention and prevention.

7. Complexity spectrum quality of the subject's cognitive limitations. The understanding, deciphering, and mastery of complexity, especially the "severe" complexity, mainly rely on the subject's grasp of complexity information, particularly complexity knowledge. The more comprehensive, complete, and profound the subject's understanding of complexity, the more the difficulties caused by the

complexity spectrum phenomena in complex systems management can be alleviated. Because of the intricacy inherent to complexity, it is not feasible for us to possess comprehensive knowledge of all information related to complexity. In general, regarding complexity, we can know some things and not know others. Furthermore, we know what we know and know what we don't know. Regardless of who it is, there are certain limitations, and it is impossible to know "infinitely many" things. It is precisely because of this that there are things we don't know about what we don't know, which creates a cognitive blind spot in the face of complexity. Often, it is precisely this last point that leads to us losing our direction or losing our ability to control in the face of complexity.

We summarize the above content with the following diagram (Fig. 9.1): The straight line in the diagram represents the increasing severity of complexity from left to right, which corresponds to the increasing intensity of complexity, or what is referred to as the complexity spectrum in this section. In plain language, it means that complexity is becoming "more severe" or "more intricate."

Below the spectrum, we list various types of attributes closely related to the degree of complexity, indicating their main characteristics and how these characteristics change. From this diagram, we can see how the overall degree of complexity is closely associated with various attributes.

The attributes closely associated with the degree of complexity are not limited to the few mentioned above; those are just a few representative examples. When complexity is concretized, according to the principle of uniqueness contextualization, there will be additional attributes that adequately reflect the characteristics of the real world to enrich the complexity spectrum.

9.4.3 Value of Complexity Spectrum Analysis

Complexity spectrum analysis provides us with the logical relationship between the degree of complexity and the attributes closely related to it. This actually gives us a starting point for changing the position of complexity in the spectrum and further adjusting the degree of complexity in a real system by changing the features of the elements. This idea has practical significance for how to operate complexity degradation and break the complex wholeness. The main process is as follows:

1. Firstly, clarify the specific requirements for changing (generally degrading) the degree of complexity, that is, the need to move the position of complexity in the spectrum to the left.
2. Analyze which associated elements (one or a group) are mainly involved in achieving this requirement and explore how to regulate the attribute features of the elements based on the knowledge of their relationship with the degree of complexity.
3. After multiple explorations and modifications, influence the degree of complexity by changing the attribute features of the elements.

Complexity Spectrum: Complexity increases from left to right	
Characteristics of system elements	Lower quantity → Greater quantity Homogeneity → Heterogeneity Loose connection → Close connection Single connection method → Diverse connection methods
System hierarchical structure	Single level → Multi-level
Scale of the Features	Single scale → Multi-scale
Uncertainty	Certainty → Weak uncertainty → Strong uncertainty → Inherent uncertainty
Adaptability	Stimulus-Response → Weak adaptability → Strong adaptability → Adaptability that creates complexity
Knowledge and Capability of Subjects	Fewer knowledge limits and strong capability → More knowledge limits and weak capability

Fig. 9.1 Analysis of the complexity spectrum

It is not difficult to understand that in the above process, the attribute features of the elements are equivalent to factors in multi-factor mapping or state variables in control systems, while the degree of complexity is equivalent to the mapping values in multi-factor mapping or the state variables in control systems. For example, reducing the closeness of associations between elements, simplifying complex associations between elements, imposing restrictions on adaptive behaviors of subjects, and avoiding excessive refinement of multi-scale divisions can all achieve the purpose of reducing the degree of complexity to some extent.

9.5 Issue 4: Robustness Management

"Management quality" is not a commonly used concept, but it is an important topic for the management of complex systems. Firstly, "management quality" implies complex wholeness, which is the same as the attributes of complex systems management. Secondly, "management quality" has a broader conceptual connotation beyond the functions and topics of management activities. For example, in the field of decision-making, management quality refers to the quality of management decisions; in the field of risk, it becomes the quality of risk prevention and control; in the field of economy, it becomes the quality of economic development; and so on. Considering the above two points, studying the management quality of complex systems management at a higher level is more universal than studying the quality of functional management activities in specific fields.

9.5.1 Overview of Management Quality

When it comes to management activities, people often talk about the level of management, the quality of management plans, and so on. The core of these evaluations of management activities and management plans is a kind of thinking and description of management quality. This indicates that management quality is a very important and profound concept in the field of management.

The concept of quality originated in the manufacturing industry and was initially aimed at physical products manufactured, such as the physical properties of product materials, durability, and stability, perceived by people as the most direct aspects of product quality. People's initial understanding of engineering quality was also mainly within this category, measuring engineering quality based on the physical attributes of artificial engineering hard systems, such as whether the engineering is robust and whether it can withstand environmental changes.

With the expansion of human cognition about the connotation of "manufacturing," more and more products, mainly in the form of non-physical attributes, have been "manufactured." For example, management activities have "manufactured" management plans as "products." Because the functionality of management plans is no longer based on physical attributes but on their non-physical attributes, namely the specific regulatory aspects of the management plan being reflected through the effectiveness and efficacy of specific provisions of management plans in solving practical problems, this gives rise to implications such as whether the management plan is reasonable, effective, can remain robust throughout its entire lifespan, whether it satisfies people, and the level of satisfaction. This actually forms the basic meaning of management quality as follows.

Firstly, management indeed has quality attributes. Firstly, management is an activity that serves a certain practice, such as production and construction practices. Secondly, management is a service activity with ultimate products, and its ultimate product is the management plan chosen and provided by the management subject. Thus, the quality of management plans can be directly regarded as the quality of management activities. It needs to be pointed out that since the management plan is the final product of the management process, the quality of management activities should not only refer to the quality of management plans but also include the quality of the process of creating management plans. In other words, the quality of management activities should be viewed and analyzed from the perspective of the entire management process. These two aspects are summarized and collectively referred to as management quality.

Secondly, when evaluating the quality of management, it is important to first respect its objective role and effectiveness in solving practical problems. The satisfaction of the management subject can only be considered within the premise of respecting its objective role and effectiveness while also taking into account the subject's preferences and values. The subjective value preferences of the subject should never override the objective effectiveness of the management plan.

Thirdly, it is necessary to be cautious of falling into cognitive misconceptions regarding the concept of management quality, for example, using the physical quality of a material product as a substitute for management (plan) quality. The physical quality of a product primarily describes the attributes of its physical performance, including the continuity of physical functions (such as product lifespan) and the reliability of physical indicators (such as pressure resistance and load-bearing capacity). On the other hand, management quality mainly considers whether the management plan can continuously and robustly deliver established functions and effects throughout the entire life cycle of production and construction, especially when significant situational changes occur. Therefore, there is a significant difference between the physical quality of products (engineering) and management quality, and the two should not be confused.

The Three Gorges Dam project in China, constructed on the Yellow River, which carries the highest sediment load in the world, is the first reservoir built on this river. It has proven its engineering quality in terms of resistance to high-sediment water flow erosion for turbines and hydraulic structures, reservoir operation, and maintenance, among other aspects. However, due to the incorrect estimation of the project's role in soil and water conservation, the design did not account for the need for sediment discharge, resulting in severe sediment accumulation. This has transformed the Wei River into a suspended river, leading to the peculiar phenomenon of "small water, big disaster" in the Wei River basin, constantly threatening the lives and property of local residents. This reflects a serious error in the management decision-making of the Three Gorges Dam project at a macro level. This case demonstrates that, in practical situations, the final quality of a major project needs to be evaluated through the comprehensive assessment of both management (decision-making) quality and the physical quality of the project itself. Different situations may arise, such as high decision-making quality but low engineering quality, low decision-making quality and low engineering quality, low decision-making quality but high engineering quality, or high decision-making quality and high engineering quality. This case illustrates that any instance of poor quality in either aspect can result in significant losses. Therefore, what we should strive for and recognize is the pursuit of "high quality" in both management and engineering aspects.

Certainly, if a certain practical activity and its corresponding management activity are relatively simple, both the design of the management plan and its implementation are relatively easy. As a result, the ultimate realization of this practical activity is more certain. In such cases, the synchronization and consistency between management quality and physical quality are easier to achieve. However, for complex systems management, achieving this is not easy. Therefore, it should not be taken lightly to equate the management quality of complex systems with the physical quality of complex man-made systems.

9.5.2 Management Scenario Robustness

From the previous section, we have experienced that adaptability is a quality attribute of management plans. Based on this, an important idea arises: how do we measure adaptability? To answer this question and avoid ambiguity, we cannot directly use adaptability to measure adaptability. Instead, we need to introduce a new concept, which is the "scenario robustness" introduced in this section.

Chapter 7 of this book points out that the natural, social, economic, and large-scale temporal and spatial evolution of the environment are important reasons for the various uncertainties in complex systems management activities. This characteristic has the greatest impact on the management quality in management activities. This is mainly because the management plan must maintain effectiveness and robustness throughout its entire life cycle, and the various scenario evolutions generated by the environmental system and the management-environmental composite system during this period can potentially have a serious impact on the effectiveness and robustness of the management plan. If such impacts lead to a decrease or failure in the effectiveness and robustness of the management plan, it means that the management quality is not high or there are problems with management quality. This viewpoint expresses our thinking paradigm of measuring and judging management quality from the perspective of adaptability to scenario changes.

Now, let's analyze this paradigm in detail. Firstly, when designing and demonstrating a management plan, the management subject generally takes into full consideration the influence of various changes in the natural, social, economic, and natural environment on management activities. At this time, the management subject considers the environment as the background of the management activities and thinks about how to respond to these influences in the design of the management plan in order to ensure the coordination between management quality and the environment. However, regardless of the approach, the coordination in real-world complex systems management activities can only be limited and relatively insufficient, resulting in possible "shortcomings" and "deficiencies" in management quality.

Furthermore, once the management plan is formed, a new complex management-environment composite system is formed as well. The new system may generate emerging scenarios that could affect the effectiveness and robustness of the management plan, thereby reducing its quality. In severe cases, it can even lead to the failure of the management plan.

In any situation, changes and evolutions in scenarios can have various possible impacts on the quality of the management plan. The quality attribute of the management plan is characterized by its ability to maintain the normal release of effectiveness and maintain robustness in response to different influences. This quality attribute is referred to as the scenario robustness of management (activities or plans). Generally, the former is called the first-class scenario robustness, while the latter is called the second-class scenario robustness. This understanding has been abstracted by Sheng (2017) as the fundamental concept for evaluating and measuring the quality of complex systems management, known as management scenario robustness [5].

The robustness and adaptability of management scenarios have different meanings. Management robustness refers to the measure of the sustained functional stability of management plans, while adaptability refers to the ability to make timely adjustments in the face of environmental changes and various uncertainties to maintain the functionality of the plans.

For complex systems management problems with inherent uncertainty, the robustness of management scenarios is of particular significance. We are usually most concerned about the impact of environmental changes on management plans, that is, whether the plans can maintain their functionality in a changing environment. This is referred to as the first type of management scenario issue [5]. At the same time, the effectiveness and impact of management plans can span time and space. Especially during the life cycle of management plans on a large temporal and spatial scale, the complex systems management-environmental composite system will undergo new changes as the functionality of the plans is implemented. These new changes will give rise to new scenarios that affect the functionality of the plans. At this point, the question is whether the management plans can still maintain their functionality in the face of new changes in the complex systems management-environmental composite system. This is referred to as the second type of management scenario issue.

To distinguish these two types of issues, the concepts of explicit robustness and implicit robustness of management plans can be used for study. For the first type of management scenario issue, the uncertainty that allows management plans to achieve the specified performance level is called explicit robustness. For the second type of management scenario issue, the uncertainty that allows management plans to achieve the specified performance level is called implicit robustness. In summary, whether the environment is changing on its own or due to changes caused by management plans, we are concerned with how the management plans respond to environmental changes. The study of scenario robustness includes both types of robustness, but implicit robustness is particularly important and should be given attention and precaution in the complex systems management with deep uncertainty to proactively avoid new scenario risks.

Thus, we have formed a new understanding of the quality of complex systems management: (1) Under inherent uncertainty, traditional management theories and methods are no longer applicable to complex systems management. Improving, optimizing, and creating new theories and methods of complex systems management are necessary to measure and improve the quality of management plans. (2) A high-quality complex systems management plan should exhibit robustness to various scenario changes throughout its entire life cycle. Robustness can be further categorized into two types, and, therefore, scenario robustness is the core feature (attribute) of the quality of complex systems management plans. It is an objective attribute that measures the effectiveness of a management plan relative to environmental changes in terms of stability or alignment.

9.5.3 Robustness Management

Based on the basic concept of management quality and the complex wholeness properties of complex systems management, this section introduces the concept of scenario robustness to measure and evaluate the quality of complex systems management. The basic idea is that the more a management plan can demonstrate scenario robustness, the higher the quality of management, and it becomes the goal of high-quality management behavior. In a paradigmatic sense, the behavior of the subject will constitute a management mode based on scenario robustness (with scenario robustness as the core goal), which we refer to as scenario robustness management.

In practical thinking, complex systems management activities exhibit various forms and patterns. However, at the theoretical level, we need to start from the basic characteristics of different activities. Of course, we can abstract and understand these characteristics from different perspectives. For example, due to the multi-scale nature of complex systems management goals, it can be considered as a type of multi-scale management. From the perspective of the "iterative" path of management plan formation, it can be seen as a type of "iterative" management, and so on. However, the phenomenon that best reflects the unique attributes of complex systems management should be as follows: The management entity must develop a management plan within a relatively brief time frame that can consistently ensure normalcy and resilience over an extended period. During this relatively long period, the management environment will experience various complex scenarios and scenario evolutions due to its inherent uncertainty, which will impact the effectiveness of the management plan. Therefore, the scenario robustness of complex systems management plans becomes crucial. Without this quality, the effectiveness of the management plan may be compromised or even fail during its lifespan, directly affecting the achievement of the management subject's expected goals and the realization of the management's own value.

In general management theory, the concept of optimal management plans is often discussed, which means that based on certain goals, the management subject selects a plan that is superior to other feasible alternative plans through certain methods, such as establishing a structured mathematical model. This is known as the optimal plan. However, for complex wholeness management problems with inherent environmental uncertainty, it is difficult for us to accurately and comprehensively propose management goals at a large temporal and spatial scale and find such optimal plans. We can only choose plans that are insensitive to damages caused by uncertain scenario changes as acceptable "high-quality" plans while taking into account the inherent uncertainty and complexity of the system.

Therefore, this approach differs significantly from traditional optimal solutions, as it overall better reflects the consideration of the subject's perception of environmental uncertainty and the complexity of managing changing scenarios. This consideration inspires us to view scenario robustness as the "optimality" of complex systems management plans to some extent, and scenario robustness management can be seen as a type of "optimality" management within complex systems management.

As a mode of operation, scenario robustness management embodies the basic principles of complex systems management in its basic process. For example, based on the principle of complex wholeness cracking, after establishing the goals and functional design of the management plan, some wholeness management issues can be decomposed into several relatively independent subproblems using appropriate decomposition methods. Individual management sub-plans can be proposed for each subproblem. Furthermore, under the principles of adaptive selection and "iterative" generation, an overall plan compatible with these sub-plans can be directly formed, or an integrated management plan can be created by adjusting some of the sub-plans (Fig. 9.2).

In addition, the most crucial aspect of scenario robustness management is to perform scenario forecasting. Through scenario adaptability analysis, phased management plans can be proposed, and scenario robustness can be assessed and measured. If the quality satisfaction of the plan is not yet recognized, the scenario robustness can be measured under the premise of scenario assumptions and reconstruction. Further assessment and recognition can be conducted based on this

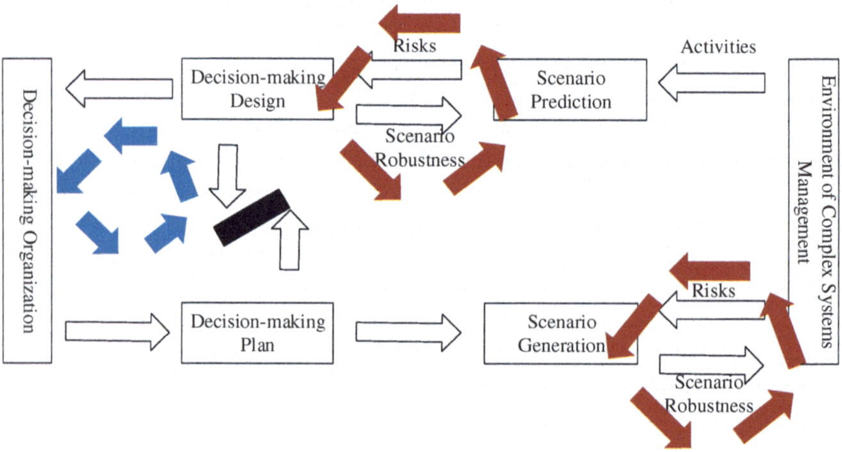

Fig. 9.2 Flowchart of Scenario Robustness Management

measurement, and this iterative feedback process continues until a satisfactory plan is obtained (see Fig. 8.3).

9.6 Issue 5: Essential Management

9.6.1 Overview

In theoretical terms, complex systems management refers to the management of complex wholeness problems within complex socio-economic systems. Specifically, the concept of complex wholeness and irreducible wholeness, particularly in relation to the irreversibility of reductionism, represents an abstraction and condensation of the objective essential characteristics of a certain subject. Any attribute must have a specific carrier. Therefore, at the practical level, complex wholeness problems, which are the objects of complex systems management, serve as the concrete manifestations of these attributes. The study of pure complex wholeness, on the other hand, is the task of complex systems management theory.

The seven fundamental principles of complex systems management presented in Chapter 8 of this book are, in fact, a summary and refinement of the thinking principles and behavioral norms for dealing with complex wholeness problems. For example, although complex wholeness problems are "indivisible" due to the irreversibility of reductionism, we can establish a relative thinking approach to decipher complex wholeness and enhance a deep understanding of the complexity of the problem by uniqueness contextualization. Furthermore, by employing a physical-system-management chain progressive paradigm, we can achieve an orderly transformation of complex wholeness problems from the physical space to the system space and then to the management space. In the process of constructing feasible management solutions, management behavior paradigms such as multi-scale management, adaptive selection, and iterative generation are employed to ensure the robustness of the management plan in various scenarios.

It is evident that, based on the core concepts in Chap. 7, the seven fundamental principles in Chap. 8 form a relatively complete methodological system for understanding, analyzing, and solving complex wholeness problems in complex systems management. From this, one can also perceive that the entire methodology system is guided by dialectical thinking, not an absolutist approach but a relative perspective in examining, analyzing, and addressing the "indivisibility" of issues. For instance, it's not about "not dividing at all" but rather about dividing with a sense of measure, dividing without causing harm, initially refraining from division, and then engaging in exchanges or compensation within the dimensions of time and space when necessary, and so forth. It should be noted that management ideas, principles, and methods cannot rely solely on rigid rules. Instead, they require imaginative and wise thinking to design rules and guide behavior. This philosophical thinking is crucial and valuable in dealing with the complex wholeness issues in complex systems management.

For example, in general, systems management, the ability of the subject to cook with rice (having resources), is considered a basic skill, and it's understandable and forgivable if they can't cook without rice (when resources are scarce). However, in complex systems management, the ability of the subject to cook without rice (to manage in resource-scarce situations) is a common skill, while cooking well without rice (effectively managing in such situations) is regarded as a mark of management "expert."

The new challenge arises in that the management of complex systems is "complex," and we cannot expect to solve all types of complex wholeness problems simply by applying the aforementioned mode of thinking and methodological paradigm. For instance, in holistic thinking, sometimes or in some scenarios, we need to face the complex wholeness of problems directly and completely rather than adopting relative approximations or decompositions. We refer to the management that directly confronts the essential attributes of complex problems (complex wholeness, irreversible reductionism) as Essential Management. Clearly, Essential Management still pertains to the management of complex wholeness problems, but its uniqueness lies in directly and completely addressing the essential attributes of problems rather than dealing with them in a relative manner.

It goes without saying that essential management, in terms of the universality of behavior and the integrity of processes, will distinctly demonstrate that complex wholeness is the origin of the problem's formative characteristics. It views, contemplates, and analyzes problems from the perspective of complex wholeness and uses complex wholeness as the "entry point" for solving the difficult aspects of problems.

Thus, essential management is not only a scientific issue of fundamental significance in complex systems management but can also be considered as a paradigm of complex systems management.

9.6.2 Theoretical Analysis of Essential Management

For the practice of complex systems management activities, the questions of "how to do it" and "how to do it well" cannot be solely accomplished and mastered based on personal, limited experience or individual wisdom that comes to mind. Instead, it requires the establishment of guiding principles and behavioral guidelines for management practices, the formulation of specific plans and processes, and the selection of appropriate technical approaches, all of which must be premised on a profound understanding of the essential attributes of the management entity and the objective laws of management. The deeper, more accurate, and more comprehensive this understanding is, the stronger the scientific, operational, executive, and affective aspects of complex systems management practice will be, leading to higher overall quality in management practice. This implies that high-quality complex systems management directly addresses the inherent complexity of the problem, and the ability to analyze the causes, evolution, and "entry points" for solving problems directly from this intrinsic property constitutes the concept and paradigm of essential management.

As revealed earlier, the theory of complex systems management is not primarily about discussing general principles of knowledge in management activities, nor is it mainly about studying specific technical methods in management activities. Instead, it is primarily about designing basic principles for management operations and behaviors based on the essential attributes of the management entity and the objective laws of management activities. Therefore, essential management includes both the essential attributes of the management entity and the phenomena and methodology of management practice. It is proposed on a "semi-virtual, semi-real" level to guide how to start from the intrinsic attributes of complex systems, discover and explore the roots of the generation, formation, and development of management problems, propose solutions from the roots, and strive to achieve long-term effectiveness and stability in management outcomes. In short, essential management in complex systems management emphasizes fundamental and source-oriented management.

In general, complex systems management problems first manifest as various physical forms perceived in human senses, referred to as the "appearance" of the problem. Different individuals, due to their varying experiences, even when faced with the same problem, may generate inconsistent appearances. Moreover, the essential attributes of problems, such as their characteristics, formation mechanisms, and evolutionary mechanisms, cannot all be manifested as visible appearances. This tells us that it is insufficient to deeply and comprehensively understand the roots of problems and to fundamentally solve problems solely based on the appearance of problems. In other words, for complex systems management, we cannot only see the phenomenon without understanding the essence or "treat the head when it hurts and the foot when it hurts."

Of course, we must not assume that essential management can always find the cause-and-effect relationship and the mechanistic "root" of the problem because, according to the theory of complex systems management mentioned earlier, the formation and evolution paths, forms, and features of complex systemic problems do not always have clear causal, mechanistic, or physical "roots" like structured physical systems. More often, on the one hand, problems are closely related to various changes in the external environment, profoundly influenced by various environmental changes. On the other hand, there are various explicit or implicit correlations among the elements of the problem itself. The "sum" of these correlation effects is the "root" of the complex systems management problem, and these correlation effects are widespread at the physical, systemic, and logical levels of the complex systems management scenario and problem. The modes of action include multiple levels and multiple scales and exhibit both determinism and uncertainty, even deep uncertainty. All these actions ultimately lead to various forms of complex dynamic evolution in the manifestation of complex systemic problems, such as mutation, emergence, submergence, and so on.

For example, in the overall aspect of complex systems management, not only are there numerous and varied elements, but there are also various complex relationships among elements and subsystems. When the state of one part of the elements changes, it can be transmitted to other interrelated elements, triggering the phenomenon of so-called multiple transmission. This multiple transmission often exhibits characteristics such as complex regularities, unpredictable trajectories, and difficulties in prediction. This may lead to a more extensive and harder-to-identify complex wholeness. Furthermore, if this process is gradually or sequentially magnified or upgraded under the action of certain implicit dynamic mechanisms or certain accidental perturbations, it may lead to the environmental state in management surpassing a certain threshold, resulting in the phenomenon of qualitative change.

In this sense, the overall state of management situations, due to their inherent essential uncertainties, long-range correlations, multiple transmissibility, and the influence of certain random factors, will exhibit new dynamic mechanisms that are difficult to detect in advance. Everything may appear normal, but "essential" risk issues stemming from complex wholeness are actually forming and occurring.

Since complex systems management itself is also a complex system, even if "essential" risk issues are a type of low-probability event, as Murphy's Law points out, over a large time scale, such risks may become a reality and cause significant losses. Therefore, in the practice of complex systems management, it is advisable to use essential management principles to address such risks and make preparations for preventive measures against high-probability occurrences. Thus, the essential governance principles of complex systems management can enhance our level of management cognition, explore deep-seated rules of management, and enhance the ability to handle the intrinsic complex wholeness of problems.

Therefore, it is easy to understand that essential management is not always about finding and must find a "root" of physical causality for a complex wholeness problem. On the contrary, what is more important and realistic than the physical "root" is to start from the systemic complex wholeness and, based on systemic connectivity and wholeness thinking, analyze and explore the possibilities of complex wholeness problems in terms of connectivity, structure, and logic. This includes understanding the likelihood and trends of the emergence of complex wholeness problems and why they may occur and assessing the magnitude of these possibilities. These aspects constitute the "roots" of more general and universal issues in complex systems management. In other words, the complex wholeness is the deeper "root" of problems in complex systems management. This forms the fundamental essence and starting point of essential management.

The above analysis tells us that in the practice of complex systems management, the theoretical understanding of essential management can be interpreted as follows: grasp the intrinsic attributes of the problem and simultaneously integrate the principles of essential management with the specificity, reality, and uniqueness of management activities. This integration forms the implementation plans, processes, methods, and techniques of management practices, enhancing the practical execution and quality of execution in the management field.

9.6.3 Typical Examples

Now, let's examine a few typical examples of essential management in different fields.

Example 9.1 In March 1979, a major accident occurred at the Unit 2 reactor of the Three Mile Island Nuclear Power Station near Harrisburg, Pennsylvania, USA. The accident involved a loss of coolant and the release of radioactive material. Following the accident, approximately 200,000 residents in the vicinity of the nuclear power plant were evacuated. The President of the United States personally inspected the accident site and organized a team of experts to investigate the incident. The investigation, led by renowned safety expert Perrow, concluded that no individual was responsible for the accident. Instead, the complexity of the nuclear power plant system was identified as the primary cause, leading to the development of the Normal Accident Theory. This theory is significant for our understanding and control of on-site risks in complex systems management. In the following discussion, we will combine the complex wholeness characteristics of complex systems management in general site to introduce the core ideas of this theory.

Complex systems management involves a complex physical-management-behavioral composite system with close connections among various heterogeneous elements. As a result, when one element fails, it can propagate the failure to other elements, leading to the occurrence of multiple failures. Within the physical system at the site, this propagation process can be rapid and catch people off guard, rendering them defenseless. Systems with such characteristics are referred to as strongly correlated systems, and they exhibit greater complexity in terms of system behavior and characteristics. For example, strongly correlated systems often "escalate" failures from the micro level to the macro level, where minor local failures transform into global risks and result in significant losses. In this sense, the occurrence of risk in a system is "normal," as such risks fundamentally arise from the system's complex wholeness, meaning that the strong correlations between system elements lead to the propagation of element failures and evolution into system-level accidents. It's worth noting that this evolutionary trend often follows unpredictable paths and modes, imbued with inherent uncertainty, thus manifesting distinct characteristics of essential risk. When such risks are formed or emerge, people often struggle to understand their causes and mechanisms. The irreversibility of risks prevents us from stopping this trend or restoring the system's original state, resulting in either chaotic responses or normal operations that exacerbate the situation.

The above understanding is of great significance for applying complexity thinking to examine, understand, and analyze the origins and evolution of risks originating from interdependencies in complex systems management. According to the "Normal Accident Theory," the control mode and methods for such inherent risks should focus on the system as a whole rather than individual elements. Furthermore, they should be designed and organized based on a mindset of complexity thinking and strong interdependencies.

Example 9.2 In early 2006, an unexpected accident occurred at the construction site of the north approach bridge of the Sutong Bridge in China. The accident involved the collapse of a 50-m-long box girder left-span movable module support. The direct cause of the accident was the fracture of a connecting component of the movable module, which led to the separation of the upper support frame and the pushing device. As the support frame fell, it caused the inner pushing device to fall as well due to its association with the chain connected to the pushing device. The falling pushing device then hit the support frame, damaging the lateral pushing cylinder connection plate, ultimately resulting in property damage and casualties.

After the accident, through the scanning electron microscope analysis of the fracture surface and metallographic examination, it was found that the equipment design company, in order to prevent the shaft from rotating with the connected parts below during operation, welded the lower workpiece and the shaft together. However, due to defects in the welding process, the hardness of the weld joint did not meet the specified requirements, leading to this accident.

The sequence of unexpected events for this accident is as follows: preventing equipment rotation-increasing welding process-process defects leading to excessive hardness in the heat-affected zone of the weld-the appearance of microscopic fractures-fracture expansion-support frame fall-inner propulsion device fall-causing the horizontal propulsion cylinder to pop out.

A slight quality issue in the equipment manufacturing process ultimately resulted in a severe, unexpected accident. Throughout the life cycle of the movable scaffolding, all the procedures were normal, and no violations occurred. However, a minor accidental "event," amplified by the strong interdependencies within the system, eventually transformed into a major accident. This is a typical example of the "normal accident theory."

After the accident, the Sutong Bridge promptly initiated emergency plans for unexpected risk events and conducted a thorough analysis of the causes. The safety monitoring system was expanded to include overseas equipment manufacturers, minimizing the small probability of "normal accidents" that could lead to significant risks.

Example 9.3 The traditional concept of lake water environmental safety is based on a certain premise, presenting a set of threshold values for state attributes, using these to determine whether the water environment meets safety standards by not exceeding (or not falling below) these thresholds. Therefore, traditional water environmental safety standards are generally rigid and static. However, if our object of study is the complex lake water environment—a colossal composite system of regional socioeconomic dynamics—then our understanding of water environmental safety acquires new implications.

In general, there are close interdependencies among various elements within complex mega-systems. When the state of one element changes, it can propagate to other elements due to their tight interconnections, leading to changes in the states of those elements as well. This propagation process exhibits characteristics such as complex patterns, random trajectories, hidden forms, and unpredictability. Consequently, the original complex system may experience more severe and inherent complexity as a result of multiple state changes propagating within the system. If this propagation phenomenon undergoes amplification and escalation through positive feedback under certain dynamic mechanisms or chance disturbances, it can lead to the overall water environmental security of a lake system crossing the safety threshold, resulting in unsafe phenomena.

In this sense, everything in the original complex system may still appear "normal," and the occurrence of unsafe phenomena should mainly be attributed to the complexity of the system. Moreover, due to the endogenous mechanisms of complex systems often lacking stable structures and exhibiting long chains of correlations, deep hiddenness, and inherent uncertainty, the concept of "complexity-induced safety risks" arises from situations that seem normal but are actually complicated due to their inherent complexity.

In the context of lake water environmental governance, the occurrence of complexity-induced safety risks resulting from multiple state changes propagating is a relatively low-probability event within a specific time frame. However, as stated by Murphy's Law, when viewed over a longer time scale, such low-probability events can occur and have significant risk implications. Therefore, we should employ probabilistic thinking to address these low-probability events, pay attention to potential risks under "normal" system conditions, and not relax our monitoring, early warning, and prevention efforts. Consequently, complexity safety risks are an essential emerging issue in the management objectives of complex lake systems, both in theory and practice.

Beyond this, a new implication of "safety" in the goals of lake complex systems management is what's known as inherent safety. This mainly refers to safety issues caused by the complex wholeness of the colossal complex system.

According to complex systems thinking, the management activities of complex lake systems are characterized as a holistic representation of the system, representing a macro phenomenon formed at the system's overall level by lake complex systems management activities, the evolution of this phenomenon, and the possible paths forming this phenomenon. It is a comprehensive reflection containing all the detailed information on human water environmental management activities and processes within a certain water environmental phase space. Thus, the safety scenario of lake complex systems requires considering all elements related to safety attributes, such as people, objects, technology, laws, and systems, and their overall conditions to be always in a safe and controlled state, gradually approaching the goals of inherent and permanent safety of the water environment.

Inherent safety in the water environment is committed to thinking about safety issues at the system's overall level and the depth mechanism level, like conducting system-wide panoramic coverage and tracking of safety risks, improvements across all scenarios and processes, emphasizing using the whole system scenario as a platform, and grasping the essential elements and mechanisms affecting the realization of water environment safety goals through the complex surface and local phenomena. In this process, inherent safety holds a prerequisite, guiding, and foundational position.

The essence of "Inherent Safety" is the panoramic robustness and resilience of the entire system's safety performance, reflecting a reconsideration of the entire scenario of "complex safety risk." For instance, from a localized perspective, a specific water management project might be effective in enhancing the safety levels of the surrounding water environment. However, on a larger system scale, the local structural changes brought about by the project might render the entire system vulnerable to certain external disturbances on a larger spatial–temporal scale, and such vulnerabilities are latent and implicit. In this way, the construction of a water safety project, while achieving localized safety "increments," inadvertently plants new safety hazards. When the system conditions touch upon some vulnerability "triggers," a sensitive "butterfly effect" might occur, passing through a series of uncertain multistate changes and eventually becoming complex wholeness safety risks.

In the safety management of complex artificial systems, such contradictions between local and global appearance and essence often exist. For instance, to enhance safety, different components are forcibly interlinked to strengthen robustness, but this increases the latent "vulnerability" between nodes, reducing the system's anti-interference or reliability and essentially lowering the inherent safety. Therefore, inherent water safety requires us to deeply understand and handle water environment safety issues from the essential attribute of complex wholeness. Additionally, any changes in the physical structure, functional structure, and topological structure of the connected regions, especially the potentially connected regions upstream, can alter the safety conditions of the water source areas. Therefore, considering the essential impacts of the changes in these connected regions' system structures on the safety of water source areas is crucial.

Based on the above points, in the concept of water environment safety, complex wholeness gives new implications to the management goals of lake governance, providing us profound insights: the management of water environment in complex systems indeed requires us to establish inherent management thinking based on complex systems management.

Example 9.4 Cost overrun in major construction projects is a challenging issue faced by the engineering community worldwide. For a long time, people around the world have been unable to find effective solutions to this problem, and there is a lack of comprehensive theories to analyze it thoroughly. However, exploring the root causes of this problem from a perspective of complex wholeness can provide us with some new insights.

In general, key engineering construction decisions often involve more political factors than ordinary projects, which means that political thinking and political interests from various aspects are incorporated into the decision-making and management process. For instance, decision-makers may exaggerate the value and prospects of the project or forcefully implement the subjective will of leaders and hastily approve key engineering projects in order to build the government's image and satisfy the personal achievements of officials. They may also use political power to change the project design, interfere with the normal construction process, or distort the bidding procedures, leading to underestimation of the project costs and resulting in cost overruns.

Furthermore, the large scale, technical complexity, and long life cycle of major construction projects all "contribute to" exacerbating cost overruns. For example, taking a key transportation project in China as an example, if we calculate based on a project life cycle of 100 years, a vertical angle error of no more than 1/100 would be sufficient to meet the requirements. However, the expected project lifespan is now 120 years. Through analysis, it is found that the allowable limit for vertical angle error has been increased from 1/100 to 1/250. This change leads to a 20% increase in project lifespan but a 2.5-fold increase in technical standards, directly causing nonlinear growth in project costs.

Additionally, factors such as rising material prices, equipment failures, and rework due to substandard quality can also contribute to cost overruns in major construction projects.

However, all of the above can be classified as routine, structural, predictable, and causally clear cost overrun phenomena in key engineering projects. They are referred to as routine cost overruns, and the corresponding risks are known as routine cost overrun risks. Reality has shown that even if we try to clarify these causes as much as possible, it is still difficult to fully understand many phenomena related to cost overruns in major construction projects. If we view major project construction as a complex system with corresponding inputs and outputs, there are complex transformation mechanisms between inputs and outputs that do not follow clear, deterministic, and structured rules. Some complex and integral aspects are not only difficult to structure and regulate but also difficult for people to comprehend, discover, and pay attention to.

If we ignore these elements in addressing the issue of cost overruns, it would be difficult to find the "entry point" to understand the root causes of cost overruns in key engineering projects at a fundamental or deep level.

Therefore, it is necessary for us to deeply reveal the causes of cost overruns triggered by the complex wholeness of key engineering projects. In other words, no matter how detailed the structured system of cost overrun factors may be, it can only describe and summarize a category of constitutive and procedural activities and the corresponding cost overrun phenomena in the process of engineering creation. However, the complex wholeness of key engineering projects will also give rise to a set of unpredictable, uncertain, and uncontrollable new constructive processes and modules that go beyond the original cost budget, making cost overruns inevitable. We refer to this type of cost overrun caused by the inherent complexity attributes of key projects as complexity overruns, and the uncertainty and potential harm resulting from the complexity overruns as the risks of complexity overruns in key construction projects.

Therefore, for major construction projects:

$$\text{Cost overrun} = \text{Regularity overrun} + \text{Complexity overrun}$$

$$\text{Cost overrun risk} = \text{Regularity overrun risk} + \text{Complexity overrun risk}$$

Complexity overrun is not just an abstract theoretical concept. Depending on the different forms of complexity in key engineering projects, complexity overrun risks manifest in different physical forms and management implications, such as

(1) Complexity overrun risk caused by complex project environments;
(2) Complexity overrun risk caused by inherent uncertainty;
(3) Complexity overrun risk caused by the evolution and emergent phenomena of the project construction process;
(4) Complexity overrun risk caused by the game of interests among different project stakeholders;
(5) Complexity overrun risk caused by project innovation;
(6) Other combinations or variations of the above points.

In conclusion, the complex nature of key engineering projects is indeed an important root cause of cost overruns, and these causes exhibit characteristics of non-structuredness, inherent uncertainty, systemic evolution, and emergence. Therefore, they cannot be addressed through reductionism or structured research frameworks.

9.6.4 Essential Management Strategy

The essential management strategy refers to the general methodology and technical approach of the subject regarding the essential management paradigm. In Chap. 8, we studied the general behavioral guidelines for management complex systems, focusing on how to decipher the complex wholeness through concepts such as inseparability, multi-scale, and iteration, gradually reducing the complexity of the problem. However, essential management directly targets the irreversibility of reductionism, which means that the complex wholeness of the problem is perceived as "wholeness" by the subject. Hence, it's imperative to approach and address issues based more on essentiality and wholeness.

Recalling Sect. 8.4 of this book, we pointed out that uniqueness is an important channel for creating complex wholeness. From the four examples provided in this section, it is evident that while the nature of complexity problems is uniform, the expressions of their essence can greatly differ. These differences not only manifest in the real plots and details of the essential form but are also deeply embedded within the mechanisms of essence. Therefore, studying the uniqueness of complexity scenarios and problems is, to some extent, studying essential management strategies.

The academic concepts mentioned above illustrate that by exploring the non-normativity or uniqueness inherent in complex wholeness, and by unveiling the distinctiveness within such complexity through detailed descriptions and contexts, the management activities and behavioral guidelines thus formulated will furnish us with the corresponding entry point for essential management.

We should revisit the mechanisms and preventive measures for routine accidents at the construction site, as illustrated in Examples 9.1 and 9.2 mentioned earlier.

In major construction sites, various personnel, equipment, and raw materials gradually form a material-based complex system under complex technological specifications and management procedures. This system is a composite system composed of resources such as "human" and "materials." Within the "material" hard system, it embodies the combined effects of various scientific and technological knowledge domains, making the "human" system even more complex. Therefore, the risk condition of the system depends not only on the state of engineering equipment and materials but also on the psychological and behavioral conditions of personnel, as well as the management systems and procedures. As a result, the physical complexity manifested by the strong correlation of major construction sites is generally much greater than the virtual complexity estimated and predicted by engineering design. This is mainly due to the following reasons [6]:

1. The actual correlation between elements in the construction site is greatly enhanced.
2. The cause-effect relationship of phenomena is no longer direct and evident. Complexity makes many problems in construction sites unpredictable.
3. Small impacts generated by certain factors on site can be diffused and amplified, changing from local to global, from minor to catastrophic. In other words, the strong path dependence of construction site accidents can cause minor differences in initial conditions and states to evolve into major accidents.
4. Engineering accidents often have unforeseen and sudden characteristics. Accidents often occur in unconventional ways and procedures, making it difficult for people to recognize them at the time.
5. In the face of accidents, behaviors that adhere to normal operations may not only fail to mitigate the severity of the accidents but even make them worse.
6. People tend to strengthen the correlation to reinforce the system and increase its safety. However, everything has two sides. After enhancing the correlation of the system, the complexity increases, thereby increasing the possibility of accidents.

These essential characteristics indicate that in major construction sites, even if the quality of every equipment, the perfection of every process, the technical proficiency of every staff member, and the rigor of every management procedure are maximized, there is still a possibility of errors occurring at some point, as pointed out by Murphy's Law. Particularly, any small accidental event related to engineering safety, whether it is related to equipment, personnel, management, or environmental factors, or a combination of several factors mentioned above, as long as it leads to even a very minor failure, the strong correlation within the engineering site can potentially amplify the fault and enhance the interaction between faults, resulting in a systemic accident. All these factors far exceed the experience of on-site personnel and the expectations of engineering design, ultimately making the originally normal operation the cause of a more significant accident [7].

Based on the above analysis, we must recognize that excellent material and equipment quality, outstanding personnel qualifications, and comprehensive management procedures undoubtedly have significant implications for preventing risks at construction sites. However, we should pay more attention to the influence of complex wholeness on the risks at the construction site. It is essential to avoid using a passive risk control mode that focuses only on specific incidents and instead establish an active risk prevention system. This system should start with all factors related to engineering accidents, with a focus on prioritizing prevention in risk management [8].

Furthermore, in the risk control model for construction sites, based on the conditions at the site, it can generally be classified into the following three categories:

(1) When there is a strong and close relationship among the relevant elements at the construction site, but their mutual influence is low, a centralized risk control model is advisable.
(2) When there is a weak and close relationship among the relevant elements, but their mutual influence is high, a decentralized risk control model is advisable.
(3) When there is a strong relationship and mutual influence among the relevant elements, a comprehensive risk control model that combines centralized and decentralized approaches is advisable.

In addition, when unexpected accidents occur at construction sites, it is not appropriate to solely focus on finding the "responsible person." The person responsible for major construction site risk control may not be an individual, and it is possible that no one bears the responsibility. The inherent complexity of the system makes the "responsible person" a result of system complexity. This requires us to thoroughly investigate the causes of risks and the mechanisms behind major accidents caused by these causes. This mechanism often represents the essence of safety accidents, providing us with extremely valuable cases and samples [6].

Taking the example mentioned above as an illustration, by studying the essential attributes of complex lake water environments, the following can be clearly understood:

(1) The appearance of lake water environmental problems is referred to as "lake disease," but their essence is "social disease." Lake management is not only about treating the lakes, but, more importantly, it is about managing people. Humans are the originators of the "social disease" of water environments and are also the doctors who treat "social disease." However, people often use simplistic thinking or general systematic thinking to view, analyze, and deal with the complexity in lake water environments and their management, or they simply use reductionist methodology to handle irreducible complex wholeness problems in water environment management. This leads to a lack and misalignment of genuine principles in governance thinking.
(2) With the advancement of people's understanding of environmental governance, a thinking evolution path has been formed in lake governance, progressing from simple systems to complex systems and ultimately to socio-ecological complex systems. Specifically, it should be recognized that lake water environments are complex giant systems consisting of human, natural, and social elements, forming a socio-ecological symbiotic complex system. The essential attribute of the giant system is socio-ecological complexity, characterized by symbiosis

between nature and society, decentralization, value sharing, and orderly evolution. Therefore, people themselves need to employ socio-ecological complexity thinking to deal with the complexity of lake water environments.

(3) Superficially, "pollutants" in water environments are actually a type of "negative" product of the entire watershed socio-economic system, transported to lake water bodies through specific supply chains and logistics networks. This entails complex natural, ecological, social, economic, and cultural implications. Therefore, water environment governance needs to consider not only traditional mechanisms such as water quantity, water quality, biology, chemistry, and meteorology but also the close logical connections with "pollutant" manufacturers, suppliers, logistics networks, water resource transactions, value conversion, and ultimate water body governance. Thus, water environment "disease" is not only an "ecological disease" but also a "social disease." In addition to the allocation of administrative power and the authority of the rule of law, it also requires a market-oriented allocation mode that conforms to market mechanisms and supply chain management rules. This includes mechanisms such as cost accounting and compensation for water environment governance, water environment data resource ownership, effective flow, and transactions. It necessitates the construction of an open, shared, fair, and trustworthy circulation platform within the governance system. Otherwise, the water environment cannot form a value Internet, and modern technologies such as big data and blockchain cannot be socialized. Therefore, the metaphor of lake water quality based on the supply chain network of water pollutants will deepen our understanding of the attributes of water environment governance in lake basins. It requires a shift from "ecological-type" governance to comprehensive governance encompassing "ecological," "social," and "economic" aspects, incorporating governance thinking and mechanisms of markets, symbiosis, trust, valuation, transactions, and alliances.

(4) The complex wholeness of lake governance are core issues in paradigm shifts because they answer how to perceive and discover new things about the research subject within the new paradigm of complex systems thinking. The transformation of Taihu Lake governance primarily addresses the deep-rooted governance problems within the social-ecological symbiotic complex giant system of the Taihu Lake basin. Obviously, if these types of problems can be decomposed and solved one by one using reductionist methods, they would no longer be "persistent" problems. Therefore, these governance problems are inevitably complex and irreducible. Although general systemic thinking may be effective in dealing with these complex wholeness problems, it often only partially works or lacks strong efficacy. To address these issues, it is necessary to adopt a comprehensive risk control model that integrates various approaches, such as centralized and decentralized methods, and incorporates the principles of socio-ecological complexity.

Through the analysis of the aforementioned Example 9.4, the following new theoretical viewpoints and strategies can be formed regarding the issue of significant project cost overruns:

1. The cost overruns (risks) in significant projects consist of regularity overruns (risks) and complexity overruns (risks).
2. Regularity overrun risks and complexity overrun risks are two different types of inherent attributes of cost overruns (risks). The former belongs to the category of general systemic overruns (risks) in engineering, while the latter belongs to the category of complex systemic overruns (risks).
3. The main factor responsible for the complexity cost overruns in significant projects is the complex wholeness of the project. In this sense, complexity overruns are similar to "normal unexpected" accidents. They are a type of "normal unexpected" overruns that arise from the complexity of the system. It should not be explained solely by inadequate capabilities or optimistic attitudes of the involved parties, nor should it be analyzed solely by intuitively listing the phenomena and causes of overruns. Such approaches only reflect a perceptual understanding and reductionist thinking and lack the depth of essential thinking in significant complex engineering systems.
4. In the face of complexity cost overruns in significant projects, we must approach, analyze, and address them based on complexity thinking. For example:

   (1) Based on the "iterative" generation principle of the main behaviors in major projects, we intend to enhance the understanding and ability to manage complexity through self-learning in the "first level" of iterative behavior. This will help us identify and control the phenomenon of complexity overruns.
   (2) Based on the principles of "second-level" iterative behaviors, we scientifically design the organizational structure and processes of cost estimation in engineering.
   (3) Based on the principles of "third-level" iterative behaviors, we compare and iterate the generation sequence of consensus and non-consensus in cost estimation proposals, continuously reducing budget errors and approaching the ultimate "true value" of costs.

5. The risk of complexity cost overruns in significant projects is an essential form of risk within the risk system of significant projects, and it has two levels of connotations: the first level is the general complexity inherent in significant projects, and the second level is the specific uniqueness inherent in individual projects. Therefore, we should not study the specific risk of complexity cost overruns in significant projects using general sampling statistical methods or general risk source system evaluation methods. Instead, we can only approach complexity overrun risks by thinking similarly to how we deal with "normal unexpected" accidents.

Here, we observe a thought-provoking phenomenon, where the recognition of the "normal unexpected" risk of complexity cost overruns in significant projects based on complex wholeness is proposed. This association between the complexity risks at the project site and the cost overruns in engineering, which were originally "unrelated," establishes a paradigm of essential thinking. It liberates us from seeking the causes of significant project cost overruns in a perceptual and intuitive manner and establishes a new theoretical thinking regarding the issue. This not only enhances our fundamental understanding of significant project cost overruns but also improves our ability to analyze and control such overruns in practice. This example vividly illustrates the scientific significance and practical value of essential management in complex systems management.

From the above, it is evident that we should not expect a one-size-fits-all essence management strategy for complex systems management that can adapt to various scenarios and apply to all types of problems. Instead, we can only formulate adaptive strategies by fully considering the unique context and real details of the specific scenarios and characteristics of complex wholeness under the thinking paradigm of complex wholeness.

9.7 Issue 6: Dynamics of Management Organization Models

The so-called complex systems management organization is a complex system with human beings as the main entities, consisting of management subject groups, capable of implementing management functions on complex wholeness problems. The management organization model refers to the composition of the main bodies of the management organization, organizational structure, management process, allocation of responsibilities, management support, the formation of overall organizational behavior mechanisms, etc. The study of complex systems management organizations and organizational models is a fundamental scientific issue in the theory of complex systems management.

9.7.1 *Overview of Complex Systems Management Organizational Models*

1. **Overview**

In Sect. 7.7 of this book, it is mentioned that complex systems management activities require the necessary environment and conditions for the formation of management organizations to address the complex wholeness of problems. At the holistic level, these organizations possess the ability to manage complex wholeness. This actually represents the functional platform form of complex systems management organizations and our understanding of their essential attributes.

As complex systems, complex systems management organizations do not follow the simple system principle of "more people, more power." Instead, they are oriented toward constructing and enhancing the overall behavioral ability of the organization, selecting and optimizing subjects, including the completeness and synergy of the subjects' levels, responsibilities, specializations, associations, abilities, and knowledge. This primarily reflects the "other-organizing" behavior of complex systems management organizations.

Additionally, as complex systems, management organizations, through their interconnected internal structure, mechanisms, and processes, can exhibit emergent abilities that are not only superior to individual entities but also exceed the simplistic additive abilities of multiple entities. This primarily reflects the "self-organizing" behavior of complex systems management organizations. The overall behavioral ability of management organizations is a comprehensive result of the combined action of "other-organizing" and "self-organizing." Among them, the mechanism of "self-organization" plays an especially important role.

To be specific, the emergence of macro functions in management organizations reflects a mechanism where the elements and structures at the micro level of the organization integrate upwards and form new overall structures and abilities at the macro level. Even if we cannot or have not yet ascertained what the mechanism forming the emergence is, we usually can perceive the phenomenon of the emergence of such new structures and abilities. The model of complex systems management organizations is precisely the propeller and driving force forming this kind of emergence. In a sense, designing and optimizing organizational models is designing the dynamics mechanism of complex systems management organizations.

2. **Scrutiny**

The internal management processes, authority allocation, and conversion of various management resources within a complex systems management organization operate similarly to the functioning of a normal working machine, known as the operational mechanism. It is through this mechanism that the management organization can develop the capability to handle complex wholeness problems. This is essentially the pattern of complex systems management organizations.

The model refers to principles, rules, and methods that either exist for things or act upon them to maintain certain stable states while the model refers to the abstract principles, rules, and methods that allow these types of actions to function normally.

In other words, the so-called complex systems management organization model refers to the rules and principles by which complex systems management organizations form stable holistic functions based on individual qualities, behavioral norms, and resource allocation, and according to certain processes and information transformation methods. This is essentially the mechanism for the formation of the stable holistic functions of the organization. Among them, the mechanisms that can activate the adaptive and self-organizing abilities of management organizations, thus making the organization exhibit holistic abilities, are especially critical and important.

3. Fundamental Force System

According to system attributes, a complex systems management organization is a complex system, formed with humans as basic elements, that has basic structures and functions and can manifest the capability to control complex holistic management problems.

Organizations formed by stakeholders of different levels and attributes with the aforementioned functions must have their fundamental mechanisms. By examining the features of complex systems management, we know that these fundamental mechanisms will enable us to grasp the general laws of the mechanisms of complex systems management organizations and design feasible and effective management organization models.

From a unified perspective of cognition in physics, biology, and sociology, people have discovered that the concept of "force" can describe the generalized relationships between elements within a system. For example, Ouyang Yingzhi in "Foundations of Complex-System Theories" (1999) said that for a collection to successfully form a system, the bonding forces between its elements must be strong… If the bonding force is greater than the force separating the elements or being attracted externally, this collection can form a system with a stable structure.

Based on this idea, whether originating externally or within themselves, such "adhesives" must exist inside complex systems management organizations; otherwise, each subject would undoubtedly be independent or in a free state.

Therefore, in a general sense, we can use the concept of "force" to describe the interrelations between individuals within complex systems management organizations. Since the individuals in organizations are "people," the "force" here must comply with the basic principles of social sciences and humanities and reflect the unique connotations of complex systems management organizations, not simply applying concepts such as mechanical force and atomic force from physics. Specifically, the "force" between individuals in complex systems management organizations must obey the principles of delegated agency relationships, management behavior, and psychology under the basic rules of political science, sociology, economics, psychology, etc., that is, a "force system" profoundly affected by multiple types of forces.

Introducing the concept of "force" in the study of organizations and organizational models is entirely a vivid depiction of the internal connections of complex systems management organizations because, in actual management, various interrelations, influences, and effects between subjects in management organizations cannot be measured and quantified by instruments like physical forces and are difficult for people to perceive and define intuitively. For this reason, using the concept of "force" is merely leveraging the basic discourse of physics, through visual thinking, to help us construct the discourse form of complex system organization management, facilitating the understanding of the basic behavioral mechanisms of individuals and between individuals at the micro level of management organizations.

To summarize the above content, the following three points are especially important:

(1) A complex systems management organization with a certain stable form has strong and stable forces between its subjects; otherwise, the interaction of forces between subjects cannot maintain their relationships and give the organization a stable structure.
(2) The force between subjects in complex systems management organizations is manifested not only in strength but also in the diversity of types, characteristics, and forms of force, and their dynamic variations. Such diversity allows management organizations to display different functions and the evolution of these functions.
(3) The function of complex systems management organizations is the behavioral representation of the organization in a holistic sense. It is closely related to the interactions, types, characteristics, and forms of forces between subjects within the organization, as well as to the external environment and internal self-organization. This indicates that the functions at the macro level of management organizations must be formed at the macro level through the multi-level progressive effects of individual states and behaviors at the micro level, and patterns and mechanisms at the meso level, namely the dynamic principles inside the organization.

9.7.2 Dynamics of Complex Systems Management Organizations

1. The Complex Forms of Forces System

Simply pointing out the "force" effect between the subjects of complex systems management organizations is not enough. More importantly, it is crucial to understand how the complex forms of force manifested between subjects and the self-organization based on complex forms enable the emergence of complex behaviors and functions at the holistic level of complex systems management organizations.

The term "form" refers to the external shape, internal structure, and overall demeanor of things, an objective attribute that varies among different entities. Typically, the form of a thing is constant; even when it changes due to some reasons, it usually involves partial alterations rather than comprehensive qualitative transformations. However, there are cases where the force forms of the subjects do not change significantly, but adaptive behaviors and self-organization phenomena occur, inducing significant overall changes in the system.

Now, let's analyze the complex forms of force among the subjects of complex systems management organizations.

Every complex systems management organization is an artificial complex system formed under certain environmental conditions and is comprised of various types of subjects. This environment, more importantly than the natural one, is the political, social, economic, and cultural context of a country or region.

Thus, the environment provides legal, administrative, and cultural foundations and sources for the subjects of complex systems management organizations. This provision includes both support and restrictions for the subjects' actions. Even economic and contractual forces are deeply affected by the political and social environment.

According to the general principle of delegated agency, there exist so-called sequential subjects in complex systems management organizations, which hold significant decision-making, discourse, and operational rights on many occasions. These subjects, whether representing an individual or an organization, all have social identities. Since every social being is firstly an individual, these sequential subjects will inevitably exhibit their independent psychologies, behaviors, cultural values, and preferences, such as memory, knowledge, information, perception, cognition, judgment, learning, innovation, and adaptability to the environment.

Thus, in the actual management activities of complex systems management organizations, sequential subjects will not only possess their forms of force but, more importantly, integrate these with the legal, contractual, and cultural forces endowed by the environment, exerting guiding, normative, restrictive, or catalytic effects on the overall group of subjects.

As for other subjects within management organizations, they mostly act as general social beings (economic beings) and establish relationships with other subjects mainly through economic and contractual forces. Due to varying legal forces, contract spirits, and cultural values, each subject, although connected by economic and contractual forces, will exhibit different forms and variations of their forces under the combined effect of legal and cultural forces.

All subjects within management organizations will also demonstrate various diversities and variabilities in the comprehensive forms of force, and, hence, the comprehensive forms and effects of each subject's force, including the nature, intensity, and scope of the comprehensive effect of forces, are not static or immutable. This is because each subject, under different circumstances, will experience adaptive and organizational changes in their behaviors, psychologies, and social characteristics.

1. **The Dynamics of Organizational Behavior**

In the study of complex systems management and organizational research, the most challenging aspect is understanding how macro-level organizational behavior emerges from micro-level individual behaviors. It is crucial to clarify the meso-level dynamic mechanisms, that is, how organizational models operate and function. The previous analysis of the mechanics between management organizational entities has constructed a theoretical framework for this purpose. However, it is important to note that this is merely an attempt and exploration, and alternative concepts and principles can be employed to analyze the relationships and structures among complex systems management organizational entities and conduct research on complex systems management organizational models.

Furthermore, even with the construction of this theoretical framework, it is extremely difficult to understand how micro-level individual behaviors, meso-level model mechanisms, and macro-level overall functions are coherent and interconnected. This is mainly because the interactions among these three levels of behavior are no longer simple transformations of behavioral states of the same attributes. Instead, they involve the emergence and disappearance of functional levels in complex systems management organizations, based on inherent uncertainty and multi-level self-organization and adaptability.

Next, we will study this problem by analyzing two mechanisms:

The collective form and dynamic evolution of the power system within the management organization of a complex system, stemming from the natural, behavioral, social, and relational attributes of each subject, embody the comprehensive internal dynamics of the management organization. The interaction between the environment and the management organization generates comprehensive external dynamics. Under the combined influence of these two dynamics, the macroscopic organizational behavior and functions of the management organization are formed. This represents the dynamic mechanism from the complex forms of individual micro-level forces to the macro-level organizational behavior and functions.

Specifically, the above mechanism consists of the following two basic mechanisms:

First mechanism: Self-organization mechanism. This mechanism refers to the active formation of an organized structure within the management organization by the entities themselves, according to mutually agreed-upon rules of force correlation driven by goals. Each subject fulfills its responsibilities and coordinates with others to form an orderly structure within the organization. This process is driven by the spontaneous behavior of the entities within the management organization and leads to a transformation from disorder to order, resulting in overall behavior and cooperation among entities. This process, known as the self-organization mechanism of the management organization of a complex system, is driven by the spontaneous behavior of individuals. The adaptability of the organizational structure and functions of the complex systems management organization is largely based on this mechanism.

Second mechanism: the generalized evolution mechanism. This mechanism refers to the evolutionary process that occurs due to external constraints, such as resource limitations, at the macroscopic level. For example, in the self-organization process where local interactions among organizational entities expand to the global level, the organizational structure and operation methods continuously self-improve in response to changes in the external environment, aiming to enhance their adaptability to the environment.

The formation and evolution of complex systems management organization's overall behavior and functions depend on the combined effects of the two aforementioned mechanisms, as well as the changes in the complex forms of individual forces and the interactions between them. Ultimately, this leads to the evolution path of the macro structure and overall behavior of management organizations.

From this perspective, there exists an interactive process between the micro-level individual entities and the macro-level organizational entities in complex systems management organizations. This process, based on the complex forms of various forces, constitutes a meso-level dynamic principle between the micro and macro levels. This meso-level dynamic principle is the mechanism behind the formation of macro-level behavior and functions in complex systems management organizations and is also the organizational model of complex systems management organizations.

It is necessary to analyze the internal operation of this important process. Firstly, each subject, as an individual, possesses internal force systems and external force systems due to its social and functional roles. The internal force system represents the various forces among the entities within the organization, while the external force system represents the various forces between the subject and the external environment. These two types of force systems are coupled and undergo dynamic and evolving changes. The term "form" here refers not only to the types of forces but also to the "intensity," "direction," and the feedback and transmission methods between forces. The overall form and evolution of the interactions among these forces represent the operational mechanism of a well-ordered machine. The overall behavior and functions of complex systems management organizations are the results of the operation and functioning of this machine.

Ideally, when designing and constructing complex systems to manage organizations, there are clear purposes and predefined functional requirements. Therefore, there are inevitable regulations on various types of systemic forces. Initially, the focus was on selecting organizational entities, designing the basic structure of the organization, and arranging its fundamental functions. These aspects pertain to the "other organizations" involved in managing the organization. Under the influence of complex systemic forces within the organization, the interaction between the entities within the organization can spontaneously generate new emergent behaviors and functions beyond the initial design. This is known as the "self-organization" of complex systems management organizations. The overall behaviors and functions of a complex systems management organization are the combined result of the aforementioned "other organizations" and self-organization. This is referred to as the emergence of overall behaviors at the macro level of complex systems management organizations. Emergence refers to the part of the macro-level behaviors and functions of complex systems management organizations that cannot be directly or simply obtained by summing up the micro-level systemic forces. The main theoretical viewpoints on the emergence of overall behaviors in complex systems management organizations are as follows:

1. Emergence is the manifestation of new behaviors and phenomena at the macro and overall levels of management organizations, which do not exist at the micro level of individual entities. In other words, there is a "gap" between macro-level emergence and micro-level individual systemic forces. The complexity of emergent phenomena suggests that we cannot establish a direct causal relationship between certain micro-level entities and macro-level emergence based on traditional thinking paradigms. Instead, we should consider how the coordination among numerous micro-level entities (including their dynamic changes) may form a mechanistic function at the meso level, which can explain the possibility of new emergent phenomena at the macro level.

2. The generation of emergent phenomena in organizational systems often does not require the appearance of new entities within the organization. It is mainly due to changes in the systemic attributes of the existing entities themselves or the emergence of new ways of interaction among the overall systemic forces, or both. As a result, the macro-level structure of the management organization may change, or the behavior of a subset of entities may change, leading to changes in the behavior of the majority of entities within the organization. This series of self-organizing behaviors among entities, under the influence of the operational mechanisms at the meso level of the organization, gradually amplifies and eventually evolves into unpredictable overall behaviors and functions, giving rise to new forms and functions of the management organization.
3. The emergence of organizational phenomena occurs during the dynamic evolution process of the organization. In this stage, the complex forms of systemic forces differ from the previous stage and are often different from the next stage as well. In the next stage, new and unforeseen dynamic phenomena may emerge, whether within the systemic forces themselves or in the interactions between the systemic forces. These phenomena may disrupt certain existing relationships between entities or create new relationships. This indicates that in the process of forming overall behaviors and functions of complex systems management organizations, both the forms of systemic forces within the entities and the directly observable relationships between the entities at the macro level undergo a series of complex and profound evolutions.
4. The aforementioned circumstances make it generally difficult to predict emergent phenomena in advance. However, the causes and origins of observed emergent phenomena can generally be explained after the fact. In other words, after observing emergent phenomena, we can provide some explanations within the macro-level cognitive framework by considering the relationships between the forms of micro-level systemic forces and the mechanisms at the meso level. In other words, we cannot directly deduce the relationships between micro-level systemic forces in advance, but we can introduce new ideas and concepts at the macro level to provide appropriate explanations.

From the aforementioned, it is apparent that to understand and analyze the dynamics of complex systems management organizational models, the principal technical route begins with examining the complex forms of systemic forces at the micro level. Subsequently, it progresses to exploring the interactions and evolutions of systemic forces at the meso level and culminates in the revelation of the emergence of complex behaviors and functions at the macro level of the organization.

The above is merely the basic logic of understanding organizational emergence, and, in general, it can be applied to theorize about emergence phenomena in various complex systems. For a complex system, the details of the system are primarily determined by individual behaviors at the microscopic level. The operational mechanisms and patterns of the system are mainly determined by individual associations and structures at the mesoscopic level. Consequently, at the mesoscopic level, many degrees of freedom and characteristics of individual behaviors will be submerged, emphasizing causality, correlation, and logic of associations and structures. This often leads to the fact that behaviors and forms at the mesoscopic level of the system cannot be fully described by individual behaviors and forms at the microscopic level. That is, there emerges a new phenomenon and meaning at the mesoscopic level that cannot be entirely explained by behaviors or mechanisms at the microscopic level. Furthermore, the combined effects of micro and meso levels in a system will also give rise to certain new and perceptible holistic characteristics and functions at the macro level. Because these holistic characteristics and functions integrate all the effects from the micro and meso levels, it is highly likely that phenomena and features may emerge at the macro level that have not been observed at the micro level or even at the intuitive level. Naturally, these phenomena and features cannot be fully explained by behaviors or mechanisms at the micro or mesoscopic levels of the system.

Furthermore, even if there exists some objective mechanism behind these macro-level phenomena, due to the mechanism spanning multiple levels in complex systems and demonstrating high complexity, or because our level of understanding and technological means are not yet sufficient to help us reveal this mechanism, we often can only perceive this phenomenon at the macro level. The complex causes and evolutionary paths behind this phenomenon are uncertain, incomplete, and often largely unknown. If influenced by traditional reductionism and deterministic causal thinking, our attempts to interpret macro characteristics and functions with meso or micro behaviors usually end in failure. Consequently, these phenomena at the macro level in complex systems, which are perceptible but not present at the micro or meso levels, become a class of perplexing and confusing phenomena. The concept of "emergence" thus becomes the most vivid metaphor and linguistic symbol for this phenomenon.

The above analysis contributes to a deeper understanding of the essence of organizational emergence phenomena.

9.8 Issue 7: Knowledge Forms and Acquisition Paths

Knowledge is indispensable for the management of complex systems: solving real-world problems requires knowledge; academic research is a process that uses knowledge as a resource to produce new knowledge; theory is the most core part of all learning, where knowledge serves as the fundamental elements or units of a theoretical system, and the theoretical system is the systematization and logical organization of related knowledge.

From the perspective of human cognitive patterns, knowledge is a unique man-made logical system. The knowledge of complex systems management, in terms of its form, characteristics, formation methods, and acquisition paths, not only follows the general rules of human knowledge production but is also influenced by the thought paradigm of complex systems. Therefore, as the theory of complex systems management is still in its infancy today, paying attention to some related issues of knowledge in complex systems management naturally holds foundational significance.

9.8.1 Overview of Knowledge Forms

"Knowledge" is described in the entry on "Knowledge" in the *Chinese Encyclopedia of Education* as follows: "*Knowledge, in terms of its content, is the reflection of the attributes and connections of objective things, the subjective image of the objective world in the human brain. In terms of its reflective activity form, it sometimes manifests as the subject's sensory perception or representation of things, belonging to sensory knowledge, and sometimes manifests as concepts or laws about things, belonging to rational knowledge.*" From this definition, we can see that knowledge originates from the external world, so knowledge is objective. However, knowledge itself is not objective reality but the reflection of the characteristics and connections of objective things in the human brain [9].

Knowledge has its form, that is, its formation and style. In human history, the form of knowledge has evolved. Initially, people gained understanding and insights into practical activities through brain thinking during their practice. Over time, they formed insights and experiences regarding certain aspects of practical activities. With more such cases, these insights and experiences gradually became richer and more profound. In the future, when encountering similar situations, people will consciously or unconsciously apply these insights and experiences to subsequent practice and further expand and improve them based on multiple successes. The insights and experiences gained through personal experience or observation are referred to as personal knowledge. Experience can be cognitive understanding of a particular problem or phenomenon, or it can be a method or technique for problem-solving. In specific practical contexts at different times and places, people often repeatedly generate similar experiences, which are more easily recognized and solidified by people. Solidified experiences, after undergoing multiple tests, become more reliable, valuable, and have stronger guiding significance for practice. This is the embryonic form of knowledge.

In short, experience can be seen as the "primary" form of knowledge. However, due to the limitations of personal practice, experience, and especially subjective thinking of the human brain, personal knowledge inevitably has issues such as fragmentation, shallowness, and errors. But through repeated testing and modification of countless individual knowledge in different contexts, the correctness, depth, and universality of knowledge as a consensus among humans will greatly improve.

The knowledge formed through experience and the solidification of experience mentioned above, besides having a slow formation process, also has limitations in "quality." In the agricultural civilization or natural economy era, due to the relatively simple requirements for the profoundness of the reflection of objective things and the guidance of practice, knowledge forms and values based on experiential forms can generally meet the needs of objective practice. However, as human society has gone through the sprouting of European capitalism, followed by large-scale industrial revolutions, especially the rapid development of information technology and social changes in recent years, not only has there been a global social transformation, but there has also been a high demand for high-quality knowledge. This has led to a significant turning point in the forms and evolution logic of knowledge.

Today, in the rich soil of practical application in complex systems management, the basic approach of perceiving personal experience and gradually solidifying it into knowledge still has a certain validity. However, from the perspective of quality standards such as the universality and depth of knowledge, it is clearly inadequate and unsuitable. Firstly, as the complexity of management practices in various fields such as human society, economy, and technology continues to increase, merely knowing partial and superficial phenomena is far from sufficient. Experiences primarily stem from people's perception and summarization of the superficial aspects and external aspects of things, making it difficult to uncover and reveal the essence and underlying laws of things. In addition, the limitations of personal practice in terms of time and space greatly restrict the universality of experience. Therefore, experiential knowledge forms are increasingly unable to meet the needs of understanding complex systems, holistic issues, and complexity management. People increasingly require knowledge in the form of principles that deeply reveal the essence and underlying laws of complex systems management.

The knowledge of principle morphology is the understanding of the essence and laws of things in the real world through rational thinking of the brain, which is accurate, in-depth, and not merely empirical and superficial, forming a judgmental statement. The important role of principle morphology knowledge is that the foundational and formalized judgments it extracts can help people establish behavioral guidelines and operational rules at the operational level. Clearly, this type of principle morphology knowledge is much deeper and more reliable than empirical knowledge. It not only has broader applicability but also has more robust operability. This means that the knowledge morphology of complex systems management has undergone an important transition.

Furthermore, practice has shown that with the progress of human society, knowledge is increasingly cross-fertilizing not only in terms of depth but also in terms of breadth. In particular, modern information technology provides people with various advanced tools, supporting the integration and fusion of knowledge from different disciplines to form new forms of principle morphology and even knowledge with theoretical breakthroughs. This is the arrival of the "Big Science" era as referred to by the scientific community. The multidisciplinary composition of the "Big Science" in the era of Big Science becomes a new cross-form knowledge generation and development platform, where many significant discoveries and improvements are made. This reality tells us that the knowledge morphology and generation path of human beings are undergoing an evolutionary trend from empirical form to principle form to Big Science form. We intend to contemplate the knowledge morphology and generation path of complex systems management in such a historical context.

9.8.2 Knowledge Morphology of Complex Systems Management

Overall, the knowledge of complex systems management is the subjective image of the objective reality attributes and connections of complex systems management in the human brain. In different contexts, knowledge can be divided into knowledge that represents human social consensus and personal subjective knowledge in individual minds. The former is mostly represented by concepts, principles, paradigms, propositions, etc., related to complex systems management. Here, we mainly discuss this type of knowledge related to complex systems management.

In the field of complex systems management, although existing knowledge from other fields can be directly borrowed to explain and solve many specific problems, since complex systems management is a complexity management field formed after the transfer of complex systems thinking paradigm, this field cannot be without a knowledge system that adequately reflects and profoundly reveals its own complex wholeness nature. It should also have a theoretical chain and basic path to form its own knowledge system of principle morphology. Especially because complex systems management is still a new field, the construction of academic and theoretical systems is still in its "primary stage." Therefore, there is a lack of mature experiences and existing resources, and a simple "importing" approach cannot be adopted. Thus, in the current early stage of creating knowledge in complex systems management, the characteristics of "creating from scratch" are highlighted, and attention should be paid to following the norms of knowledge formation and connotation. The following points are worth considering and focusing on in the current process of creating knowledge in complex systems management:

1. The theoretical suitability of core concepts in complex systems management should be treated with caution. Concepts are the linguistic expressions of people's understanding of the essential attributes of objective things in the knowledge system. In this sense, the core concepts of complex systems management need to profoundly reveal and embody the essence and connotation of complex systems management and must be expressed in scientific terms rather than "plain language." [10] For instance, the foremost issue is the basic concept of "complex systems management." If we were to define it solely based on its natural meaning, it would translate literally to "management of complex systems," where the term "complex" imparts a strong subjective nuance. Consequently, the term "complex systems management" can become ambiguous. To address this, it's essential to standardize the connotations of "complex systems" and the combination of "complex systems + management." This has prompted an evolution in the scientific community's understanding of "complexity," "complexity vocabulary," "complexity science," and "complex systems" over the past century. By selectively adopting various thought processes related to "complex systems," we can understand why this book dedicated an entire chapter (Chap. 3) to clarifying the specific style of "complex systems" that would merge with "management" to form the distinct "complex systems management." Imagine, if not for this

meticulous approach, for every perception of "complex systems," there would be a corresponding concept of "complex systems management." Could there then be a definite and unified principle and scientific essence for complex systems management?

2. Coherence of principles. Knowledge creation should not primarily rely on the formation of knowledge through artificially created "new terms." Chapter 3 of this book, "Complexity and Complex Systems," provides a detailed introduction and distinction between the complexity-based view of complex systems and the methodology-based view of complex systems. After a comprehensive comparison, the methodology-based view of complex systems is established as the thinking principle to establish the fundamental principles and core concepts of complex systems management. It specifically discusses how "complex systems + management" becomes the adaptability and versatility of "complex systems management." The purpose is to ensure that the knowledge generation of complex systems management has a clear starting point and a solid theoretical foundation.
3. Moreover, the knowledge value of the concept of complex systems management is not only manifested in its individual connotations but also, more importantly, with concepts as a basis or through the combination of concepts, it can analyze phenomena in complex systems management practice by combining experience. It strives to provide reasonable explanations for the logical relationships, causal relationships, and associative relationships in the phenomena and extract new concepts as well as formulate principles in the theoretical framework. A principle represents a relatively independent unit of knowledge. This reflects the coherence of the basic concepts of complex systems management. In terms of logic, foundational knowledge should have the ability to expand and generate new knowledge.

As an example, this book, based on the refined complexity attributes of complex systems, integrates complexity with various attributes of management activities and problems, constructing several important new concepts based on complexity in different contexts. Additionally, in Chap. 5, it is pointed out that "complexity," as a core concept in the field of complex systems management, is a condensation of the attributes of intricate management activities, scenarios, and problems, and thus holds significant consistency and expansibility. In the specific context of complex systems management, allowing "complexity" to combine with other discourses can express richer connotative meanings, such as "complexity problems," referring to management problems that cannot or should not be addressed with reductionism, "complexity thinking" referring to the complex systems thinking paradigm adopted by subjects, "managing complexity" indicating the various forms of complexity (complex holism) inherent in management activities, and "complexity management" signifying a category of management models based on the shift to complex systems thinking paradigms.

Moreover, this book establishes the logical correlations between knowledge of basic principles such as scenario orientation, decoding of complex wholeness, progression from physics to systems to management, contextualization in unique scenarios, indivisible relativity, multi-scale management, adaptive selection, and iterative generation. Also, the concept of scenario not only serves as a core concept in complex systems management and scenario orientation as a fundamental principle, but also scenario-based robust management is a foundational scientific topic. Scenario modeling, a key methodology in managing complex systems, has evolved into an interdisciplinary technique for big data-driven scenario modeling of complex systems when combined with big data-driven approaches. All these elements together form a relatively complete theoretical chain of knowledge formation in complex systems management: delving deeper into complex systems management could start from a "problem-oriented" principle to propose a theoretical chain, consisting of "scenario-oriented" principle (an innovation in management thought)—based on scenarios in complex systems management (a core new concept)—scenario robustness (a new fundamental principle)—robustness management (a new scientific topic)—complex systems scenario modeling (a methodology)—big data-driven scenario modeling technology (a new key technology).

From this case, it is evident that starting with innovation in management thought, as long as the content of knowledge units in complex systems management has strong logical consistency and integrability, it not only forms multiple mutually communicative new knowledge contents theoretically but also, significantly, can construct a relatively complete chain of theoretical innovation. This endows the academic and theoretical system of complex systems management with robust self-organizing and self-growing capabilities.

The above discussion and examples tell us from multiple perspectives that, in the process of constructing the knowledge system of complex systems management, it is essential for the knowledge that serves as the principles of complex systems management to have distinct theoretical connotations and linguistic carriers. It should neither replace scientific discourse with "plain language" nor simply use terms from other fields. Without good, autonomous, and influential concepts and principles, there will be no place and opportunity for academic ideas, theoretical viewpoints, and knowledge claims to be narrated and expressed through discourse. Academic research would be like breathing in a vacuum, which is impossible.

In summary, regarding the quality of complex systems management knowledge, attention should be paid to the transformation and sublimation from experiential knowledge to principled knowledge and integrated knowledge. It is especially important not to stay at the level of natural concepts or replace the inherent connotations of knowledge with fancy "new terms."

Finally, it should be pointed out that currently, as the formation of the complex systems knowledge system is still in its early stages, particular attention should be given to guarding against the following possible tendencies:

(1) Neglecting the research on foundational and principle-based knowledge. Complex systems management research has breakthroughs in thinking paradigms and academic logic and requires the support of original foundational theories. Therefore, researchers should not overlook or avoid research on foundational and principle-based knowledge solely for the sake of quickly achieving

research results or simply following a ready-made technical method system. Otherwise, not only will it make it difficult for complex systems management itself to achieve true academic breakthroughs, but it will also be challenging to achieve foundational theoretical support for the research of complex systems management techniques and methods.

(2) Fuzzy logic in research questions. Both complex systems and complex systems management based on natural concepts are fuzzy concepts, which bring practical difficulties to complex systems management theoretical research. From a basic logical perspective, the main research object of complex systems management is a type of complex wholeness problem in social and economic complex systems. Therefore, complex natural system problems that solely follow natural laws and lack social and humanistic characteristics should not be included in the scope of complex systems management problems. It is also not appropriate to treat complex wholeness management problems as complex mechanical systems or automatic control systems and solve them using purely engineering technical methods. Furthermore, the study of complex human behavior norms and phenomena in complex systems management should not rely solely on traditional stimulus/response-based "control" and "governance" thinking and means. Instead, it requires the application of modern thinking and new research paradigm shifts, such as sharing, integration, co-governance, multi-centricity, and adaptability. It is important to note that the reductionist thinking of studying complex systems management with simple systems should be avoided.

(3) Tendency toward mathematical tooling. Mathematical formalization of management problems has played and will continue to play an important role in management science research. However, we must also be fully aware that the complexity and integrated methodology system of complex systems management tells us that mathematics not only does not have the ability to fully describe and analyze complex problems, but often, due to the complexity of reality, it has to "drastically" reduce complexity to "adapt" to the constraints of mathematical paradigms. This will inevitably "damage" the essence of the realistic scenarios and the complexity of human behavior in complex systems management. Thus, even if mathematical models are novel and technically skillful, they are merely replacing genuinely complex wholeness problems with severely "distorted" and simplified "complex problems." This tendency easily deviates from the reality of complex systems management.

9.8.3 Knowledge Acquisition Paths for Complex Systems Management

The term "knowledge acquisition paths" here clearly does not refer to the ways of obtaining knowledge from books or learning in classrooms. Instead, it refers to the channels and means of creating firsthand knowledge. For example, throughout history, through the scientific advancements of the Middle Ages and the early modern scientific revolution, many fundamental features and paradigms of modern science began to take shape. These include the use of sophisticated instruments and mathematical methods, the description, prediction, and explanation of natural phenomena through observational and experimental evidence, which led to the acquisition of new knowledge in various disciplines. This involves the selection and optimization of knowledge acquisition paths [11].

It is not difficult to understand that different fields of study and research objects require different channels and means of knowledge acquisition. Whether in the natural sciences, social sciences, or humanities, the knowledge acquisition paths differ. For example, within the natural sciences, disciplines like mathematics and chemistry have different paths for acquiring knowledge, such as inorganic chemistry and computational chemistry. However, when summarized and generalized from a methodological perspective, they typically involve theoretical analysis, experimentation, empirical evidence, mathematical models, and so on.

This section specifically refers to knowledge acquisition paths in complex systems management. However, it is not appropriate to discuss it from the perspective of methodological generalizations and principles. Instead, it should be approached from the essential attributes of complex systems management and the classification of related knowledge types, with an emphasis on identifying the most common channels and means of knowledge acquisition.

First and foremost, fundamentally speaking, the knowledge of the principles and forms of complex systems management is based on the complexity and integrity of real-world problems. The corresponding knowledge is a condensed and abstract representation of the concerns and responses to these objective problems. Therefore, knowledge obtained through practices that are detached from reality relies solely on assumptions without practical validation, and often lacks true theoretical depth and practical significance. Such knowledge has limited long-term value for the construction of knowledge on the principles and forms of complex systems management.

Secondly, due to the rich content of complex systems management as a practical field, it inevitably possesses abundant domain-specific knowledge. Each knowledge domain is discovered, created, and improved along specific research (technological) paths for studying and solving certain types of real (or theoretical) problems. Therefore, it is not feasible to list the means of knowledge acquisition for each specific problem. Firstly, the number of problems is vast and endless. Secondly, lacking distinctive attributes, it is difficult to organize them systematically. What we need is a comprehensive categorization of knowledge that reflects the essential attributes of complex systems management and the basic paths of knowledge acquisition, providing methodological insights for theoretical research in complex systems management.

As mentioned earlier, in terms of information attributes, the knowledge related to complex systems management activities and processes can be broadly divided into three categories:

The first category includes mechanistic cognitive information about the management activities and processes of complex wholeness based on prior knowledge.

This category embodies the scientific rigor and normativity inherent in complex subject matters, displaying features like standardization, determinism, causality, and measurability. This type of information is frequently conveyed through the causal relationships existing between distinct states and characteristic quantities.

The second category includes cognitive knowledge about the relevance of management activities and processes of complex wholeness based on various principles. This category reflects the cross-level and cross-scale correlations within complex wholeness and is characterized by understandability and perceptibility. Such information is often expressed through relationships between different states and characteristic quantities.

The third category includes cognitive knowledge about the unique characteristics of management activities and processes of complex wholeness based on certain real-world factors. This category reflects the contextuality and contingency within complex wholeness and is characterized by uniqueness, singularity, and irreproducibility.

In this way, it is not difficult to find the corresponding knowledge acquisition paths. Specifically, for the first category, relevant knowledge is generally obtained through methods and means similar to those used in natural sciences and engineering technology sciences. Mechanistic causality is often used to describe them, such as quantitative descriptions or mathematical models. For the second category, relationships between different states and characteristic quantities are often used to describe them, for example, traditional statistical methods based on sample collection and modern big data-driven methods. For the third category, specific contextual language is often used to describe them.

The complexity of complex systems management arises because there is a scarcity of the first type of rule-based knowledge, a prevalence of the second type of correlational knowledge, and a significant but unique and scarce characteristic of the third type. This uniqueness and scarcity make it impossible to extract statistical regularities from the parent population of the same type of scenario as representatives. In particular, for the unique aspects of scenarios, it is necessary to closely approach realistic scenarios and pay full attention to the details and variations of the scenes in order to discern the "subtleties" of the complex wholeness of the scenarios. By analyzing the intricate details of complex wholeness and the details related to complexity, we can discern the microscopic elements that make the macroscopic manifestation of complex wholeness "possible." This approach contrasts with seeking to uncover knowledge that renders macroscopic complexity inevitable under specific microscopic dynamic mechanisms. To achieve this, it is possible to establish various constituent elements of scenario details in realistic scenarios to reflect the social, humanistic, and contingent aspects of the complexity of the overall integrity. In terms of methodology, this can be understood as the pathway to acquiring comprehensive knowledge about the complex wholeness of the entire scenario.

Another crucial method for acquiring common knowledge is prevalent in today's era of "big science." In this trend, many new forms of knowledge, including technologies and methods, display interdisciplinary traits and are developed on the multidisciplinary platforms inherent to "big science." A typical example is the complex systems scenario modeling technique as part of complex systems management knowledge. This technique employs an integrated methodology system including natural language, quantitative analysis, panoramic qualitative analysis, generalized algorithms, new symbol systems, big data-driven approaches, and artificial intelligence. The core of this system is the thinking paradigm for modeling the complex wholeness, deducing the microscopic mechanisms of scenario processes and scene autonomy, and hybrid systems, demonstrating and sampling scenario language in terms of scenario chronology, such as who/what/when/where/why/how and the outcomes, to obtain local and micro-level complexity cognition (the third type of information). From a holistic perspective, it relies on demonstration and big data-driven methods to form scenario-associative logic (the second type of information). Considering the scarcity of overall scenario samples, statistical concepts are generalized into the "scenario cultivation" method, employing computer simulation techniques to generate scenario "sliced-flow" at a certain scale and granularity, which is then merged at multiple levels to form the macroscopic emergent and evolving panorama of the management process and field (panoramic approximation and process reconstruction).

In summary, the knowledge form and acquisition pathway of complex systems management represent a crucial scientific topic in the field. "Knowledge form" is the fundamental connotation and main content of academic research in complex systems management, while the "acquisition pathway" ensures the ability of complex systems management to solve specific problems and provide practical value in terms of technical methods.

9.9 The Logical Correlations of Scientific Issues in Complex Systems Management

The scientific issues in complex systems management are primarily fundamental and comprehensive problems in both practical management and academic research. It can be observed that

1. With the development of complex systems management and in-depth academic research, the scope and depth of scientific issues will become broader and more profound.
2. From the perspectives of both practical and theoretical thinking in management, scientific issues encompass academic problems at the theoretical level as well as operational problems at the practical level.

Therefore, it is not appropriate to confine scientific issues in complex systems management to a few abstract theoretical research problems. It is necessary to consider both academic and operational aspects. Even for predominantly academic issues, they can be approached from multiple research perspectives, including operational considerations. Once purely abstract scientific problems are combined with practical applications and can solve real-world problems, they often require appropriate contextualization of the "pure abstraction." This may seem to "diminish" the "scientific nature," but it precisely demonstrates the practical vitality of scientific rigor.

In light of the understanding mentioned above, this chapter adheres to a coherent epistemological and methodological logic in the domain of complex systems management. It begins by addressing the most fundamental and immediate scientific issue: the "deciphering of complex wholeness." This discussion underscores the vital attribute of "complex wholeness" that qualitatively cannot be "disrupted."

However, it's crucial to clarify that this represents a form of "virtual" cognition at the theoretical thinking level. In practical terms, the concept of "indivisibility" must be contextualized, and various techniques aimed at "deciphering" integrity should be deployed. By doing so, theoretical and practical thinking can engage in a dialogue and fulfill their integrated roles effectively.

Using the aforementioned topics as the "trailblazers," it becomes possible to identify issues of shared value that are perhaps of a slightly lower level but are approached from diverse perspectives, such as.

1. Developing new paradigms for qualitative transfer.
2. Analyzing the complexity spectrum.

Moreover, in order to determine the "entry points" for the study of complex systems management, it is suggested that management functions and qualities be integrated through two pivotal topics: robustness management and essential management. These topics initially distill the means by which to transcend conventional thought and formulate complex systems thinking within the realm of complex wholeness. They consolidate multiple management principles and seize the central theme of management activities. They also contemplate the methodologies for formulating core guidelines for subjective behavior in the midst of intricate and unpredictable environments. In doing so, at the principled level, this approach modifies the traditional method of managing complexity, which relies on subjective experiential knowledge, and evolves it into leveraging complexity thinking and foundational knowledge to oversee the complex wholeness of problems.

Considering the execution of complex systems management activities, there is a need for a defined and stable management organizational model at the macro level, and the subjects need to possess the necessary comprehensive capabilities at the micro level. These capabilities stem from the breadth and depth of the subjects' relevant knowledge and the effectiveness of knowledge acquisition pathways. Therefore, the dynamics of management organizational models in complex systems management, the knowledge forms and acquisition pathways of complex systems management, become two fundamental and platform-level scientific issues (Fig. 9.3).

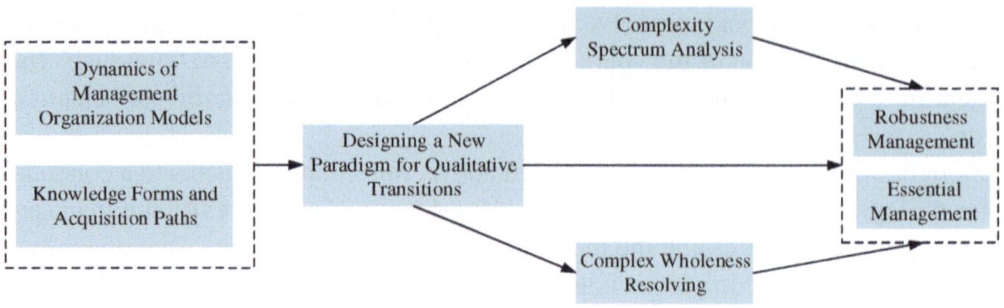

Fig. 9.3 The Logical correlations Diagram of Basic Scientific Issues

References

1. Oppenheim R, Putnam, H. Unity of science as a working hypothesis [M]. In H Feigl, M Scriven, G Maxwell. Minnesota Studies in the Philosophy of Science[M]. University of Minnesota Press, 1958.
2. Nagel E. The structure of science: problems in the logic of scientific explanation [J]. Philos Rev. 1964;73(1):104–6.
3. Tao Y, Zhaohan Sheng. A panoramic decision-making on environmental responsibility for major projects: a case study of the protection of Chinese white dolphins on the Hong Kong-Zhuhai-Macao Bridge [J]. Environ Protect. 2020;48(23):56–61.
4. Pan Q. What economics can do in the face of "fundamental uncertainty". [N]. Shanghai Securities News, 2018.
5. Ye Q. On the self-organization mechanism of the system: the interpretation of the dissipative structure mechanism diagram [J]. J Syst Dialectics. 1994;02:57–63.
6. Zhaohan S (2009) A preliminary study on the comprehensive integrated management model of large-scale and complex projects: Theoretical thinking on the management of sutong bridge project [J]. Construct Econ. 2009;(5):20–2.
7. Wang Q, Cheng S. System complexity studies for large-scale projects [J]. Scientific Decision-Making. 2009;01:11–7.
8. Xu T, Sheng Z, Li J. Research on complex engineering management system based on comprehensive integration [J]. Complex Syst Complexity Sci. 2008;03:48–54.
9. Jiao Y. Encyclopedia of China: Education volume [J]. Educ Stud. 1985;10:1.
10. Wu L. Talk about terminology and the systematization of terminology systems [J]. Res Scientific Techn Terms. 2005;02:44–8.
11. Sheng Z, Huo H, Chen X et al. Moving forward and innovating more than others: a review, reflection and prospect of 70 years of management science and engineering in China [J]. Manag World. 2021;37(2):185–202+13+13.

Chapter 10
Introduction to Scenario Modeling in Complex Systems

The preceding chapters of this book provided an overall introduction to the principles and basic theories of complex systems management. In this chapter, we are prepared to progressively introduce the key techniques of scenario modeling in the complex systems management methodology over the span of four chapters. The sequence of these chapters is as follows: this chapter is an introduction to scenario modeling in complex systems, Chap. 11 covers the basic principles of big data-driven modeling, Chap. 12 discusses scenario computability and big data-driven scenario modeling, and Chap. 13 focuses on the key techniques and processes of big data-driven scenario modeling.

Before delving into the content of this chapter, let us summarize a few key points from the preceding chapters that are relevant to this chapter, serving as the logical starting point for our research:

1. Scenarios fully embody the complex wholeness of a system and are a central concept in the theory of complex systems management. Scenario orientation is a fundamental principle of complex systems management, and thus scenarios hold a special foundational and significant role within the complex systems management theory. As scenarios themselves are complex systems, scenarios in complex systems management are generally referred to as complex system scenarios or simply scenarios.
2. The knowledge of complex systems management principles includes techniques for understanding, analyzing, and describing phenomena and problems related to complex systems management activities, including relevant modeling techniques.
3. When addressing practical problems, specialized techniques are required to decipher and manage complex wholeness. Scenarios play a crucial role in this regard since they are important carriers of complex wholeness.

Z. Sheng, *Outline of Complex Systems Management Theory—Based on Irreversibility of Reductionism Thinking*, International Series in Operations Research & Management Science 360, https://doi.org/10.1007/978-981-97-3361-3_10

Based on these three points, it is evident that scenario modeling in complex systems management is a key technique within the complex systems management methodology.

10.1 Introduction to Modeling Complexity

As pointed out earlier, in today's complexity management activities, there has emerged a category of problems where reductionism is irreversible. The inherent attributes of such problems will lead to a series of significant transformations in corresponding management activities, issues, and theories. Therefore, adaptive responses are needed in management methodology and method systems. For example, in terms of modeling, it is necessary to construct modeling techniques aimed at complex wholeness. Of course, complex wholeness is an abstract attribute; in reality, modeling techniques targeting complex wholeness need to be implemented as modeling technologies for phenomena, entities, issues, etc., that possess the attribute of complex wholeness.

To effectively address complexity through modeling techniques, we need to respect and embody the objectivity and logical rigor of modeling within the corresponding paradigm shift in thinking, academic thoughts on modeling, principles, and tools.

British physicist Andrew Lyddiard posits that significant challenges in the universe can be categorized into three tiers of complexity: Level C comprises problems that have discernible solutions; Level B includes problems that have theoretical explanations but lack empirical evidence; while Level A represents problems that lack even a foundational starting point. He argues that defining facts belongs to the Level A difficulty because even though scientific tools allow us to understand the laws governing phenomena, we cannot explain in a short period of time why these mechanisms exist [1]. Lyddiard's viewpoint undoubtedly serves as an important reminder for us during the process of modeling complex wholeness: how to define the modeling object, how to represent the uniqueness, singularity, and uncertainty manifested by this difficulty, and how to prudently handle the complex wholeness hidden behind the phenomena of complex management problems.

Due to these considerations, in practice, the process of modeling complex systems management activities or problems is inevitably a cognitive and recognition process of the subject(s) regarding the complex wholeness of complexity management facts. This process cannot be a one-time "overnight" action but rather a comprehensive process of the modeling subject(s) constantly comparing, providing feedback, trial and error, iteration, approximation, convergence, and ultimately forming a holistic view of the final results of their cognitive process.

In this process, the complex wholeness of management activities and problems and the intelligent adaptability of the modeling subjects are mutually coupled, forming a complex system of subject-object modeling. This system's orderly approximation of the abstraction, description, and representation of complex wholeness of the modeling object not only fully reflects the inherent organization and self-organizing behavior of the modeling object but also represents the modeling subject's profound understanding and adaptability in modeling the complex wholeness of management.

It is evident that the modeling process of complex wholeness in complex systems management encompasses not only the complex wholeness of the object itself but also the complex wholeness of the modeling subject's behavior during the modeling process. Furthermore, it includes the complex wholeness of the higher-level composite system formed by the collaboration between the subject and object in modeling.

For example, in the modeling process, the subject not only expands its perception of the complex wholeness of management facts synchronously with the continuous enrichment of complex wholeness information but also deepens its understanding of the complex wholeness. Furthermore, these two forms of complex wholeness are integrated into the complex wholeness of the modeling process. In other words, the modeling entity, through its self-organizing and adaptive capabilities, not only gradually refines and improves the originally generated complex systems management model based on the abundant new information and data emerging from management facts, but also further explores more "known facts," "knowable facts," and even "unknowable facts" in the process. Thus, the complex systems management modeling sequence is continuously orderly iterated, approximated, and converged. This academic thought and technical approach go beyond the traditional one-time interactive complete model cognition between the modeling entity and the modeling object (facts). Instead, it involves multiple interactions, feedback loops, and iterations between both parties to gradually achieve the modeling of complex systems management. Clearly, this fully embodies the paradigm shift in complex system thinking and forms a corresponding new technical pathway at the methodological level.

10.2 The Concept and Characteristics of Scenario Modeling

This section mainly discusses the fundamental ideas behind modeling and constructing scenarios with complex wholeness. To begin with, let's summarize several important viewpoints regarding the concept of scenarios that have significant foundational significance for the content of this chapter:

1. Research on complex system management is better expressed as "scenario-oriented principles" rather than "problem-oriented principles."
2. Scenarios are the most comprehensive perception of complex wholeness attributes in complex system management activities. They possess both rational abstraction and complex wholeness, effectively linking the two together as an entity.

Therefore, although a specific modeling process may involve phenomena, activities, or problems, as a methodology, complex system management should further abstract the "collective" nature of complex wholeness, which is referred to as a scenario. In other words, at the methodology level of complex system management, modeling complex wholeness can be understood as scenario modeling in complex systems management, also known as scenario-based modeling.

In the process of scenario modeling, special attention should be given to important connotations of scenario complex wholeness, such as:

1. Scenarios are generated and dynamically evolve in the present, past, or future of management activities. Scenarios constitute an ordered and coherent process that cannot be severed between the "future" and "present" or "past" of a scenario. The "future" encompasses the "past" but is not entirely determined by the "present" and "past."
2. The composition of scenarios includes diverse elements such as physical, managerial, behavioral, cultural, and psychological components. Scenarios reflect both the structural components that reflect objective natural laws, the semi-structural components that reflect social laws related to behavior, and the non-structural components that reflect cultural and value orientations. They also encompass the integration of these three aspects.
3. In the context of complex wholeness, scenarios consist of both components designed and specified by humans, components generated based on the interrelationships among system elements, and components that emerge through the self-organization of system complexity.
4. Within the framework of complex wholeness, the scenario space and its evolutionary trajectories frequently display traits like structural multi-levelness, multidimensionality, multi-scale, multi-granularity, and varying degrees of uncertainty.

In traditional management and decision-making research, modeling approaches mainly include qualitative modeling, structural modeling, process modeling, and simulation modeling, resulting in qualitative, quantitative, and various experimental modeling techniques. It can also be seen as a structured, fixed, or discretized modeling based on static cross-sections. However, in complex system management, the complex wholeness, dynamic evolution, self-organization, emergence, submergence, and abrupt changes exhibited by scenarios, which serve as the modeling objects, cannot be achieved solely through these traditional modeling approaches. For example, applying traditional statistical models to predict future scenario states is based on the assumption that historical data and information fully contain future states or that historical operational dynamics will be fully valid in the future. However, the complex wholeness of scenarios tells us that the future scenarios contain the present and the past but are not entirely contained within them, and the path from the present to the future is essentially uncertain. Cognitively, there exists a blind spot where the subject is unaware of what they don't know, making it impossible to grasp unrevealed scenarios regarding the future. This requires us to apply new thinking, methods, and approaches to abstractly model complex system scenarios using a comprehensive symbolic system. Therefore, the core idea of complex system scenario modeling is guided by the basic form of scenario complex wholeness, breaking through the reductionist and simplistic thinking of traditional modeling, integrating multidisciplinary approaches from various domains, and constructing a symbolic system capable of reflecting the complex wholeness of scenarios. On this computable and visualizable symbolic system platform, the reconstruction, reproduction, and prediction of scenario behaviors and evolutions can be achieved.

The value of complex system scenario modeling technology lies in its ability to generate abstract representations of scenarios, encompassing the past, present, or future scenarios with complex wholeness, as well as their dynamic evolution. This enables the analysis of complexity management problems and decision-making activities based on scenarios. All of these are difficult to achieve under traditional modeling methodologies, thus demonstrating the fundamental connection between complex system scenario modeling technology and complex system management.

To summarize the previous content, the following points need to be further emphasized in scenario modeling:

1. The complexity of the scenario involves a complex "multi perceives," meaning that the complexity of a scenario encompasses the complexity of scenario elements and their relationships, scenario structure, scenario forms, scenario evolution, and so on. It is reflected in all aspects of the scenario, such as its spatiotemporal aspects, levels, dimensions, scales, dynamic changes, and evolution. Therefore, scenario modeling is a "panoramic" modeling of the scenario, not just modeling the qualitative aspects, architecture, a specific state, or a static form of the scenario.
2. Since the understanding of the complex wholeness of the scenario is a process of deepening, approximation, and convergence, the process of modeling complex system scenarios will be an iterative closed-loop process of trial and error, testing, feedback, correction, and approximation. It often exposes uncertainties and even the loss of control of the modeling purpose. Therefore, in the process of modeling complex system scenarios, it is necessary to pay full attention to the comprehensive balance of model abstraction granularity and resolution. Emphasizing high-resolution or fine-grained modeling without considering other factors such as modeling purpose, cost, and feasibility can lead to problems. A more cautious approach is to balance modeling requirements, modeling techniques, and modeling costs based on specific goals of managing complex systems, striving to achieve a "just right" modeling process.
3. The holistic attributes of the scenario require the modeling process to be as comprehensive as possible. For example, it should consider the coupling between scenario features in different dimensions or scales, how the micro-to-meso-level actions in complex management activities give rise to macro-level holistic scenarios, and how the "strong correlation" between scenario parts elevates the complexity of the scenario from local to global, for instance, how a fault scenario gradually evolves into a local accident and further escalates into a global disaster, or how management activities in a complex system-environment composite system give rise to new holistic scenarios.

4. When considering "wholeness" in the process of modeling complex system scenarios, it should include the wholeness of the scenario itself and the wholeness of the scenario-object modeling subject composite system formed by the mutual influence of the modeling subject and the adaptiveness in modeling. It is not about separating the modeling subject as one entity and the scenario as another entity.

10.3 Basic Connotations of Scenario Modeling for Complex Systems

The analysis above tells us that since scenarios are the perception of the complex wholeness of the structure, functions, environment, and evolution of complex system management activities or problems, it can be considered that modeling scenarios are, to some extent, a means of realizing the modeling of complex wholeness.

In terms of its connotation, modeling complex system scenarios involves the holistic abstraction and description of scenario-specific language and modeling symbols, such as when, where, who, what, why, and how the application of a fully defined and clearly delineated scenario will occur. The basic points of modeling complex system scenarios include:

1. Defining and describing the time and space scope of a specific scenario's occurrence and development.
2. Defining and describing the occurrence and development of important plotlines and events in a specific scenario, especially the core plotlines and events closely related to the research problem, as well as the corresponding basic dynamics and mechanisms.
3. Defining and describing the characteristic information of different entities involved in the occurrence and development of a specific scenario, such as individual psychology, behavior, habits, goals, preferences, and responsibilities, with a focus on establishing and exploring the causal or associative relationships between individual micro-behaviors, behavioral interactions, and the complex wholeness of the scenario.
4. Defining and describing the information of different organizations involved in the occurrence and development of a specific scenario, including goals, interests, functions, structures, processes, and behaviors, with a focus on describing and abstracting the causal or associative relationships between different organizational behaviors, behavior interconnections, and the complex wholeness of the scenario.
5. Defining and describing the content, expectations, conflicts, and constraints of important plotlines and events in a specific scenario, with a focus on describing and abstracting the causality and correlation between specific spatiotemporal backgrounds, dynamics, individual behaviors, and organizational functions.
6. Defining and describing the environment in which a specific scenario occurs and develops, including natural, social, economic, political, and cultural aspects.

Analyzing the interaction between a specific scenario and its environment and making expected definitions and descriptions of new scenarios in the complex system management-environment composite system.

7. The task involves detailing and abstracting the physical forms and dynamic alterations of a specific scenario, primarily concentrating on organizing and analyzing the relationships and structural characteristics (such as richness, spatiotemporal attributes, transmission and transformation patterns, and operational modes) of the scenario's physical elements.

In summary, the academic essence of the process of modeling scenarios can be further simplified as follows: the modeling agent captures "slices" of the complex integrity of the scenario at different key temporal and spatial points. Furthermore, they utilize basic principles of scenario evolution to obtain a "slice flow" composed of ordered scenario slices that describe the evolution of the scenario. This approach allows for an overall approximation of a specific panoramic scenario.

For modeling complex system scenarios, the key is to determine which scenario elements the model should include and how to describe and abstract individual behaviors, events, and plots. Since most complex system management problems are described and abstracted based on scenario elements (temporal and spatial aspects, tasks, individuals, organizations, problems, and society), this facilitates a unified modeling approach for complex system management modelers and enables them to analyze problems and develop solutions for complex problems in the later stages of the modeling process.

10.4 Complex System Scenario Modeling Methodology

In scientific research, the design and selection of methodology are fundamentally determined by the characteristics and features of the problems to be solved. Firstly, as previously noted, modeling predominantly involves devising and defining representative behavior and techniques reflective of reality, whereas modelization refers to the specific process or workflow involved in modeling. Therefore, having introduced the overview and essence of complex system scenario modeling in the preceding sections, this section logically extends to a deeper discussion on the methodology of complex system scenario modeling and the principal issues related to it.

Complex system scenario modeling is the abstract description of specific complex system scenarios using normative and procedural scenario language and symbol systems under the paradigm of complex system thinking. Secondly, from a theoretical perspective, a complex system scenario itself is a complex system. Therefore, complex system scenario modeling is the holistic abstraction representation of the scenario elements, structure, form, and evolution of specific complex scenario systems using normative and procedural scenario language and modeling symbol systems under the paradigm of complex system thinking. Therefore, designing the methodology of complex system scenario modeling requires the organic integration of the specific modeling goals with the essential attributes of the scenario and provides corresponding new paradigms in a methodological sense.

The transfer of the scenario concept based on the paradigm of complex system thinking has positive implications for the innovation of scenario modeling methodology.

From a methodological perspective, in management research, researchers need to use a symbolic system as a medium to abstractly describe management activities and processes. In 1988, Ostrom proposed three "symbol systems" for use by social scientists. He pointed out that in addition to the qualitative methods familiar to us and certified by natural language and the quantitative methods described in mathematical language, there is a third research method: computers and procedural languages [2]. We can use standardized and procedural languages of computers to describe our thoughts and discuss the past, analyze the present, and predict the future through computer assistance. This methodological principle tells us that we can achieve the modeling of complex holistic scenarios with the help of computer technology based on traditional qualitative and quantitative methods.

Specifically, any scenario, whether at a specific point in time and space or within a certain spatiotemporal range, as the overall form and complete dynamic evolution process of a complex system, regardless of the complexity it exhibits, has its fundamental deterministic mechanisms and certain causal regularities, known as dynamic laws. These laws mostly reflect the objectivity and naturalness of the scenario and exhibit characteristics such as norms, determinism, causality, and measurability. For scenarios, the important function of these laws is similar to constructing a building and maintaining a stable and specific style as the fundamental architecture or guiding and constraining traffic rules that ensure the safe progress of cars on highways. Thus, the most basic aspect of complex system scenario modeling is to demonstrate the fundamental architecture and dynamical mechanisms of scenario regulations, behaviors, patterns, characteristics, and attributes in a holistic manner.

Therefore, scenario modeling primarily involves condensing the dynamic mechanisms of the scenario and accurately reflecting the essential attributes, overall structure, spatiotemporal distribution, and evolutionary laws of the scenario through the representation and description of these mechanisms. In this stage, the knowledge about the scenario, especially the fundamental knowledge, plays an important and fundamental role. It may also involve the computer simulation and experimentation of a large amount of scenario information data to explore this type of knowledge. This methodological idea effectively addresses the challenge of explicitly structuring complex, holistic, dynamic mathematical models of scenarios. In recent years, the development of this idea has become the big data-driven scenario modeling method introduced in the last two chapters of this book.

Furthermore, if the entire or major part of the complex system scenario exhibits the deterministic and causal dynamic laws mentioned above, then the scenario may not be as complex. Scenarios also involve a significant amount of uncertainty (randomness, fuzziness) expressed in various forms. Correlation is also a kind of mechanism that reflects the possibilities of scenario forms more than the actual laws. It often exhibits characteristics such as comprehensibility, perceptibility, and imaginability. In many cases, the complexity of the scenario is precisely caused by these correlations. If these correlations occur between different levels and scales of the scenario or manifest as implicit, lagging, and long-chain characteristics, they contribute to the complexity of the scenario. Therefore, complex system scenario modeling requires specific methodological approaches, such as using knowledge of the correlation between different states and characteristic quantities to represent such architectures and rules.

Moreover, complex system scenarios are not only complex in a general sense but often possess unique and scarce complexities. Therefore, it is generally not feasible to extract statistical regularities from the same type of scenario population. In particular, for the unique components of the complex wholeness of panoramic scenarios, it is necessary to closely align with realistic scenarios, paying full attention to the details and variations of the scene. Hence, in the process of modeling complex system scenarios, it is necessary to reveal what unique micro-level details of the complexity morphology and complexity correlations will enable the macro-level behavior to exhibit complex wholeness as a "possibility" rather than attempting to reveal how the macro-level complex wholeness becomes "inevitable" under certain micro-level dynamic mechanisms.

To accomplish this, the data (information) sources that support the description of details must be extensive. This includes observational research data conducted for specific purposes as well as non-research-related real-world data, such as various data related to the problem collected and recorded by multiple institutions. It encompasses real-time large-scale data generated from practical management processes as well as small-scale data collected from specialized management processes. Based on data collection, various elements of scenario details can be set in realistic scenarios to reflect the social, humanistic, and contingent aspects within the complex wholeness. From a methodological perspective, this should be understood as achieving the panoramic complexity of the scenario through social and humanistic aspects and finding standardized entry points to address the non-normative unique components in scenario modeling.

In conclusion, the methodology of modeling complex system scenarios not only involves exploring and representing the fundamental dynamic elements of the scenario that reflect its essential properties, overall structure, temporal and spatial distribution, and evolutionary patterns, but also relies on valuable and unique scenario samples or clues. Utilizing computer systems as the "soil," these samples and clues are treated as "seeds" to be sown, nurtured, and allowed to grow. Eventually, scenario "fruits" with consistent underlying dynamic mechanisms are obtained. Building upon these basic forms of "fruits," a diverse "real world" of scenarios is formed through the contextualization of the unique characteristics of complex system scenarios. This represents the core technological roadmap of the methodology of modeling complex system scenarios.

References

1. Xinhua. Western media take stock of eight unsolved mysteries in the world today. Xinhua;2018.
2. Zhang J. The role of computational experiments in social science research. Res Explor Lab. 2009; 28(06):75–8+90.

Chapter 11
Fundamentals of Big Data-Driven Principles

11.1 Overview of Big Data

After understanding the basic content, paradigms, and methodologies of complex system scenario modeling, our focus naturally shifts to the fundamental principles and operational techniques for designing specific and practical complex system scenario modeling at the operational level. This is essentially a question of consistency between epistemology and methodology. Before directly addressing this question, we will first spend some time in this chapter to provide some preliminary knowledge preparation, such as the concept of big data and the basic principles of big data-driven approaches. Then, in the next chapter, we will provide a detailed introduction to the complex system scenario modeling techniques driven by big data.

1. **Concept of Big Data**

In the present era, modern information technology has rapidly developed and become a powerful force driving social progress. Firstly, the emergence of the Internet has brought about significant changes in global socio-economic activities. The Internet, as a globally extensive computer network comprising various types and scales of independently operated and managed networks, represents a vast global information and service resource. With the rise of smartphones, tablets, and mobile internet, the era of mobile internet, which combines the Internet and mobile communication, has begun. In 2007, cloud computing emerged. Cloud computing is an internet-based computing method that connects thousands of computers and servers into a cloud, where shared software, hardware resources, and information can be provided to computers and other devices on demand in remote data centers.

Subsequently, the concept of the Internet of Things (IoT) emerged based on the concept of the Internet. The IoT is an extended and expanded network built on the Internet. It connects all objects, including appliances, vehicles, equipment, forests, lakes, houses, etc., through information sensing devices and IP addresses and enables

Z. Sheng, *Outline of Complex Systems Management Theory—Based on Irreversibility of Reductionism Thinking*, International Series in Operations Research & Management Science 360, https://doi.org/10.1007/978-981-97-3361-3_11

information interaction even between any objects, thus achieving intelligent identification and management. In 2008, the concept of big data was proposed. Big data refers to the massive amount of data that is involved and cannot be collected, organized, and analyzed by human brains or mainstream software tools within a reasonable timeframe, thus becoming information that assists decision-making or management.

In summary, big data refers to the information resources that have stronger analytical, insight, and decision-making capabilities through the fusion and mining of large-scale, dynamic, and continuously collected data using new processing patterns, tools, and models.

Specifically, with the current rapid development of IT technology, especially internet technology, information collection has become more convenient and cost-effective. Personal computers, various sensors, and technologies that digitize non-digital data, such as voice, images, videos, files, sketches, self-media technologies, wearable smart devices, and others, have greatly expanded the channels for obtaining information. In addition, the internet technology that connects wide-area networks and interconnects them provides the technical means for transmitting and sharing these data and information, allowing various data and information to interconnect and generate new value during their flow.

2. **Characteristics of Big Data**

Based on the above introduction, it is not difficult to summarize the following four characteristics of big data, referred to as the 4Vs [1]:

(1) Volume. The first characteristic of big data is the large volume of data. This is primarily because big data is no longer solely obtained through sampling. Social networks (such as Weibo, Twitter, and Facebook), mobile networks, various intelligent tools, service tools, sensors, and the internet have all become sources of data.
(2) Velocity. Big data is generated rapidly, and the storage and processing of data (information) are fast. The generation of big data is highly accelerated, mainly through internet transmission. In daily life, everyone is constantly providing a large amount of data, as nobody can be separated from the internet. At the same time, big data imposes strict requirements on processing speed, and the abundant resources in servers can be used for data processing and computation.
(3) Variety. The broad sources of data determine the diversity of big data. Text, images, audio, video, and any other form of information can generate data. However, the causal and correlation effects of different types of data vary, so their actual value also differs. We cannot determine the significance of each data type solely based on prior knowledge, and some level of discernment is required for big data.
(4) Value. The value referred to here is not the value of individual data or a specific portion of data but the overall value of the dataset. The value of big data lies in its overall contribution to problem-solving. On the one hand, due to the characteristics of big data, such as volume, velocity, and variety, we cannot expect all acquired data to be significant. In this sense, the value of individual or partial

data seems to decrease. On the other hand, the abundance and diversity of data provide us with a comprehensive description of the problem of wholeness. Especially through in-depth analysis using machine learning or data mining methods in the methodology of big data, valuable new knowledge can be extracted from seemingly unrelated data of various types, revealing deep patterns and implicit trends in complex systems. This overall value is the essence of big data.

11.2 Big Data Perspective Based on Scenario Modeling

Based on the aforementioned fundamental concepts and characteristics of big data, combined with previous knowledge of complex systems management, we will focus on discussing the value of big data in the field of complex systems management, especially in the research of scenario modeling.

1. For convenience, let's refer to the period before the concept of big data as the "small data era." In the small data era, the research methodology centered around data can be summarized as the "sample perspective." It means that the entire data of the research subject is referred to as the population, while a portion of data actually observed or collected during the research is called a sample. People use samples as representatives of the population to study the population, and the corresponding attribute values of the samples are the sample data. The reason for studying the population through samples is that data collection techniques were difficult and costly in the small data era. Studying the population with only a "portion" of the data is incomplete. Regardless of the statistical techniques used, it is challenging to overcome the inherent lack of information about the population. If we continue to use this sample-based approach and methodology in the research of complex system scenarios, its shortcomings will be even more significant.

This is because the essential attribute of complex system scenarios is their complex wholeness. If we adopt the traditional sample perspective in scenario modeling research, we are essentially decomposing the reductionism of scenario wholeness. This not only directly damages the wholeness of the scenarios, but also makes it difficult to "capture" the "rare events" within the overall scenarios. Since these "rare events" are, by definition, infrequent, finding sample data for this part is like "finding a needle in a haystack." On the contrary, in this regard, the characteristics of big data, such as volume, velocity, and variety, effectively compensate for this significant deficiency. Although not every piece of data in big data has a significant impact, the datasets in the context of big data can approximate the details of various aspects, dimensions, and scales of the scenarios as a whole. Furthermore, it can uncover deep correlations between data that were previously considered unrelated, thus demonstrating the comprehensive objectivity of the scenarios. Therefore, for complex system scenarios, a sample-based approach based on reductionism fails to reflect and guarantee wholeness, while the characteristics of big data make it possible to approach the wholeness of scenarios, providing us with a means to model complex system scenarios.

2. In addition to exhibiting a holistic nature, complex system scenarios also manifest various types of complexity, especially those complexities that can only be revealed through implicit, deep-level, and multiple correlations. It is very challenging to construct mechanistic, structured mathematical models to reveal such complexities. However, as carriers of traces of scenario information, big data inherently contains various types of complexities. By collecting an adequate amount of diverse data and conducting effective in-depth mining and analysis, we can reveal these complexities to a certain degree of accuracy, even if it is a weak correlation. In the small data mindset of sample-based approaches, these weak correlations are often mistaken as some form of low-probability uncertainty due to the incomplete sample. It is worth emphasizing that if such weak correlations happen to be a crucial detail in complex system scenarios that is difficult for people to perceive, big data undoubtedly provides a powerful new tool for deepening our understanding of the essence of complex system scenarios.
3. The scenario of a complex system includes not only the complex overall form, but also the complex evolutionary path of the scenario. In the study of complex systems management, only studying the static state of a scenario at a certain moment, without studying the dynamic process of the scenario, often fails to fully meet the needs of analyzing and solving complex wholeness problems. This requires us to conduct a systematic and coherent study of the entire process of the past, present, and future scenarios based on the dimension of time. The coherent dynamic evolution of the scenario will inevitably be governed by some complex dynamic mechanisms, and we are incapable of fully mastering and effectively controlling such dynamic evolution.

This difficulty can be summarized as the challenge of connecting the "past," "present," and "future" of a scenario together for the subject in the absence of prior knowledge, personal experience, and cognitive understanding of structured dynamic models. It is acknowledged that the "present contains the past but is not completely contained within the past, and the future is not completely contained within the past and present." This means that complex system scenarios are not systems that can be extrapolated based on deterministic dynamics principles of history and the present.

Therefore, it is impossible to achieve what Laplace stated: "We can consider the current state of the universe as the consequence of its past and the cause of its future. If a wise person knows the forces of all natural movements and the locations of all objects constituted by nature at a certain moment, and if he is also able to analyze this data, then the movements of everything in the universe, from the largest bodies to the smallest particles, would be included in a simple formula. For this wise person, nothing would be ambiguous, and the future would appear before him just like the past."

This intellect is Laplace's demon, born from the reversibility of classical mechanics [2]. However, the essential attribute of complex systems is the irreversibility of reductionism, and complex system future scenarios are not predictions of a specific state variable of a complex system but predictions of the future complex wholeness of a complex system. To predict scenarios that have never occurred in history or the present, the capabilities required far exceed the imagination and the ability to "extrapolate based on deterministic dynamic principles." It is also not solvable using small data sample methodologies.

To summarize the previous content, complex system scenarios are the entirety of human complexity management activities and processes in the real-world space, including all the details. From the information (data) carried by these details, we try to dig out and find traces of complexity management activities and processes. Due to technological conditions today, it is possible to trace, observe, describe, and measure these traces and further quantify them as mappings or projections of complexity management activity and process scenarios in data space.

Taking a closer look at scenarios, they include natural, social, spatiotemporal, organizational, individual, and task-related scenario components. These components are composed of various types of associations based on when, where, who, what, how, why, cause, effect, and change, forming the plot and evolution of the scenario. The information characteristics of the scenario can be summarized as complex wholeness, "infinite" associated elements, heterogeneity, and non-structuredness. This has a close theoretical unity with the characteristics of big data (volume, velocity, variety, and value depth). Therefore, the methodology based on big data is suitable for scenario modeling research, especially for scenario modeling research involving scaling of three dimensions (dimension, scale, and granularity), cross-domain associations, holographic scenarios, and scenario orientation.

This provides us with a big data perspective on scenario modeling. By perceiving, cognizing, experiencing, knowing, theorizing, associating, assuming, guessing, and collecting data through multiple channels based on the perception of the complex wholeness of scenarios and using big data techniques such as independent testing of scenario-associated elements data sequences based on the principle of scenario computability, a core set of scenario elements under a certain dynamic mechanism can be formed. Then, based on big data and utilizing the spatial network, event time series, and heterogeneous data fusion technology of scenario space, along with the descriptive methods of scenario spatiotemporal states and analytical methods of scenario evolution dynamics mechanism, we leverage the infinite computing and other statistical analysis capabilities of machines to complete data pattern analysis and value discovery. Ultimately, within the scenario space, we achieve the historical reconstruction, real-world reflection, or future prediction of complex system scenarios in different temporal states.

11.3 Definition of Big Data-Driven

In the above analysis, we have explored the potential role of big data characteristics and methodologies in modeling complex system scenarios by theoretical correlation analysis at the methodological level. From a theoretical perspective, we recognize the positive significance and role of big data thinking and methodologies in complex system scenario modeling.

In the domain of complex system management, the previously mentioned theory can be synthesized into the technology of big data-driven complex system scenario (or modeling) modeling. In this section, we will initially offer an elucidation of the foundational concept of being big data-driven.

In the real world, people directly observe, collect, and perceive data. Data originates from the synchronous or lagged records or advanced predictions of human practical activities and processes, naturally containing certain information about scenarios. However, data generally does not completely encompass all attributes of a specific scenario and only represents partial characteristics and traces of the scenario. Nowadays, on the one hand, various advanced technologies help us conveniently obtain a large amount of scenario data. On the other hand, when people attempt to directly model scenarios using data, they still encounter issues such as missing data and low quality. Nevertheless, due to the increasing amount of data that can be collected and the methods available to improve data quality, it is possible to extract information about a specific scenario inherent in the data and the associative relationships between different pieces of information. Therefore, we can gradually "externalize" the potential value of raw data sets, namely the value of scenario reconstruction, simulation, and prediction regarding the wholeness of the scenario. This process, after a series of transformations, leads to the concept of big data (scenario value) thinking.

In many contexts, people refer to big data thinking collectively as "big data." Here, big data is not a narrow concept but rather a way of thinking, a resource, a tool, a means, an algorithm, a research paradigm, wisdom, and more, based on a massive amount of data. However, regardless of the definition, big data should not be understood simply as a large-scale or massive data set, nor as a large sample data set in traditional statistical methods. The value of big data is not fixed; its "value dimension" expands as people enhance their data processing and mining capabilities and clarify the purpose of data value. For example, the emergence of big data (scenario value) thinking is based on the recognition of the positive significance and role of big data in modeling complex system scenarios. Through the study of the fundamental connotations, characteristics, and basic value of big data, a specific path for externalizing the potential value of scenario modeling through big data can be clarified and designed.

Now, let's delve into the concept of "big data-driven" in more detail. Undoubtedly, "big data-driven" has become a highly used scientific term across various sectors of society. It is a compound phrase composed of "big data" and "driven," forming a modifier-noun structure, with "driven" as the central word. The complete structure of this phrase should be "A drives B using big data," which is a subject-verb-object structure. Here, A is the subject, the agent; "driven" is the action (behavior) of A, the predicate. "Big data-driven" implies that A utilizes big data as a resource, method, or tool to drive and empower big data, thereby generating some value. Therefore, big data functions as an adverb, while B represents the object or purpose of the driving process. In this light, "big data-driven" is a phrase that has been condensed, similar to how the complete meaning of "gasoline-driven" is "the driver uses gasoline to drive the engine."

From this perspective, big data-driven requires the existence of a driving entity A and a driven object B. B can be a thing, such as a solution obtained through driving, or an action, such as scenario modeling driven by big data. Therefore, in general, big data-driven is usually focused on B as the target (objective, direction) of the driving process. Once this point is clear, it is not difficult to understand that the core of big data-driven is the principle of using big data to drive toward a specific goal (objective, direction) and designing the specific driving programs and processes to achieve it.

Generally, whenever a goal is achieved, it reflects the acquisition of some form of value. Big data-driven indicates that the latent and implicit value attributes of big data gradually manifest and emerge during the driving process. It is similar to extracting oil from rapeseeds, where the oil is originally contained within the seeds, but through pressing (driving), the seeds become edible oil in reality.

In this sense, the basic meaning of big data-driven is to extract (mine) the corresponding latent value contained in big data based on a specific goal orientation, making the value of big data tangible, referred to as "externalizing" value. Therefore, the understanding of the connotation of big data-driven should revolve closely around the following two questions:

1. What potential value does big data possess to meet a specific goal?
2. How do we systematically review and articulate the principles that can meet this goal requirement and design specific paradigms and processes to "externalize" the value of big data?

Next, by briefly reviewing the path of human understanding of information (data), we can further comprehend the interpretation of big data-driven.

In the real world, people directly perceive and encounter various phenomena, behaviors, and the states and processes of various objective things. Initially, people used relatively simple methods such as sound, language, and text. With the development of human society, people have added digital, graphical, visual, and video-based methods to track, observe, collect, record, store, and digitize this information. Dataization is a process of transforming various non-data forms of information into quantifiable symbols.

Initially, people directly perceived and encountered non-quantitative information, which gradually transformed into quantified information, namely data. Further digitization provided important technological conditions for computers to process information and data. Today, there are increasingly diverse means of obtaining data, lower costs, faster frequency, and more advanced technologies for processing data and extracting useful information from it. A type of data set that was initially overlooked by people is now regarded as a new symbol set called "big data" due to its unique 4V characteristics, displaying its multifaceted and valuable nature through various technologies.

To draw two analogies, data can be likened to disorganized and chaotic gold ore, while Big Dataization[1] is akin to refining the gold ore into gold nuggets. This represents the initial work driven by big data. Meanwhile, purifying the gold nuggets and crafting them into exquisite gold crowns is equivalent to the later stages of work driven by big data. Similarly, a raw tree root represents data with certain underlying values. Splitting it and using it as firewood was once believed to be the only value of data for calculation purposes. However, an artist equipped with artistic aesthetic thinking, wisdom, necessary craftsmanship, and tools turns a tree root into an elegant root carving artwork, and its value soars. This is equivalent to the process where data is driven and deeply mined, and its potentially valuable nature is "externalized." Similarly, in this process, the artist not only possesses big data wisdom, but also serves as the driving force to externalize the latent value by utilizing data processing techniques. The fundamental difference lies in the fact that small data is about computational thinking, while big data represents computational wisdom.

In relevant research, we often come across the term "big data environment," which generally refers to the platform and conditions that provide and support the formation, collection, and mining of big data and its value, such as the Internet of Things, various sensors, analytical tools, algorithms, software, hardware, and related equipment used to extract useful value from big data.

Additionally, there is the term "big data background," which, from a temporal perspective, can be understood as the overall form, level, and characteristics of the social, economic, and technological aspects of the big data era, particularly information technology. In a more detailed understanding, it refers to the basic capabilities provided by information infrastructure and the storage and transmission technologies of data, such as the Internet, cloud computing, the Internet of Things, and mobile devices, which collectively form the platform for realizing the value of big data.

However, regardless of how they are understood, "big data environment" or "big data background" should not be confused with "big data-driven." Sometimes, we may see researchers conflating these three concepts, resulting in studies that focus on the "application of large amounts of data" in a big data environment or background rather than studies on "big data-driven."

11.4 The Fundamental Connotation of Big Data-Driven

This subsection mainly provides a detailed interpretation of the fundamental connotation of big data-driven based on the basic definition of big data-driven.

1. **Data-Driven Versus Big Data-Driven**

Data derives from human practical activities and processes, serving as the identification and trace records of these activities and processes. This determines that data

[1] The big data processing of data.

possesses certain attributes, characteristics, and information related to these activities and processes, thereby having potential value for us to understand and analyze them. When we extract such potential value from the data through certain techniques to serve a specific practical need, this way of thinking and behavior is known as **data-driven**. Depending on the different attributes of human activities and data, the data digitization in the real world is the mapping or projection (quantification into data) of human activities, behaviors, facts, forms, and feature information in the data space. Data-driven, on the other hand, involves processing certain datasets in the data space and reversing the value formed back to a specific activity (event) in the human real space, where it can exert its actual value.

To avoid ambiguity in the meaning of terms, we refer to the data before the emergence of the concept of big data as small data. Therefore, data-driven generally refers to small data-driven.

Due to the outdated data collection and recording methods before the era of big data, data was extremely scarce, much like the fleeting traces left by geese in the sky or the faint footprints left in the snow by swans, or like elusive spider silk and horse tracks. Additionally, there was a lack of technology and tools to deeply process and utilize the data, making the exchange and sharing of data difficult. The value of data needs to be realized through communication. Therefore, although the fundamental implication of data-driven is very simple, for a long period of time, people had to be very innovative with methods in order to extract value from data-driven approaches. For example, it was necessary to very clearly define value needs, that is, the objectives of the problem; in order to save on data collection costs, efforts were made to control the sample data volume and to obtain the best possible value to meet needs with as little sample data as possible. All of these were not easy tasks in reality.

For example, to describe a complete elephant, we collect 100 pieces of data. If these 100 data points are all from the elephant's leg, the value obtained through data-driven would only conclude that the elephant is a thick and sturdy pillar of meat. Only when these 100 data points are evenly distributed throughout the elephant's body, we can gain precise knowledge about the elephant, such as its large ears and long trunk, which provides accurate cognitive value about the elephant. Clearly, this is not an easy task.

As management evolves from systematic to complexity, the demand for uncovering the potential value of data becomes increasingly strong. Fortunately, comprehensive technologies for data collection, recording, analysis, and mining have become more advanced and sophisticated. In this context, a new methodology of big data-driven has emerged.

The basic definition of big data-driven has been introduced earlier. The difference between big data-driven and data-driven lies primarily in the richness of the connotation. With the advent of the Internet (information physical devices) and cloud computing (information processing platforms), it is now possible to collect and analyze data in the form of "all" traces related to human activities or behaviors using statistical techniques, machine learning, data mining, and visualization. This has created unprecedented quantifiable dimensions, where many "traces" (information) that were previously unmeasurable, difficult to store, analyze, and share can now be digitized, rapidly transmitted, analyzed, and utilized. Therefore, "big data" initially represents the closest digital portrait mapping to the real world. From an objective perspective, there is only data in the real world, not "big data." The latter is simply a subjective creation of human thinking.

In short, in the era of small data, as data was hard to come by, every piece of data was valuable, and people naturally focused on how to make data representative and not waste the value of each piece of data. However, original big data is abundant, of poor quality, mixed, unordered, imprecise, both abstract and concrete, variable in density, structurally diverse, and inconsistent in format. Therefore, to avoid fishing in the muddy waters, it is necessary to conduct essential "pre-processing" work on the data in the early stages of the drive to improve the quality of big data and the efficiency of uncovering the deeply "hidden" value in the data.

Furthermore, in many cases, describing a complex phenomenon or solving a complex problem solely relying on the value provided by big data-driven approaches is not sufficient. It is necessary to integrate big data-driven approaches with theoretical analysis, structured models, and computer simulations to form a more powerful methodology for managing complex systems study.

In addition, from the perspective of the subject, data-driven primarily involves accepting the value provided by data as much as possible based on specific needs. In this sense, data-driven exhibits a unidirectional relationship between the subject and the data set. However, for big data-driven, due to the high speed and frequency of data collection, the continuous exploration of new data during the driving process leads to the formation of new values. This, in turn, deepens the subject's understanding of the problem and further generates new value demands for the data set. In this sense, big data-driven exhibits a bidirectional interaction between the subject and the data set.

2. **Basic Connotations of Big Data-Driven Approach**

Broadly speaking, the situations that most need to employ this "heavyweight" methodology of big data-driven approaches are typically complex wholeness problems. This is because, for general additive wholeness problems, we can decompose the problems through reductionism and solve the relatively simpler problems post-decomposition using established qualitative methods, mathematical models, etc. Even if data is involved, the problems can essentially be resolved through sample statistical methods based on small data. Therefore, for relatively simpler problems, the functionalities provided by the traditional methods, combined with the researcher's experience, knowledge, and insight, can basically meet the research needs.

Additionally, in a relatively simple problem, there is often more determinacy and structural clarity within the relationships between elements and the overall structure of the problem. Therefore, structured mathematical models can not only clearly describe the dynamic mechanisms of the problem but also make accurate predictions about the future development trends and states of the problem. In this process, if data is involved, general sample statistical methods can usually resolve the problem.

However, as previously mentioned, to solve complex wholeness problems, it is essential first to achieve a paradigm shift in complex system thinking. This leads to the emergence of new cognitive paradigms, such as the irreversibility of reductionism, essential uncertainty, unique scenario contexts, adaptability, and iterative cognitive paths. Subsequently, it inevitably induces a paradigm shift in research methodology, for example, transitioning from methodologies driven by a priori knowledge and assumptions to those oriented toward "real-world" research methods. "Big Data," originally, is, in fact, the closest digital portrait mapping to the real world, representing the quantified mapping or projection in data space of human activities, behaviors, phenomena, and traces. Therefore, the methodology driven by big data has become one of the new feasible methodologies for researching complex wholeness problems.

Specifically, when traditional methods are insufficient to address the new essential attributes of complex wholeness problems, a new methodological system naturally emerges, integrating a combination of traditional methods and innovative methods. The various methods in this system address different needs and synergistically complement each other, resulting in a more robust overall capability to manage and decipher the complexity of complex systems. In this regard, the big data-driven approach is one of the new tools with significant empowering value.

Furthermore, based on the definitions of the big data-driven approach from multiple perspectives and the comparison with general data-driven approaches, we believe that in the sense of "driven," "big data" has a broad connotation, including thinking, resources, tools, wisdom, and paradigms. From this point of view, we can understand the cognitive connotations of a big data-driven approach from different perspectives:

(1) **Cognitive Thinking of Big Data-Driven Approach**. The essence of big data-driven methodology is not about having "more" data but whether one can extract value needs satisfying certain entities and behaviors from the dataset. For example, whether one can distill orderly patterns of human activities from seemingly "chaotic" datasets, especially when such patterns are not strongly causal and apparent; whether fragmented, discontinuous scenario data snippets can be replicated, reconstructed, and assembled into a relatively complete scenario (scene, context), essentially achieving a "restoration of the shattered mirror" based on initial scenario fragment data; and whether the "externality" of the value of big data can be converted into some originally absent but indispensable resources for certain decisions or management, supporting the subject to overcome the predicament of "a clever wife can't cook without rice."
(2) **Enhancement and Optimization Thinking of Big Data-Driven Approach**. Researchers with a clear understanding of the research need of a given problem, utilize the empowerment approaches of big data-driven techniques to enhance their research and problem-solving capabilities through the externalization and value-added nature of the big data. This reflects the fundamental thinking of a standardized big data-driven approach. The basic idea is that when a complex problem leaves researchers lacking a priori knowledge and mature theories to construct structured optimization models, big data-driven approaches serve

as a new and effective methodology to compensate for the deficiencies in optimization knowledge.

(3) **Comprehensive Integration Thinking of Big Data-Driven Approach**. The earlier discussion in this book has demonstrated the theoretical equivalence between big data and scenarios in terms of their attributes of complex wholeness. This determines the potential value of big data-driven approaches in describing and addressing complex wholeness and, under certain technical frameworks, transforming such value into realistic scenario modeling capabilities. In particular, for complex system scenario modeling, the reductionist and structured thinking in traditional modeling approaches struggle to adapt to the characteristics of complex wholeness. However, an integrated modeling methodology consisting of big data-driven approaches and other modeling techniques can better meet the externalized demands of complex wholeness scenarios. This methodology encompasses theoretical analysis, mathematical models, computer experiments, big data-driven approaches, and the unique contextualization of small data, forming a comprehensive, integrated modeling symbolic system.

(4) **Emergence of Big Data-Driven Approach and Adaptive Thinking**. The continuous influx of new high-frequency and high-velocity data in the process of big data-driven approach not only enriches and completes the real world, but also, within certain paradigms, brings forth new phenomena, characteristics, and problems that have never appeared before or cannot be explained by the existing theories and mechanisms. Among them, there are often elements that go beyond traditional cognition and imagination. At this point, the big data-driven approach serves as a "timely help" by helping us form new cognition and imagination of the original scenarios and their evolution based on adaptive thinking. This undoubtedly greatly expands our understanding and cognition of the complexity of the real world and the deep uncertainty of the future world, enabling us to make more evidence-based and robust decision plans or risk response measures.

(5) **Paradigm Shift Thinking of Big Data-Driven Approach**. Based on the complex wholeness principles of big data, the methodological nature of the big data-driven approach inherently involves a paradigm shift in complex systems thinking. For example, based on the formation logic of big data, the data in the original big data sets is characterized by the coexistence of constructiveness, generativeness, evolution, high volume, poor quality, and disorder. Therefore, it cannot provide information regarding what is "accurate" or what is "causal." However, logically speaking, due to its massive scale, big data can infinitely approximate the true entirety of a scenario or problem. The ubiquitous interrelationships among the data can be obtained by mining sufficient data information related to certain missing data or by analyzing even weakly associated data to improve the quality of important information. This implies that the strong correlation of big data implies its quality of completeness. At the same time, it cannot directly reveal the "why." It can tell us "what" on the whole and the future direction of overall development, which is of practical significance for understanding the complex world.

However, it should be noted that the correlation mainly studies the relationship between phenomena, i.e., "seeing what is visible," and mainly involves superficial thinking about problems. On the other hand, causality mainly studies the relationship between phenomena and essence, i.e., "seeing what is invisible," and mainly involves deep-level thinking about problems. Therefore, if we completely replace causality with correlation, it is possible to "shield" many deep-level mechanisms analysis and revelations of complex problems and merely analyze problems at a superficial level, leading to the simplification of complex problems into simple ones. Therefore, we should not only look at data without considering the mechanisms. It is especially important to note that the nature of big data thinking is essentially a "technical thinking" of internetization. It attempts to use the unlimited computing power of computers and the analytical capabilities of other statistical tools to achieve data pattern analysis. We should not consider technical thinking as the complex wholeness thinking for complex system management. Instead, we should understand that complex management also involves human psychology, cognition, social, political, cultural, and other factors and requires interdisciplinary and multidomain thinking to complement and compensate for the limitations of big data-driven computational thinking. Only then can we have a more complete understanding of the connotations of a big data-driven approach.

11.5 Basic Principles of Big Data-Driven Approach

This section's discussion of the so-called basic principles of big data-driven approaches is not aimed at a specific target. Rather, it is a principle understanding of the general and universal sense of "realizing certain values through driving big data," encompassing both a comprehensive methodological perspective and a practical operational paradigm.

Based on this fundamental understanding, by embedding specific characteristics and corresponding problems driven by a particular value requirement, a basic principle of "achieving specific values through data-driven approaches" is formed. Furthermore, appropriate and comprehensive technical chains, as well as specific processes, are designed to constitute a complete and orderly "data-driven" basic principle.

Before introducing the basic principles of data-driven approaches, it is once again emphasized that although big data has many advantages, it is not always necessary or suitable to apply data-driven approaches in all cases. However, when facing complex wholeness problems, combining data-driven approaches with other methods forms a multi-symbolic methodological system with unique capabilities. Here, "problems" refer not only to scientific problems, but also to tasks, phenomena, processes, etc. We abstractly refer to all of them as "problems" with specific goals and values. Therefore, in the following text, "problems" should be understood as complex wholeness activities and processes with defined objectives and values. Based on this, the basic principles of data-driven approaches can be summarized as follows:

1. **Principle of Complex Wholeness Orientation**. Data-driven methods are often needed for studying a complex wholeness problem with a clear goal (purpose); the problem could be a complex decision-making activity, the restructuring of an organizational model, the pre-planning and feasibility analysis of a major project, the formulation of an emergency response plan for a significant event, or the modeling of complex system scenarios, and so on.

For complex wholeness problems, it is important to have a comprehensive and clear understanding of the problem's environment, spatiotemporal aspects, social aspects, organizational aspects, individual aspects, tasks, events, relationships, causal relationships, and dynamic evolution. Based on this understanding, it is necessary to characterize the problem's structure, features, and attributes conceptually and logically, as well as define the problem's boundaries, environment, types of research objects, spatiotemporal characteristics, research purposes, theoretical foundations, and basic assumptions. Finally, we can form a complete structured architecture and logical system for studying the problem.

To unify the expression of the above points in terms of system language, although they represent different activities and processes in various fields, they all belong to the practice of a complex system. Under certain mechanisms, this complex system adopts certain resources as inputs, and the value of the system lies in its functionality. If data is one of the resources in the process of realizing the system's value and transforming its potential value into actual value, the externalization process of data value represents the essence of data-driven approaches. This is referred to as the **principle of complex wholeness orientation**.

2. **Principle of "Big Dataization"**. In today's context, the interplay and iteration between human-to-human, human-to-machine, human-to-network, and network-to-network interactions constitute the set of elements associated with real-world problem activities and processes. The quantified form of the traces of these activities and processes serves as a mapping or projection in the data space, and the most comprehensive data set is the raw big data set. It should be noted that the data here is not derived from a small data sense of a sample set. The small data sample method aims to answer prior research hypotheses using the information contained in the sample data as much as possible, with the key being to obtain the best possible value for research with as few samples as possible. On the other hand, the raw big data set is not a specialized sample data set designed for a specific research hypothesis but rather data collected from various sources related to problem activities and process traces. The collection channels include the internet, various sensors, wearable smart devices, personalized internet information, storage of non-digitized data, and digitization technologies.

Data represents the objective measurement and recording of the real states, characteristics, behaviors, and traces of complex wholeness problems. Therefore, to some extent, complex wholeness problems can be represented by coherent or correlated data, depicting the laws and evolutions of the physical world pertaining to these problems within specific scopes. However, data acquisition is subject to measurement methods, technological means, and the influence of the subject's cognitive thinking, level, and perspectives. Thus, in the process of depicting the physical world's landscape of complex wholeness problems, data will inevitably exhibit various quality defects, such as missing data, inconsistent standards, semantic differences, and significant impurities.

Furthermore, the increasing influx of new data pertaining to complex wholeness problems needs to be examined to determine if they can be used alongside existing data to depict the problem landscape without causing theoretical or logical conflicts. Therefore, the original data must undergo cleaning, filtering, and impurity removal to enhance the quality and value "grade" of the data. This implies that, during the initial modeling phase, some tolerance or accommodation for data imperfections may be allowed within certain limits. However, targeted measures should be taken for the raw data intended to play a valuable role in big data-driven approaches to improve data quality.

Due to the potential presence of various mixed situations in the original data, the principle of "big dataization" encompasses a wide range of content. However, the most essential characteristic of these principles can be summarized as data regularization. Data regularization involves transforming non-data format information recorded, transmitted, and stored from various information carriers into data format information through pattern recognition techniques, statistical monitoring methods, sensor perception technologies, and middleware software. It also involves converting unstructured and semi-structured data into structured data. Next is achieving data digitization, which includes.

Form normalization: Data comes in various forms, such as numerical, textual, image, audio data, etc. Different forms of data are challenging to interact with, and even entities described or monitored by different data may differ. Constructing a unified data form involves transforming data of different forms describing the same entity into a consistent format recognizable by specific entity features. Additionally, the data of the same entity should be unified in terms of precision, scale, dimension, and granularity, conforming to specific data model and format requirements.

Semantic normalization: The meaning of data is referred to as the semantics of data, and the interpretation of data actually refers to the interpretation of the semantics of data. Problems such as homonyms or synonyms may exist between different data. The construction of unified semantics representation requires mapping the semantics related to the same subject from different forms and sources through semantic mapping, converting them into a globally shareable semantic representation, enabling data to interact with consistent meaning and semantics across various localities and forms.

The purpose of data regularization is to facilitate the dataization of information content and the digitization of data. Through semantic formal unification to align entities, establishing data integration infrastructure, researching centralized management and cross-platform interoperability of environmental data, dynamic updates of entity models, data integration and standardization, the diverse forms and representations of heterogeneous information in the big data set can be deeply fused across boundaries. Only through deep fusion of information can we not only "see visible things" based on relevance, but also "see invisible things" based on causality.

In summary, data normalization, standardization, structuring, and other data transformation operations, along with data cleansing to correct errors, data sampling to reduce dimensions, and data fusion to enhance completeness, all fall under the umbrella of data regularization. Achieving an increase in data value is referred to as the principle of "big dataization."

3. **Principle of Kernel Data Extraction**. In reality, for any specific complex wholeness problem (scenario), countless pieces of data carrying certain information about the problem can be collected through various methods. If the dynamism of the problem over a certain period is taken into consideration, such data will keep increasing. Therefore, in practical terms, big data-driven modeling requires us to extract from the "infinitely many" data a "finitely many" set of data that best reflects and encapsulates the characteristics of the problem.

Any specific complex wholeness problem in reality can be seen as "one" within a class of problems. This class of problems has either fundamentally similar or highly similar dynamic mechanisms, which determine the essential attributes, overall structure, and evolutionary laws of this class of problems. The "one" problem is centered around this dynamic mechanism and is "embedded" with unique contexts or various contingencies.

Therefore, constructing a complex wholeness problem (scenario) first requires building a core that represents the basic structure, attributes, and laws of the problem class based on the dynamic mechanism of this class of complex wholeness problems. This ensures the essential determinism and stability of this class of problems. This is undoubtedly the most important and fundamental aspect of modeling complex wholeness problems (scenarios). As for modeling a specific "one" problem or scenario, it is done by incorporating unique contexts and various details based on this foundation.

The above analysis also implies that the datasets corresponding to complex wholeness problems contain data categories that characterize the dynamic mechanisms, relevancies, and uniqueness of the problem, respectively. When the mechanisms of the problem cannot be obtained or confirmed from theories or other channels, people can only propose cognitions, conjectures, and metaphors about the basic dynamic mechanisms of the problem based on their initial intuitive perception of the problem through perception, association, assumption, experience, knowledge, and the inspiration of imagination, and construct the basic structure and evolutionary laws of the problem's mechanism around relevant data as much as possible. Since the data related to dynamic mechanisms are subsets of the "complete data" set of the problem, this is a complexity reduction behavior for complex wholeness problems conducted during the big data-driven process, greatly leveraging the value of big data in constructing the foundational structure of complex wholeness problems, thus termed as the principle of complexity reduction or kernel data extraction principle.

Here, "kernel" means "core," that is, among all the "infinitely many" data of complex wholeness problem information, they are the "core" that can best characterize the dynamics and basic structure of complex wholeness problems. While the principle of kernel data extraction has not yet formed a specific, unique fine structure for a problem, it provides major demonstrations and insights into the foundational structure and behavioral attributes of a class of problems with the same or highly similar dynamic mechanisms. Based on it, continually adding or altering other relevant and unique data can yield various scenarios of complex wholeness problems with the same basic core but different characteristics.

We can also understand and grasp the principle of kernel data extraction from a different perspective. The relevant information of a complex wholeness problem is extremely rich and diverse, and each type of information has its own source or form, called a modality. For example, human cognition of a problem is a multimodal collection, including touch, hearing, visual perspectives, and so on. The carriers of information can be language, text, videos, and others. To achieve a more comprehensive understanding of complex wholeness problems, it is necessary to integrate multiple modalities. This means processing and comprehending multimodal information through various methods, enabling individuals to obtain an overall cognitive model of the problem. In this process, the training stage models multiple times using data that represents the dynamical mechanism information or basic rule-based information of the problem. The result is not a specific problem but a shared foundational model for a class of problems that have the same (or similar) dynamical essence. Constructing this foundational model is equivalent to establishing a common basic structure for a class of problems with shared dynamical characteristics. Building a model for a specific problem is then akin to making unique house "decorations" on the foundation of the basic model.

The principle of kernel data extraction consists of two main points: First, using as few dimensions of data as possible to "span" a problem essential structure that is as comprehensive as possible. Second, the problem structure composed of kernel data should not differ fundamentally from the parallel mirror image of the real problem structure in terms of overall structure, features, and evolutionary patterns. It is similar to two houses built according to the same architectural design blueprint. Their basic structures, overall styles, and core functions are generally the same, but due to the cultural and preferential differences of the homeowners, the interior decoration and color vary.

4. **Principle of Complex Wholeness Integration**. Under the principle of associative thinking, one integrates the data of a problem and the various relationships between this data to create a coherent, connected, and aggregated whole. The main focus is on how the "fragmented" data is integrated to represent the entirety of the problem. The term "wholeness integration" refers to the operational process of achieving the "wholeness" of the problem.

The process of integration is built upon the foundational framework constructed by kernel data. This foundational framework can be viewed as the systemic fundamental structure of holistic problems. The various informational attributes and characteristics inherent in big data are analogous to new materials. Integration involves incorporating these materials into the systemic basic structure, continually refining and enriching the problem scenario.

To be more specific, integration is to allow data (along with the information value they inherently possess) to form a parallel mirror image of the original real-world problem in an abstract data space based on rules and principles associated with time, space, events, associations, and evolution, all while retaining the complex holistic attributes. This requires us to primarily use the information from big data as a clue or foundation. Through experience, assumptions, imagination, predefinitions, and especially foundational knowledge like dynamic mechanisms, one continuously embeds the foundational information carried by big data and the unique information from small data into the foundational structure. Through continuous iterations, one gradually develops increasingly refined parallel mirror images.

In practice, wholeness integration technology primarily utilizes computationally feasible structured technology routes. On a micro level, it determines how to perform reductive deductions on the micro-mechanisms of autonomous and federated mixed scenarios of problem activities and processes. It mainly constitutes problem mechanisms through arguments and small data, takes "sliced" samples of process sequences with key languages, and acquires local and micro-level complexities. On a macro level, it mainly studies how to form problem association logic driven by arguments supplemented by big data. Due to the scarcity of samples for complex wholeness problems, the concept of statistics is generalized to form a "slice stream" of problem discrete "slice" sampling at a certain scale and coarse granularity using computer simulation technology. It then undergoes multi-level integration to form the macro scenarios emergence and evolution of problem activities and processes (panoramic approximation and process restoration).

In the process of wholeness integration, the subject has various understandings of the problem, the essence of data, the association between data, and the value implied by data. New relevant data continuously emerge, and the new data and original data have different objective relevancies and integrabilities, as well as the subject's subjective understandings of these problems. Therefore, the results obtained according to the principle of wholeness integration are not only not unique, but also do not have a so-called "standard" answer.

This indicates that the data set mapping or projecting a complex wholeness problem in data space could very likely be traced back to a new form of a problem that is not the original one. As the mapping or projecting data set continues to enrich, the activities and processes traced back may also become richer and more novel, so much so that the new form of the traced-back problem surpasses or even overturns people's presets or imaginations. This means that the association of the mapping or projection between real activities and processes and the original large data set is dynamic, irreversible, and not one-to-one corresponding. However, it should not be perceived as only creating troubles; on the contrary, it is precisely because of this that some rare values of big data can be unearthed in this driving process.

In summary, the value of real complex wholeness problem datasets is not just the record of activities and process traces, nor is it solely the driving force for reverse simulation and reconstruction of the original real problem. Through big data-driven approaches (or in conjunction with other methods), it extracts and mines the deep-level information and knowledge contained in the big dataset, transforming them into people's cognitive abilities and capabilities to deeply understand, analyze, and manage complex wholeness problems. Of course, all of this is not primarily accomplished through theoretical assumptions and mathematical models but through the wisdom of big data thinking and a series of empowering technologies that "externalize" the value of data.

5. **Principle of Feedback Learning Verification**. The aforementioned points describe the construction of an artificial virtual system based on big data driving, which serves as a parallel mirror of complex wholeness problem activities and processes. This mirror platform provides resource allocation and environmental support for unleashing the value of big data. However, the quality of the results driven by big data, including integrity, scalability, functionality, adaptability, and effectiveness, needs to be validated using a verification method. The feedback learning verification principle introduced in this section is used for functional training, performance testing, and evaluation. Based on the test results, adjustments are made to the various stages of the data-driven process to prevent overfitting during the training phase and improve the adaptability to new samples, ensuring the reliability of data driven by big data.

For example, due to the path dependence, irreversibility, and evolutionary uncertainty of any specific complex wholeness real-world activity or process, it is not sufficient to evaluate the "goodness" or reliability of the value driven by big data based solely on the degree of "simulation" of a particular reality. In a macro sense, any parallel mirror is a potential reality that appears in virtual space, and therefore, any virtual mirror obtained from data-driven has its own inherent value.

To evaluate or correct the universality and robustness of a specific data-driven process and its value, the parallel mirror driven by big data can be backward shifted on the timeline so that the "future" generated by the data-driven mirror is actually a real result in the real world. Then, the obtained "future" from data-driven is compared with this real result in the real world, and the comparison result is used as feedback learning to refine, improve, and enhance the functional value of data-driven. This stage mainly relies on the feedback learning approach to study whether data-driven can reflect the essential attributes of complex wholeness activities and processes. It builds a verification and evaluation system based on the consistency and compatibility of the driving process inputs, the credibility of the driving result outputs, etc. It investigates the reliability of data-driven and the termination criteria for modeling iteration, which forms the feedback learning mechanism-based verification principle for the data-driven process.

In summary, the discussion on the basic principles of data-driven can be summarized in the following key theoretical viewpoints:

1. In the real world, as in the native form, there is only data without natural big data. Data is ubiquitous, and big data is a concept generated by human thinking, which refers to data processed and endowed with specific connotations in a certain context. In different contexts, big data may be a way of thinking, a resource, a tool, a means, an algorithm, a research paradigm, and even wisdom. The process of transforming data into big data is not only the application of certain modern information technologies, but also the transformation of "calculation" into "counting" in thinking. Scholars should not simply regard "a large amount of data" as "big data."
2. Big data and complex entirety share a theoretical unity; hence, big data, as a resource supply, aligns logically with the demand to reveal complex entirety. Big data-driven approaches often constitute an important (but not the only) part of the methodological system for solving complex entirety issues.
3. "Data-driven" refers to the process where big data with potential (latent) value is driven to achieve a specific value goal and release the "externalities" of its own value during the realization of this goal.
4. "Data thinking" is essentially the internetization of "technical thinking." It seeks to utilize unlimited computing power and other statistical analysis capabilities of machines to achieve data pattern analysis and value discovery. However, a possible pitfall is the mistake of considering technical thinking as the entirety of thinking. Solving complex wholeness problems requires not only technical thinking, but also social, political, psychological, cognitive, and cultural thinking.
5. In the specific operational process, data-driven emphasizes the relationship between phenomena, i.e., "seeing the visible," while causal relationships mainly reveal the relationship from phenomena to essence, i.e., "seeing the invisible." If we simply replace causal relationships with correlations, it may lead to simplifying complex wholeness problems into simple wholeness or even non-wholeness problems. Therefore, we should not only focus on data, but also consider mechanisms and unique contexts. The notion that data science is the beginning of the "post-theory era" needs to be approached with caution. In practice, including data-driven, there is never a one-size-fits-all method.
6. As mentioned above, in a certain sense, big data-driven is a type of "mending the broken mirror" methodology, incorporating thinking, technology, and capability. For addressing complex holistic problems, a complete and illustrative metaphor for the fundamental principles of big data-driven is the core knowledge as the "pillar," big data drawing the "dragon," and small data adding the finishing touch or "bringing the dragon to life."

References

1. Chen J. Multi-role exploration of big data fusion analysis applications. Chin Informatiz. 2021;06:100–2.
2. Laplace PS. Essai philosophique sur les probabilités. Bachelier;1825.

Chapter 12
Computability of Scenarios and Big Data-Driven Scenario Modeling

Chapter 10 of this book introduced the basic theory of scenario modeling for complex systems, and Chap. 11 discussed the fundamental principles of big data-driven modeling. Building upon the big data perspective of scenario modeling, this chapter focuses on the thinking and basic principles of big data-driven modeling. However, merely understanding the theoretical aspects is not enough. It is necessary to establish the fundamental principles, specific technical approaches, key techniques, and modeling processes for big data-driven scenario modeling at an operational level. These are the main contents of this chapter and the next.

These contents can be divided into two levels. The first level is mainly concerned with integrating big data and scenario modeling, combining data-driven scenario representation and description. The visual representation can be summarized as how to "compute" scenarios using data. To answer this question, we need to first understand the computability of scenarios, which is the theoretical issue of whether scenarios can be computed. Once this issue is analyzed, we can then explain the definition, principles, and technical approaches of big data-driven modeling for complex systems. The second level involves further clarifying the key techniques and main processes of big data-driven scenario modeling.

Do not assume that the previous two chapters have already covered the basics of scenario modeling and the principles of big data-driven modeling. In fact, this chapter is where the integration of big data-driven modeling and scenario modeling begins, forming a new methodology and framework for big data-driven scenario modeling. There are new ideas and concepts that have not been clearly discussed before.

This chapter primarily introduces the first level, while the next chapter focuses on the second level. Additionally, for simplicity, "scenario for complex systems" will generally be referred to as "scenario."

Z. Sheng, *Outline of Complex Systems Management Theory—Based on Irreversibility of Reductionism Thinking*, International Series in Operations Research & Management Science 360, https://doi.org/10.1007/978-981-97-3361-3_12

12.1 Computability of Scenarios

Due to the involvement of new thinking and interdisciplinary knowledge in big data-driven scenario modeling compared to traditional conceptual modeling or mathematical modeling, let us first outline the basic ideas of big data-driven scenario modeling in a general and intuitive manner before discussing it in detail.

1. Firstly, scenario modeling is the abstract representation and description of scenarios. Any form of representation and description requires certain tools and methods. In recent years, an academic viewpoint has emerged: in scientific research, these tools and methods are defined as languages and symbol systems, which serve as media for representing and describing scenarios. Commonly used languages include natural language and mathematical language. Recently, computer programming languages have become increasingly popular for representing and describing scenarios. Therefore, computer programming languages have become a new language system for scenario modeling. This means that scenarios can be modeled using computer programming languages and specific symbol systems.
2. Secondly, if, as described above, scenarios can be obtained by a computer through a certain program in a new form of results, the scenario can be "computed" (by the computer). There is no need to be surprised by this viewpoint; it depends on how we understand the computer's "computation." The so-called computer computation essentially refers to the computer applying a set of programming languages and rules that it can understand and interpret, transforming certain elements (such as data) into a new result (computational result), or representing and describing the result using a computer language. Let's further expand this idea: as long as we use computer programming languages to transform certain inputs into specific outputs according to certain rules, or the output is represented and described by inputs, it is a form of "computation." The nature of the output, whether it remains as data or takes on another form, is not essential. It can be considered that the output has been "computed" by the computer. Therefore, if the form of the output is a scenario, then the scenario can be computed by the computer or is computable.

Interpreting the integration of data and scenario modeling based on the above idea, the data here mainly refers to the mapping, projection, and recording of scenario elements in the data space, all of which rely on the computer environment. This indicates that big data-driven scenario modeling naturally requires the use of computer technology and language, with scenario element data as input and abstract and descriptive representations of the scenario as output. This also demonstrates that scenarios can be modeled using computer programming languages, specific symbol systems, and the broad concept of computation. In simple terms, scenarios are "computable."

However, "computing" scenarios using computers are much more complex and challenging than performing general numerical computations on data. Regarding scenarios, their complex wholeness attributes make it difficult to abstract and describe them using natural language or mathematical language. For example, people often understand the evolution of scenarios and their related influencing factors but cannot use mathematical methods to describe the degree and mechanism of the influences. Even if a mathematical model regarding the influences is constructed, it cannot be used due to the inability to solve it. In such cases, it is necessary to transform unsolvable mathematical models into a computable form and repeatedly iterate through computer computation to identify critical influencing factors and analyze their magnitude and modes of influence.

Furthermore, we find that when using computer programming languages and computational simulation techniques to abstract and describe scenarios, the following aspects can demonstrate advantages:

1. Computer programming languages have strong capabilities in representing the evolution process of scenarios and can handle parameters that are difficult to quantify directly due to complexity.
2. Computer programming languages have a greater advantage in dealing with parallel processes or processes lacking rules, making them particularly suitable for evolving processes where the activity modes and sequence of scenario elements cannot be fully defined in advance.
3. When dealing with the wholeness of a scenario, computer programs can be designed as modular structures, allowing modification of certain modules without affecting others. This makes research schemes easy to modify, update, expand, and integrate with other methods.
4. Computer technology "excels" in simulating individual behaviors at the micro level, enabling the study of interactions between the micro and macro levels by establishing artificial systems composed of heterogeneous intelligent agents from a bottom-up approach.

(1) Computer technology can effectively describe uncertain and difficult-to-measure elements such as psychology and behavior within scenarios.
(2) Computer technology can handle problems that do not have explicit solutions through the application of rules and ordered iterative methods, providing a process equivalent to system evolution.

This inspires us to fully utilize computer programming languages and computational simulation techniques to describe many challenging aspects of scenarios' complex wholeness, such as:

1. Multiple independent heterogeneous subjects within a scenario with various complex associations. These subjects can optimize and reconfigure their associations to form the complex wholeness of the scenario under certain environmental and resource constraints.
2. The complexity of scenario evolution and paths is primarily caused by the adaptive behaviors of independent subjects in response to changes in the environment.

The above conditions and functions of scenarios can be equivalently transformed into formal "IF/THEN" statements that can be understood, accepted, comprehended, and executed by computers. Assuming that the behavioral capabilities of subjects are limited, their impact on the entirety of the scenario within a unit of time is also limited, and the overall form and behavior set of the scenario are limited and encodable under limited resources and a range of environmental changes. With adequate computational support, the aforementioned form and evolution of scenarios can be reconfigured, reproduced, or predicted to a certain extent. In this sense, we refer to scenarios as computable or possessing computability. In the context of computer systems, traditional numerical computation, computer experiments, and the simulation of scenario reconfiguration, reproduction, or prediction all fall under the category of scenario computability.

Since most people's initial exposure to the concept of computation comes from mathematical calculations, when discussing computation, they immediately associate it with quantitative calculations involving numbers or measurements. This creates a mindset that objective things must be quantified before they can be computed. The notion of computation here implies calculation based on (mathematical models) numbers. In fact, in computer science, "computation" has a broader meaning. The object of "computation" can be the encoding of "quantified" calculations or the encoding of concepts without "quantities," such as the processing of text using encoding, where the order of positions under certain rules may differ, such as the sorting rule for letters where "A" comes before "B."

Therefore, computation encompasses two main aspects: the describable and manipulable properties of the objects being processed and the computational description of these objects, which corresponds to the methods of data structure in computer science. The processing objects can be represented using numerical or non-numerical encoding, and the relationships between them can be represented using appropriate structures.

Correspondingly, the computational operations on the processing objects correspond to a sequence of processing procedures. Such procedures can be based on a set of mathematical equations (formulas) or a set of rules. Equation-based computation corresponds to the fundamental machine instructions in computers, namely numerical and logical operations. Rule-based computation can be represented in a normalized form close to natural language, such as "IF/THEN," and the corresponding machine instructions in computers primarily involve logical operations [1].

In this way, the processing objects that computers need to describe can be represented using numerical or encoded forms, and computations can be realized through equation-based computation or rule-based operations. Computers "compute" the simulation of scenario-based data using computer programs.

Regarding scenario computability, the following points need to be emphasized:

1. It is possible to establish a finite and discrete logical space for the activities of each part of a complex system within a scenario. The activities of the various parts in the realistic scenario can be mapped onto this logical space. Based on the required precision of the spatial mapping relationship, the logical space can be subdivided.

2. It is possible to establish a finite and discrete time sequence for the evolution processes and interactions of each part within a complex system scenario. The evolution processes of the various parts in the realistic scenario can be mapped onto this time sequence. Based on the required precision of the temporal mapping relationship, the time sequence can be subdivided.
3. Complex system scenarios and the states of their various components at each moment can be described by a set of finite and unambiguous characteristic values. By appropriately partitioning these characteristic values, the precision of the state descriptions required for the scenarios can be set. In fact, this sets the level of detail for the modeling of the scenarios.

From this, it can be discerned that a computational scenario allows for the creation of an "artificial scenario" within a computer. Here, the term "artificial" doesn't imply "artificial reality" as understood in the fields of social, economic, and management sciences but rather denotes the utilization of computer technology invented by humans to fabricate a virtual reflection possessing the fundamental structure, features, form, and evolution of the authentic scenario. This reflection incorporates an open environment, finite pertinent resources, and a compilation of rules that direct the interaction of elements within the scenario. The artificial agents embodied in this scenario may represent humans or organizations, be animate or inanimate, and may exhibit varying levels of intelligence or lack thereof.

The main idea of artificial scenarios is to extract the basic operational dynamics of a scenario and use computer technology to construct an artificial system with realistic scenario characteristics and functions. It involves studying the memory, learning, adaptation, and interaction of artificial agents in the dynamic evolution process of the scenario, as well as the allocation of relevant resources. It also involves observing, analyzing, and adjusting the complex wholeness of macroscopic forms that emerge in the scenario's environment.

In a dynamically constructed artificial scenario, each computational object has its own data structure and corresponding computational processing methods, which correspond to different levels or structures of the realistic scenario. In this process, the activities of elements at the next level are based on the changes of elements at the previous level, and the elements at the previous level, in turn, impact the elements at the next level. The self-organization and self-adaptation of artificial scenarios in structure and function give rise to the overall dynamics of artificial scenarios and the self-evolution between multiple levels. This demonstrates the bottom-up emergence and irreversibility of reductionism in the sense of scenario complex wholeness.

With the continuous advancement of computer hardware, increasing computing speed, the development of agent-based programming and network technologies in the field of computer science, and the maturation of computer graphics and visualization technologies, the use of artificial sceneries for studying complex system scenario modeling has become a reality.

The so-called scenario cultivation technique described in the later part of this book is a computer-based technique for generating artificial scenarios.

The computability of sceneries is a crucial theoretical foundation for scenario modeling. Since it is possible to represent and describe scenarios in computer systems using programming languages and computational techniques, To "compute" scenarios, turning them into interactive and immersive virtual digital situations, or to let the scenario be "pieced together again" while also ensuring a certain computational "quality," allows people to "see and hear" the virtual mirror image of real scenarios through the aid of visualization means. This solves the core technical route of scenario modeling. Furthermore, when combined with the basic principles of big data-driven modeling, it not only forms the basic principles of big data-driven scenario modeling, but also provides a complete technical route architecture. After a series of key technology designs are completed, an operational big data-driven scenario modeling process "emerges."

12.2 Interpretation of Big Data-Driven Scenario Modeling

The logical basis of big data-driven scenario modeling originates from the inherent similarity between big data and scenario attributes. This clarity of logical basis is important for scenario modeling, aiming to externalize the value of big data scenarios.

From a reductionist perspective, a scenario can be decomposed into a set of discrete elements. In this sense, the information and data sets derived from realistic scenario elements represent the "fragmented" representation of the scenario. The more abundant and complete the information in the data set, the better it can represent the scenario as a whole. Although there is not a one-to-one correspondence or mutual completeness between the scenario data set and the scenario, it is undeniable that big data sets contain abundant scenario value.

Based on this, the basic idea of big data-driven scenario modeling is to use "big data-driven" as a transformation mode, representing it as a mapping in the data space. This mapping reconstructs a "parallel mirror" of a certain scenario, thereby externalizing the scenario value contained in big data.

It should be noted that since the "big data-driven" transformation mode is not unique, the final form and quality of modeling may vary. These factors are closely related to the modeling subject's objectives, perspectives, pattern designs, and modeling methods.

The above explanation provides a logical understanding of big data-driven scenario modeling. Furthermore, to truly drive big data and externalize its scenario value, specific technical approaches, key technologies, and process designs are required. This can be seen as making "driving" a "converter" that transforms the potential value of big data scenarios into an externalized form. In other words, "driving" needs to be concretized as a paradigm for value transformation.

From today's philosophical perspective of science, the "drive" associated with "big data" is largely not just an "algorithm." It represents a method for addressing value transformation, an organized procedure, a process, and specific operational protocols. From the perspective of the co-evolutionary relationship between technology and humans, the aforementioned points suggest that the essence of "big data-driven" is that humans, through designed programs, methods, and procedures, and by utilizing necessary tool systems, enhance their ability to externalize the potential scenario value in big data. And nowadays, the foundation of such a tool system should be computers and algorithms. This further clarifies that big data-driven scenario modeling is based on modern computer technology, implementing the externalization of the value of big data scenarios through designing a complete set of "algorithms," that is, transformation methods, organized steps, processes, and specific operational procedures.

In academic research and practice, people use scientific language to express, communicate, exchange, and inherit academic ideas and theories, forming specific academic discourses. Academic discourses have gradually expanded from language, writing, and mathematical symbols to rules, logic, instructions, procedural processing, and ways of thinking. This expanded cognition helps us understand that the paradigm of academic discourse has a high degree of consistency or similarity with the "algorithm" that people have recently recognized as a philosophical universal thinking mode.

Since modern scientific philosophy considers "algorithm" as a contemporary thinking mode with philosophical universality, any set of rules, procedures, and steps that can place something into a logical program and process it using certain methods, including computation, is an "algorithm." Furthermore, natural language, experimental science, theoretical deduction, computer simulation, and data-intensive science can be integrated into a comprehensive "generalized algorithmic system," enhancing people's ability to describe, interpret, and solve problems, especially complex holistic problems. Big data-driven scenario modeling is a new "generalized algorithmic framework" technique developed in this sense, specifically designed for addressing complex holistic problems like scenario modeling.

12.3 Basic Principles of Big Data-Driven Complex System Scenario Modeling

The basic principles of big data-driven scenario modeling refer to a paradigm that has been derived by extracting the logical relationships, causal relationships, and associative relationships in the specific modeling process's thinking principles, technical routes, and process designs. The basic principles generally consist of several interrelated and independent ordered principles, where each principle represents a relatively independent unit of knowledge. It mainly focuses on the behavioral principles, operational guidelines, and the modeler's understanding of the essential attributes of modeling in a universal sense within the activities and processes of big data-driven scenario modeling.

As mentioned earlier, the computability of scenarios primarily addresses the theoretical aspect of using computer programming languages and computational techniques to "compute" scenarios. The establishment of the basic principles of big data-driven scenario modeling should be consistent with the general principles of big data-driven, which are discussed in Chap. 11 but tailored to the context of scenario modeling.

The important role of principles lies in the foundational and formalized rules and logic they imply and extract. They guide people on how to replicate the basic procedures and operational rules in the modeling process and achieve basic success. The relationships among these principles can be represented in Fig. 12.1.

Principle of Complex Wholeness Scenarios. Analyzing the characteristics and interactions between scenarios and big data features, including the elements of the scenario's environmental, spatiotemporal, social, organizational, individual, task, event, association, and causal relationships, as well as management and decision-making activities. Studying the boundaries, features, basic attributes, and core attributes of scenarios themselves.

Principle of Scenario Data Fault Tolerance. Since the data containing scenario element information is not perfect for the value of scenario modeling, it exposes inaccuracies, incompleteness, and inconsistencies in format and is not "conclusive," and even with the continuous generation of new data, the data itself is continually redefining and changing scenarios. Therefore, from the dimensions of data quality,

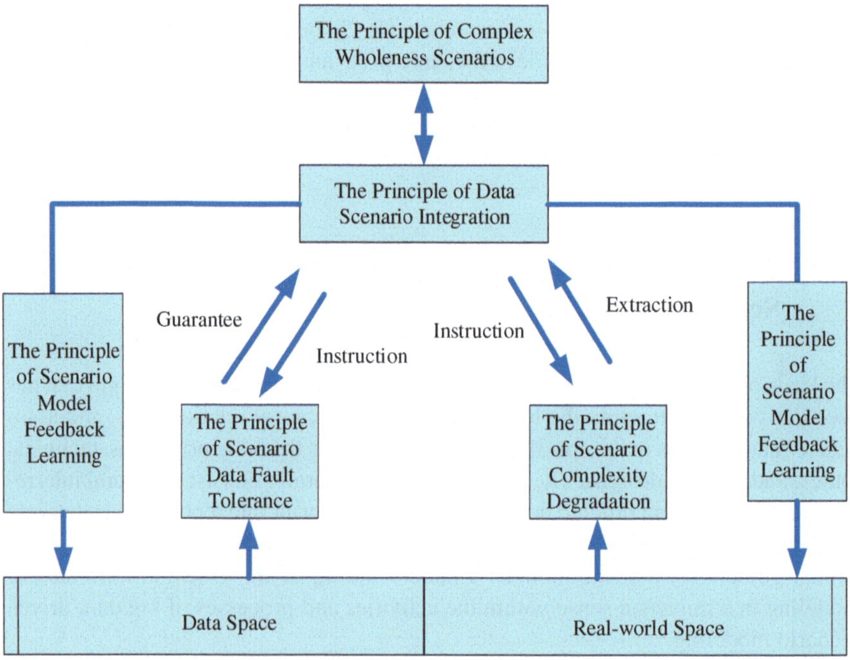

Fig. 12.1 Relationship diagram of big data-driven scenario modeling principles

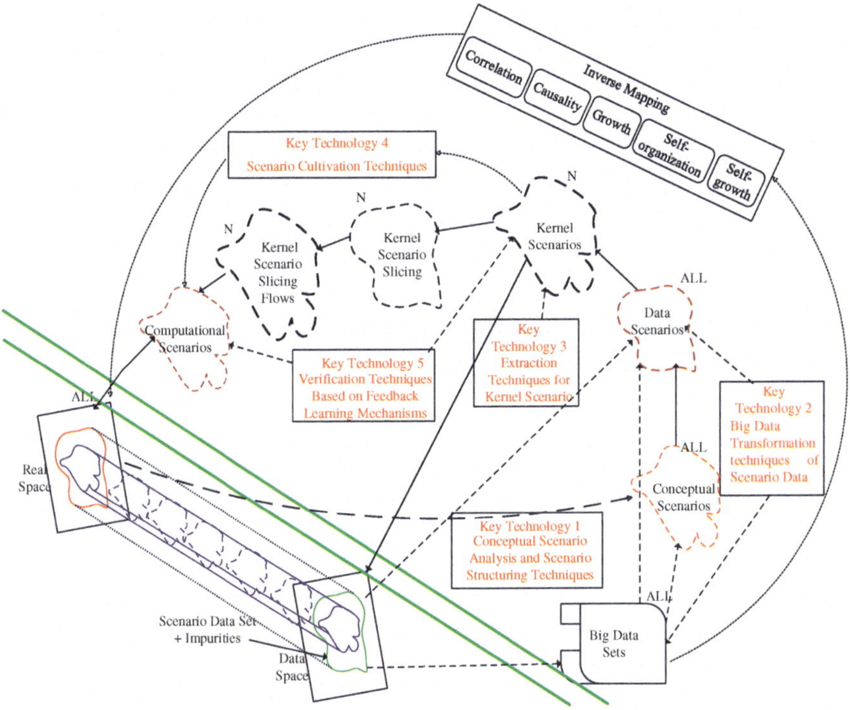

Fig. 12.2 Technical roadmap of big data-driven scenario modeling

function, and value, research is needed into methods for data interaction verification, correction, cleansing, and filtering, as well as standards for data quality and paths to improve data quality.

Principle of Scenario Complexity Degradation. In reality, one form of carrier for complex wholeness scenarios is an "infinite" dataset of scenario elements. Therefore, at the theoretical level of thinking, scenario modeling involves the transformation of this dataset along the reverse path toward the overall orderliness of the scenario. At the practical level of thinking, it involves extracting a "finite" core dataset of scenario elements from the "infinite" dataset of scenario elements while preserving the essential attributes of the scenario and reflecting the fundamental dynamical mechanisms of the overall scenario. The main focus is on researching methods, processes, steps, and stopping criteria for reducing the complexity of scenario element data.

Principle of Data Scenario Integration. Big data-driven scenario modeling integrates the potential value of scenario element data by coherently connecting and aggregating it through multiple dimensions such as time and space, focusing on the bidirectional interaction techniques of "fragmentation" and "integration" to construct a technical chain that achieves complex holistic scenarios.

Principle of Scenario Model Feedback Learning. To determine whether a big data-driven scenario model exhibits high quality in terms of entirety, functional extensibility, adaptability, effectiveness, and other aspects, it is necessary to employ the principle of feedback learning. This involves conducting functional tests based on newly generated data and adjusting the model according to the test results. Through an iterative process, a satisfactory model is formed in terms of effectiveness and error control.

12.4 Technical Roadmap for Big Data-Driven Complex System Scenario Modeling

A technical roadmap refers to the researcher's approach to achieving research goals, including the technical means, specific steps, and methods to solve key problems. It is an orderly process of preparation, initiation, progress, and results in scientific research. A well-defined technical roadmap ensures the smooth achievement of research goals. Therefore, each operational step should be clearly explained in detail. If possible, flowcharts or diagrams can be used for better understanding. The technical roadmap should be comprehensive, clearly explaining the key points of each step and its operability.

The technical roadmap for big data-driven scenario modeling is based on the unity of scenarios and big data theory, the computability of scenarios, and the basic principles of big data-driven. It involves designing the complete and ordered chain of steps for "computing" scenarios using computer programming languages and computational simulation techniques. Each step's predetermined goals and key technologies are clearly indicated. The following is the technical roadmap for big data-driven scenario modeling:

Step 1: Conceptual Scenario Analysis and Scenario Structuring. This step involves clarifying the form and boundaries of a specific scenario. It begins by conducting a top-down analysis of the perceived scenario to obtain a conceptual scenario, which is then transformed into a structured scenario. The process involves summarizing, abstracting, or generalizing scenario elements, associations, behaviors, structures, and functions. This includes considering the overall hierarchy of the scenario, as well as the levels of subjects, organizations, and elemental components.

Furthermore, through a bottom-up scenario construction, the conceptual scenario is transformed into a computationally implementable scenario. The focus is on structuring the semi-structured or even unstructured scenarios in the conceptual scenario using various methods. This process enables further abstraction and symbolization of realistic scenarios, allowing computers to "understand" the logic between the clearly defined structural and relational elements of the conceptual scenario.

Step 2: Big Dataization of Scenario Data. The product after the data undergoes "refinement." Since scenario data is limited by the original purposes of measurement, collection, and recording and is not entirely for the needs of scenario modeling, there naturally exist various problems with data quality in terms of scenario modeling, such as missing data, non-standardization, and semantic differences. These require multiple stages of cleansing, filtering, and impurity removal to improve the quality of data regarding scenario information. The data processed in this manner will enhance their value in scenario modeling. Additionally, in the process of scenario modeling, there is continuously added new data that needs to be verified to decide whether it should be used as new data for scenario modeling. All these data are no longer the original scenario data but become scenario "big data."

Step 3: Extraction of Core Scenario Data. From an "infinite" dataset of scenario elements, a limited number of elements are extracted that best reflect and preserve the essential attributes, overall structure, spatiotemporal distribution, and evolutionary patterns of the scenario. These extracted scenario elements, along with their interrelated information, form the basic framework and evolutionary rules of the scenario foundational model. These elements are referred to as core scenario data, which are crucial in the scenario modeling process. The ability to extract core scenario data relies on the knowledge possessed by the subject about the scenario.

The extraction of core scenario data follows two main principles: first, using as few types and quantities of elements as possible to encompass a complete scenario space, serving as the fundamental basis for the integration of the scenario model; second, during the modeling process, the parallel mirror image of the scenario model constructed using core scenario data should align with the realistic scenario in terms of overall structure, features, and evolutionary patterns.

Therefore, this section emphasizes the acquisition of core knowledge and important assumptions about the scenario, extracting core elements that embody the essential attributes of the scenario, as well as typical environments, plots, and evolutionary paths related to the core scenario.

The significance of core scenario data lies in its ability to establish fundamental scenarios that are highly consistent with a class of dynamical mechanisms and essential attributes.

Step 4: Scenario Cultivation. From a theoretical perspective, the operation of scenario modeling driven by big data involves using large datasets that carry scenario information as basic elements. These elements, along with their inherent scenario information value, form a parallel mirror image of a scenario in an abstract virtual scenario space based on the rules of time, space, events, associations, and evolution. The goal is to preserve the complex holistic attributes of the scenario. However, due to the limited information available, we can only use certain scenario element information from the data as clues or foundations. Through our experience, knowledge, assumptions, imagination, predefinitions, and especially the core scenarios, we construct a class of foundational scenarios with dynamic mechanisms. Then, we embed the unique information contained in the scenario big data into the foundational scenario and cultivate new scenarios through continuous cultivation.

In the scenario modeling process, we can consider the foundational scenario as the "parent plant," and the information attributes and features carried by the big data act as new mutation techniques to cultivate new varieties of scenarios. Obviously, in this process, big data fully demonstrates its foundational and driving role, hence the term "big data-driven." The more abundant the scenario big data, the more varieties of new scenarios can be cultivated. In summary, this is referred to as the big data-driven scenario cultivation technique in scenario modeling.

Step 5: Verification based on feedback learning mechanism. Big data-driven scenario modeling essentially constructs an artificial virtual scenario system as a parallel mirror image of the realistic scenario system and uses this mirror image as a platform to study relevant scientific issues. However, this raises concerns about its reliability. Due to the path dependence, irreversibility, and evolutionary uncertainty of any specific, realistic scenario, it is not sufficient to evaluate the "goodness" or "badness" of big data-driven modeling simply based on the degree of "simulation" with a particularly realistic scenario. In a macro sense, any parallel mirror image is a possible scenario in the virtual scenario space, and in terms of possibility, any virtual scenario obtained through scenario cultivation is "real."

However, when it comes to the analysis and predictive capabilities of a specific virtual scenario model, the model can be shifted backward on the timeline and analyzed against the background of realistic scenarios that have occurred up to that point. By comparing the future results generated by the mirror image with real phenomena, the comparison results can be used to refine and improve the model through feedback learning, thereby enhancing the quality of the model's functionality. This is the verification technique based on the feedback learning mechanism. Therefore, in this step, the focus is on utilizing the feedback learning approach to investigate whether the scenario model can reflect the essential attributes of the scenario. The verification technique system is built based on model input consistency, model compatibility, and the credibility of result outputs, aiming to establish criteria for terminating the modeling iteration.

Based on this, the complete technical roadmap of big data-driven scenario modeling can be characterized as follows: in the process of scenario modeling, "big data-driven" is treated as a clear thinking principle and generation mode. It involves fully exploring and utilizing the scenario value embedded in the data and designing an orderly transformation system with paradigmatic significance to effectively and stably convert the inherent value into the real value that constitutes the overall form of a specific scenario.

It should be noted that this "panoramic" scenario generation process is achieved through multiple iterations, and therefore, the restored scenarios have different quality attributes. Sometimes, it may be limited "panoramas" or probabilistic restorations. The technical route is not based primarily on theoretical assumptions but relies on data mining to obtain complex holistic scenarios.

Specifically, since data originates from the synchronous or lagged records of people's practical activities and process scenarios, it naturally contains partial information about the scenarios. However, on the other hand, data cannot encompass all attributes of a scenario but only represents some pulses and traces of the scenario. Therefore, big data-driven scenario modeling can only start with as comprehensive data as possible and use various methods to uncover the incomplete information and flawed associative information inherent in the data itself. People can only work within these objective constraints, carefully calculating and strategizing to study how to make the best use of big data and gradually externalize the value of scenario information along the orderly transformation and iterative process.

Reference

1. Zhang J. The artificial social method is used to study the evolution of social behavior. Comput Simul. 2007; 1:277–80.

Chapter 13
Key Technologies and Processes of Big Data-Driven Scenario Modeling

In the previous chapter, we introduced the technical roadmap for big data-driven scenario modeling. Therefore, it is easy to understand that the technical roadmap for big data-driven scenario modeling includes various types of technologies. Each modeling step often has its own specific key technologies. A step and one or several key technologies are closely tied together, meaning that implementing certain technologies completes a step. As for unique and specialized technologies, even those tailor-made for big data-driven scenario modeling, they naturally become key technologies within this roadmap.

Key technologies for big data-driven scenario modeling do not refer to a single technology or a simple collection of technologies. Instead, they consist of a chain of multiple ordered and logically related technologies, which are relatively independent. The overall significance of this chain is that it completes the entire process of modeling scenarios driven by big data along this technological chain.

Based on the above logic, the key technologies (chain) for big data-driven scenario modeling are highly consistent with the basic principles discussed earlier and the modeling process that follows. This also indicates that such key technologies (chains) do not have to be unique as long as the functional requirements of scenario modeling are met. As an example, a set of modeling key technology chains will be introduced.

13.1 Key Technology Section One: Conversion Technology Between Real Scenarios, Conceptual Scenarios, and Structured Scenarios

Scenario modeling generally needs to be oriented by a real scenario with a definite physical form. Therefore, it's necessary to appropriately define the specific connotation and physical boundaries of this real scenario according to the research purpose

Z. Sheng, *Outline of Complex Systems Management Theory—Based on Irreversibility of Reductionism Thinking*, International Series in Operations Research & Management Science 360, https://doi.org/10.1007/978-981-97-3361-3_13

so the scope of the scenario is neither too large to manage nor too small to encompass the essence of the problem. We consider this phase as the first section of the big data scenario modeling process. It manifests as a complete chain of orderly conversion among real scenarios, conceptual scenarios, and structured scenarios. Moreover, specific key technologies are needed for each conversion between the two. The details are as follows.

13.1.1 Techniques for Defining Realistic Scenarios

1. Identify the problem or phenomenon that serves as the guiding focus for scenario modeling, with an emphasis on its complex wholeness, such as dynamic evolutionary processes and emergent macro-level phenomena that cannot be explained by reductionism.
2. Choose the perspective and entry point for scenario modeling, such as predicting scenario structures and trends, revealing the emergence of macro-level features, and analyzing the adaptability of subject behaviors.
3. Define the spatiotemporal scale of the scenario. Every scenario is a "living subject" within a specific spatiotemporal context. Different spatiotemporal characteristics have a significant influence on the scenario and its forms. Therefore, it is necessary to define the temporal and spatial specifications and features of the scenario.
4. In addition to being influenced by modeling techniques and tools, scenario modeling is greatly influenced by the modeling objectives of different modeling subjects. These objectives may include uncovering the underlying mechanisms behind scenarios, validating a theoretical viewpoint or hypothesis, predicting future trends, exploring extreme scenario states, and designing control measures for scenario complexity.

The above points mainly involve clarifying the purpose of scenario modeling and defining and organizing the modeling of realistic scenarios appropriately. The techniques and tools used in this process mostly rely on the direct perception of the modeling subject toward the realistic scenario and the knowledge acquired through various means.

13.1.2 Transformation Techniques from Realistic Scenarios to Conceptual Scenarios

As mentioned earlier, realistic scenarios primarily consist of physical entities, and scenario modeling requires elevating the intuitive perception of physical scenarios to rational cognition. This is achieved by representing and describing realistic scenarios using an abstract symbolic system, which involves abstracting realistic scenarios into conceptual scenarios at the conceptual level. Conceptual scenarios are scenario models constructed based on abstract concepts and logical relationships derived from realistic scenarios.

Considering that scenarios are complex systems, the transformation from conceptual scenarios to realistic scenarios needs to be realized within the system space. Therefore, during the transformation process, it is necessary to have a comprehensive understanding of the complex wholeness of the scenario based on scenario intuition and rational thinking. Qualitative analysis techniques, qualitative research methods, and systems analysis techniques together form a panoramic qualitative analysis method, which serves as a key technology for transforming realistic scenarios into conceptual scenarios [1].

The formation of conceptual scenarios mainly involves abstraction and logical deduction, focusing on extracting and summarizing the uniqueness of the scenarios. This process often involves the use of qualitative methods such as language, textual description, and exposition in the context of realistic scenarios. Qualitative research generally employs open and unstructured means to collect diverse information, such as images, videos, sounds, and documents, or adopts qualitative methodological approaches using language, textual description, and exposition in the context of realistic scenarios, which aligns well with the abstraction of perceiving realistic scenarios [2].

For conceptual scenarios, whether it is scenario reconstruction, implementation, or prediction, it is essential to reveal the complex wholeness of the scenario. This is the core essence of the panoramic qualitative analysis method. The term "panoramic" in this technique refers to the wholeness of the scenario, referring to the overall scenario, while "systems analysis" addresses the complexity of the scenario, including its details. The combination of these two approaches aligns well with the complex holistic characteristics of scenarios.

13.1.3 Transformation Techniques from Conceptual Scenarios to Structured Scenarios

In the previous chapter, we mentioned that the essence of "big data-driven" in big data-driven scenario modeling, from the perspective of the co-evolution of technology and humans, lies in humans enhancing their ability to externalize the potential value of big data through specially designed methods and steps and leveraging appropriate tools. In today's context, these tools primarily include computers and networks. This further clarifies that big data-driven scenario modeling is based on modern computer technology and realizes the externalization of the value of big data scenarios through the design of a complete set of "algorithms," which are methods of transformation, ordered steps, and specific operational procedures.

This implies that once the transformation from realistic scenarios to conceptual scenarios is completed, the conceptual scenarios include not only the structured parts of the original scenarios, but also certain semi-structured and unstructured parts. However, for general-purpose computers, it is challenging to "read" and "understand" these parts. Therefore, to further enable computers to play a role in scenario modeling, it is necessary to structure the conceptual scenarios by expressing and describing them using symbols, rules, formulas, logical languages, and so on. This is the origin of the transformation from conceptual scenarios to structured scenarios [3].

"Structured scenarios" should not be equated with structured mathematical models. It is a broader concept than structured mathematical models, as certain logical relationships or rules and principles can also be considered as forms of structuring. This process allows researchers to further abstract and symbolize the subjects, behaviors, structures, associations, and rules within the "conceptual scenarios," enabling computers to "understand" the structure and logical relationships of scenarios in complex systems management.

Specifically, the structuring of scenarios can be divided into scenario object structuring, system environment structuring, scenario structuring operations, and scenario structuring presentation, according to the modeling process.

1. **Structuring Scenario Objects**

Structuring scenario objects involves encoding the subjects in the scenario using numerical or non-numerical representations, such as private attributes like gender, age, and expertise, as well as social attributes like position, role, and family roles. The structured description of scenario objects can be achieved through numerical or non-numerical encoding using the "classification + state" method. Additionally, appropriate structure trees or tables can be used to represent the relationships between these objects.

Structuring scenario objects is one of the core aspects of scenario structuring. The design of structured models for scenario objects can be approached from two perspectives: induction and inference. The structured model of scenario objects can be derived directly from human perception and abstraction of general features of the subjects in the scenario. It can also be obtained through extensive data analysis, mining, literature review, and theoretical research. In object-oriented modeling, objects are abstracted and characterized by their attributes, methods, and relationships. Therefore, the modeling of subjects in a scenario can be based on object modeling principles.

Furthermore, the subject objects can be viewed as simple nodes or complex intelligent robots in cyberspace. It is important to note that a scenario model is not and cannot be a complete replica of the subjects in the scenario but rather selectively extracts their features, attributes, capabilities, behaviors, and relationships with other subjects or the environment. Hence, the selection of content for attributes, methods, and relational associations in subject-object modeling should be determined based on specific modeling purposes. Additionally, the creation, updating, or evolution of subjects, subject attributes, and relationships should be considered. Finally, in large-scale complex scenario modeling, the design of scenario subject templates should be taken into account to create a large number of subjects through template instantiation.

2. **Structuring Scenario Environment**

Scenario structuring not only considers the structure, features, and patterns of the subjects (objects) within the scenario and their expression methods, but also takes into account the influence of the scenario environment. Generally, the natural and social environment surrounding the scenario subjects are the main environmental elements that affect their decision-making and behavior. To successfully achieve structured modeling of the scenario environment, two aspects need to be considered: the expression method of the scenario environment and the rules governing the relationship between subjects and the environment. The expression of the scenario environment can be a simple global variable or a system structure with its own internal structure and functionality. The specific way to express the scenario environment should be determined based on the specific environment.

3. **Structured Operations of Scenarios**

Structured operations in scenarios primarily refer to the operations involved in scenario structuring. It involves using a program to express the dynamic changes and sequencing of scenario operations. Such a program can be based on a set of mathematical equations or a set of rules. Calculations based on equations represent numerical and logical operations using basic machine instructions corresponding to the fundamental machine instructions of computers. On the other hand, rule-based calculations can be represented in a normalized form close to natural language, such as "IF/THEN" statements, primarily involving logical operations in computers. For example, the negotiation process of iterative bidding between an engineering supplier and an engineering contractor can be represented as a series of bids, decisions, counteroffers, and so on.

As complex system scenarios involve interactions among various scenario elements, sub-scenarios, and scenario hierarchies, the overall scenario exhibits dynamic evolutionary phenomena and continuously presents new scenario characteristics. In the process of dynamic evolution of the scenario environment, the scenario subjects also exhibit different forms of evolution behavior. Therefore, only through in-depth research on the evolution content, rules, and underlying mechanisms of scenario subjects can we better understand the wholeness characteristics of complex system scenarios. The core of data-driven scenario modeling lies in taking the evolution of scenario subjects' behavior as the starting point and considering the three main aspects of behavior evolution, rules, and expression methods in scenario composition and generation. Specifically, by systematically analyzing the "internal factors" and "external factors" of subject evolution rules, a structured model of subject evolution rules can be formed. This involves a significant amount of rule model computation and data processing. The commonly used representation of subject evolution involves using data lists to record and update the dynamic changes of key state parameters of the subject or using various types of graphs (2D or multidimensional) to express the dynamic and real-time evolution process of the subject.

4. **Structured Presentation of Scenarios**

Structured presentation of scenarios involves using graphical or chart-based methods to express key attributes and states of scenario subjects, the network status of subject relationships, and the environmental status of the system. For example, using a two-dimensional spatial coordinate system to represent the intimacy of relationships between subjects or using the quantity of a certain aquatic organism over a large-scale time frame to depict the pollution level of a lake.

From this, it is clear that the process of scenario structuring refers to the conversion of conceptual scenarios into computable models through certain means and methods. However, it's crucial to note that not all scenarios can be structured. For instance, structuring cultural aspects and value systems poses significant challenges and is often hard to achieve, representing a difficult point in this phase of processing.

In the process of scenario structuring, certain logical relationships or rules and procedures can be considered as a form of structuring. This process allows researchers to further abstract and symbolize realistic scenarios and extract the subjects, behaviors, structures, relationships, and rules within the structured scenarios.

Scenario structuring can be divided into macro level, meso level, and micro level. The macro level focuses on the overall structure of the system, its functions, and macro-evolution trends. The meso level focuses on organizations, networks, and multi-level and multi-scale relationships. The micro level focuses on individual behaviors, emotions, preferences, and so on. The purpose of scenario structuring is to approach and achieve scenario modeling through a progressively accurate sequence of structured conceptual scenarios.

13.2 Key Technology Section Two: Big Dataization Techniques

13.2.1 Formation of Big Data Sets Techniques

As mentioned above, the structured presentation of scenarios involves decomposing and organizing the elements, relationships, structures, and functions of scenarios in the context of system analysis. Depending on the modeling purposes, scenario information (data) sets of varying granularities are obtained. Since the content, boundaries, and uniqueness of a defined scenario are not inherent but determined by the research objectives, it means that the scenario information (data), including the data for modeling, has specific requirements. However, at the beginning of the modeling process, we faced two challenges. First, we are uncertain about which data are essential for scenario modeling and which data may be redundant. Second, because the scenario being modeled is defined by humans, it is not possible to prepare all the necessary and appropriate information (data) beforehand. Therefore, we either collect data temporarily based on certain judgments or use as much existing data as possible, even though these data may not have been specifically prepared for scenario modeling.

For example, to study the environmental governance of a river basin and model the water environment scenario, a wide range of data from various fields is needed.

This includes river water quality monitoring, water source quality monitoring, non-point source monitoring, industrial wastewater monitoring, satellite remote sensing images obtained through IoT, hydrological monitoring, meteorological data, data on industrial emissions, waste disposal, and wastewater treatment; various social development data such as population, employment, education level, technological level, laws and regulations, various standard specifications; various economic development data such as GDP, industrial categories, and industrial scale data; internet data such as search engines, portals, social networks, media reports, and so on. These data can also belong to different domains and dimensions, such as spatial, temporal, natural, social, economic, humanistic, behavioral, etc., and can be classified as structured, semi-structured, and unstructured data.

If we look at it from two dimensions, vertical and horizontal, in the vertical dimension, scenario data refers to the observed results of state indicators of complex system scenarios. These data are arranged in the order of their occurrence time with the same or different time intervals. These data include time series data, text data, image data, video data, etc., reflecting the changing patterns of complex system scenario phenomena or multiple phenomena. In the horizontal dimension, scenario data can also represent the overall state of complex system scenarios at a specific point in time, reflecting the overall form formed based on a specific time, place, and internal relationships of complex systems. Therefore, from the perspective of complex systems, scenario data contains the overall representation of complex systems at different times (locations, conditions, etc.) [4].

Complex system scenarios are dynamically changing over time and have different characteristics and mechanisms at different stages, making it difficult to uniformly describe and measure them. However, in each small stage, the scenario exhibits relatively stable operational mechanisms and forms. Thus, a specific point in time (segment) of the scenario corresponds to a slice of the scenario (a static snapshot), while the overall dynamic evolution of the scenario corresponds to a flow of scenario slices (a continuous video). The scenario element information (data) represents the comprehensive representation of the internal mechanisms and external forms of each slice (or the time series of scenario data). Due to the complexity of scenarios, scenario data, and scenario data, time series are not only infinite (in a class), but also diverse and heterogeneous, including time series data, text data, image data, video data, etc. The data often contain noise and useless information, requiring further noise removal processing.

Furthermore, when faced with this diverse range of data from various categories, we don't fully know which data are important and useful for scenario modeling, which data are irrelevant, or even completely useless. We also don't know which data we are still missing, and we may not even know what we don't know. Generally, we tend to believe that "having more modeling data is better than having less," or we hesitate to "discard" any data, or we are unsure which data should be supplemented, and so on. Thus, in essence, after the structured transformation of scenarios based on certain cognition and knowledge, we obtain a large data set according to a specific objective and a vague sense. At this point, we should consider this data set as ordinary data. As for the "big data set" driving scenario modeling, it is a data set that externalizes its own potential value under specific scenarios and value guidance. According to this understanding, the original data set is merely a "big data candidate set."

13.2.2 Big Dataization Techniques

So, can we directly start big data-driven scenario modeling from the "big data candidate set"? Not yet. There are two main reasons:

1. Due to the inevitable "defects" in the data quality of the big data candidate set, such as data missing, inconsistent standards, semantic differences, and significant impurities, various techniques need to be used to clean, repair, organize, and select the data in the candidate set. This is done to eliminate data that lacks scenario value or improve the quality of data with low scenario value.
2. The presentation of information in the big data candidate set can be in various forms, such as text, images, videos, and audio. Non-digitized information expressed in different ways on different data carriers needs to be transformed into digitized information.

Therefore, in the early stage of big data-driven scenario modeling, various preprocessing techniques are applied to the data in the big data candidate set. The goal is to enhance the value of big data contained in the data as a whole. This stage employs techniques known as principles (techniques) of big data transformation.

The specific big dataization techniques used in the process of big data-driven scenario modeling depend on the data types, characteristics, and the overall complexity and features of the scenarios in the big data candidate set. In general, the following common techniques can be used as big dataization techniques to address and handle common and routine problems, including:

1. Data transformation of multi-source heterogeneous data in the big data candidate set to meet data format requirements. This includes techniques such as data denoising, data aggregation, data conceptualization, data standardization, inductive processing, and attribute construction. By merging data, data that cannot capture the key features of complex system scenarios is eliminated, and similar types of data are merged to obtain a refined data set that adequately describes complex system scenarios.
2. Techniques that unify data with different precision, scales, dimensions, and granularities based on the computer's logical processing principles (computability). This involves transforming different forms of data that describe the same entity into the same data format that represents identifiable entity features, achieving formal uniformity; building global entities and establishing a semantic sharing mode for global entities; studying semantic mappings from local entities to global entities, achieving semantic normalization; establishing a data integration architecture based on aligned data and real-world entities; studying data centralization management and cross-platform interoperability to enable dynamic updates of entity models and achieve semantic-based data integration applications.

3. Based on the understanding and analysis of complex system scenarios, techniques that reduce data volume and improve data quality while preserving the core functionality of the data. This includes data dimensionality reduction, normalization, standardization, transformation of semi-structured data into tabular data, transformation of text data into unique identifier codes, verification of outlier data, etc.
4. Once scenario data is digitized, it can be represented as a scenario time series with added noise. Noise can obscure the true nature of the underlying mechanisms within the actual scenarios, leading to a decrease in modeling quality. Therefore, techniques are required to handle noise in scenario time series. There are two main types of noise in scenario time series: measurement noise and system noise. Measurement noise refers to noise unrelated to the scenario itself, caused by measurement instruments or observation methods during the process of observing information contaminated by errors. Assuming the scenario state z_n satisfies a condition $z_{n+1} = F(z_n)$, the measured data is $x_n = h(z_n) + \omega_n$, where $h(\cdot)$ is a smooth function that maps the scenario state variables to measurement values, ω_n is a random number, and the sequence $\{\omega_n\}$ is called measurement noise. In this case, the noise is superimposed on the scenario time series during measurement and is also known as external noise, which can generally be removed.

System noise refers to the interference caused by small random fluctuations within the scenario at each stage, resulting from the propagation of minor random fluctuations within the main scenario. It may also be caused by random events occurring within the scenario, directly affecting the evolution of the scenario. In this case, the scenario satisfies a condition $z_{n+1} = F(z_n + \eta_n)$. Among them, η_n represents scenario noise. Here, the noise is independent of the external environment and enters the scenario as an input, influencing its evolution. This type of noise is also known as internal noise and cannot be easily removed through simple filtering techniques. It requires the utilization of the scenario's evolutionary characteristics for processing.

13.3 Key Technology Section Three: Extraction Techniques for Kernel Scenario Elements

13.3.1 Basic Ideas of Kernel Scenario Formation Techniques

Let's imagine that when a structured scenario data set is transformed into big data, the retained data set for realistic scenarios is never complete and always has missing and incomplete parts. Furthermore, the data in the dataset belongs to different types, such as dynamics, correlations, and uniqueness. It is easy to understand that among these three types of data, dynamic data can best reflect and carry essential information about scenario attributes, overall structure, interrelationships, spatiotemporal distribution, and evolution. Therefore, if we use this type of data as much as possible to drive the basic architecture and evolution rules of the scenario model, we should obtain a high-quality foundational scenario model. We refer to these elements as the core elements in the process of scenario modeling, and the scenarios they construct are referred to as kernel scenarios. The corresponding element information (data) is called kernel scenario information (data). Kernel scenario data contains rich information about the dynamic behavior of the scenario system. It is the closest digital representation of real complex system scenarios and reflects the essence of actual complex system scenarios.

The significance of kernel scenarios is to determine a category of scenarios (types) that have the same foundational dynamic mechanisms and essential properties by using the "kernel" of the scenario. Kernel scenarios represent the mechanisms, laws, and essential properties of this category of scenarios. Based on the foundation of kernel scenarios, by adding additional information that describes uniqueness and details, a complex wholeness scenario with specific dynamics and distinct uniqueness is formed. This includes static or transient scenarios, or the dynamic evolution of static or transient scenarios, which can be considered as a wholeness model of complex system scenarios.

Extraction techniques for kernel scenarios in scenario modeling primarily involve extracting a finite subset of elements that best reflect and preserve the essential properties, overall structure, spatiotemporal distribution, and evolutionary laws of the scenario from the "infinite class" of real scenario elements. This subset can best embody the basic dynamic mechanisms and behavioral rules of the scenario fundamentally and contains the essential features of the scenario corresponding to certain parameters.

As mentioned above, extracting kernel scenarios that can reflect the essential attributes and fundamental dynamics of the scenarios and constructing a foundational scenario model based on them are critical steps in the big data-driven scenario modeling process. The methods for extracting kernel scenarios are the key techniques in this stage.

The extraction principles for kernel scenario elements mainly consist of two points: first, using "as few as possible" elements to form a relatively complete scenario space to achieve the integration of the scenario model. Second, the scenario model constructed from kernel scenario elements should not have essential differences from the parallel mirror image of the realistic scenario in terms of the overall structure, characteristics, and evolution laws. To avoid excessive specialization, the following provides a brief qualitative introduction to this academic viewpoint. Interested readers can refer to other relevant books for a more in-depth understanding.

1. **Basic Ideas of Dimensionality Reduction Techniques for Scenario Elements**. Earlier, we mentioned that kernel scenario elements are a limited number of elements from the "infinite class" of scenario elements that can reflect the essential attributes, overall structure, spatiotemporal distribution, and evolution laws of scenarios. We can use qualitative and quantitative methods to select a set of elements as "candidate" kernel scenario elements. To avoid missing any important elements, we often select more elements than necessary. However, this may

result in element "redundancy." For example, the selected scenario elements may come from the same dimension or different dimensions, but they essentially reflect the same attribute features of the scenario, indicating a strong correlation. This redundancy needs to be reduced or eliminated through dimensionality reduction or removal of such redundant data.

Consider the correlation between two sets of scenario data time series. Let X and Y be two subsystems of a complex system at a certain stage and let $\{x_n\}_{n=1}^{N_1}$ and $\{y_n\}_{n=1}^{N_2}$ be the scenario data sequences generated through observation or experimentation. If they are statistically interdependent, subsystems X and Y are said to be coupled, meaning that some information from subsystem X can be obtained from subsystem Y and vice versa. The most fundamental method to measure statistical dependence between two stationary sequences is the method of orthogonal cross-correlation. The advantage of this method is its simplicity in computation, but it can only measure linear dependence. It cannot measure nonlinear dependence. Although higher-order orthogonal functions can also be used to measure nonlinear dependence, the computations become much more complex [5]. Here, we primarily illustrate the basic principles.

Since scenario data sequences are nonlinear, mutual information is a commonly used method for measuring statistical dependency between nonlinear data sequences. When X and Y are coupled, they share common information, and the amount of this shared information can be measured by the mutual information between $\{x_n\}_{n=1}^{N_1}$ and $\{y_n\}_{n=1}^{N_2}$. When the subsystems X and Y are not coupled, it means that there is no shared information between them, i.e., the mutual information is zero. In statistical terms, this indicates that the two data sequences are independent and provide no information about each other. When the subsystems X and Y are maximally coupled, it means that there is a deterministic relationship between the two subsystems. In this case, almost all functions of $\{y_n\}_{n=1}^{N_2}$ are functions of $\{x_n\}_{n=1}^{N_1}$, and vice versa. In this situation, it indicates that one of the scenario data sequences is redundant. Therefore, in the entire complex system scenario, when two subsystems have maximum coupling, we only need to consider one of them, and the other can be removed as redundant, achieving the goal of dimensionality reduction [6].

Thus, mutual information can be regarded as a measure of the coupling strength between the two subsystems in a scenario. It can effectively reflect the coupling between the two systems. However, its calculation is complicated and requires the use of histogram-based methods to estimate the probability density functions. The generalized mutual information can solve this problem by measuring statistical dependency without assuming consistency in one of the data sequences, although it requires complex transformations to be used if one of the data sequences does not satisfy the consistency assumption. Consequently, a statistical dependency detection method that does not require a prior assumption of consistency in the data sequence was developed.

2. **Basic Idea of Orthogonality Technique for Kernel Scenario Elements**. As complex system scenarios evolve over time and develop, in the data processing and analysis process of complex system scenario modeling, based on the time

dimension and the real situation of the complex system, the complex system scenarios can be classified into primary dimensions such as spatial dimension, behavioral dimension, problem dimension, and functional dimension (in practical problems, the division of primary dimensions such as spatial dimension, behavioral dimension, problem dimension, and functional dimension can be adjusted based on the different complex system objects). These primary dimensions form an N-dimensional orthogonal space to describe the complex system. Further subdivision is done for each dimension in the N-dimensional space to find orthogonal spaces that describe secondary dimensions. This process continues, forming new lower-level orthogonal spaces as needed. The number of subdivision dimensions and subdivision levels for each hierarchical level can vary based on the characteristics of the actual complex system scenario.

Figure 13.1 shows the multi-level orthogonal space of kernel scenario elements for water environmental ecological governance scenarios. The first-level nodes form the smallest dimensional space, and each second-level node forms the smallest dimensional space. This process continues, forming new lower-level dimensional spaces as needed.

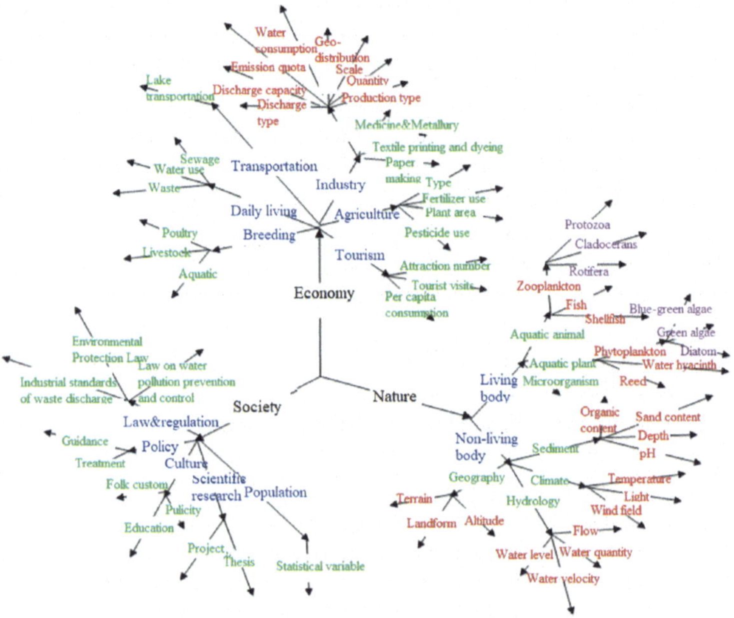

Fig. 13.1 Refinement of the orthogonalized kernal scenarios of water environment

13.3.2 Extraction Techniques for Kernel Scenario Elements

Based on the above ideas, the observed or experimentally obtained K group scenario data sequences are first divided into L sets of scenario data sequences based on dimensions such as time, space, problem, function, or behavior. Each set of scenario data sequences contains several groups of scenario data sequences, which represent the presentation form of the features of complex system scenarios in certain dimensions. For each set of scenario data sequences, the independence between any two groups of scenario data sequences is examined as described in Sect. 13.3.1. If there is statistical dependency, one group of scenario data sequences is removed. Then, the statistical dependency between any group of scenario data sequences and the remaining data sequences is examined. If there is statistical dependency, that group of scenario data sequences is deleted. After multiple iterations of the above process, no time series in each set of scenario data sequences depends on any other $I_l - 1$ ($l = 1,2, \cdots, L$) scenario time series. From each set of scenario time series, one group of scenario data sequences is selected, and the statistical dependency between this group and other sets of scenario data sequences is checked. If there is statistical dependency, that group of scenario data sequences is deleted. Through this process, a set of scenario data sequences $I_1 + I_2 + \cdots + I_L$ is finally formed, constituting the kernel scenario data sequences.

13.4 Key Technology Section Four: Big Data-Driven Scenario Cultivation Techniques

This section is the core part of the entire process of big data-driven scenario modeling. Building upon the three previous sections, this section combines the principles of big data-driven scenarios with the theory of scenario computability. It follows a technical approach based on generalized algorithmic thinking and the multi-symbol system of modeling complex wholes. Computers will "calculate" the parallel mirror image of realistic scenarios in the data space. Consequently, the value of big data-driven scenarios is externalized in a computer environment as interactive and immersive virtual digital scenarios. Scenario cultivation is the key technology to achieve this value.

Now, let's provide a brief introduction to what scenario cultivation is. As we all know, any complex system scenario is not only complex, but also often unique. Therefore, it is generally not possible to extract statistical regularities of new scenarios from a large number of known scenario samples. Instead, we can only use a small amount of valuable realistic scenario element data or clues as "seeds" in the computer system's "laboratory" to sow and cultivate, allowing them to grow. Ultimately, we obtain a class of foundational scenarios, which can be further enriched with various key element information to produce different scenario "fruits." Through the dynamic evolution process of these "fruits" and analysis, prediction, and reconstruction of their types and characteristics, we can gain knowledge and rules about complex system scenarios. This technology, in which scenarios are "computed" by computers, is called scenario cultivation.

It should be noted that scenario cultivation is significantly different from general computer simulation methods. Computer simulation requires a real system as a benchmark, aiming for realism. However, scenario cultivation simulates complex system scenarios. Because specific scenario phenomena have uncertainties or mutations such as path dependence, irreversibility, and evolution, we cannot judge the "good" or "bad" or the "true" or "false" of scenario cultivation results based on a specific scenario we have seen or recognized as the "true scenario." We cannot predict which scenario evolution path is acceptable or optimal. Scenario cultivation results should be understood as a possible scenario and a possible evolutionary path formed in the scenario space under certain assumptions and rules through the self-organization, emergence, and the mutual coupling of complex systems management and the environment-complex system composite system. In terms of "diversity" and "multiplicity," any result of scenario cultivation should be considered as an uncertain "reality."

In this section, key technologies focus on how to organize and integrate the scenario information contained in the data space into a complete scenario whole. Figuratively speaking, it is about how to achieve the "reconstruction of the shattered mirror" of scenarios through the data carriers of scenario fragments.

13.4.1 Integration Techniques for Heterogeneous Scenario Data

Scenario element information exhibits heterogeneity, and achieving a "reunion" of scenarios requires their integration into a complete scenario, which calls for integration techniques for heterogeneous scenario data. Heterogeneous data integration techniques can be categorized into three types: stage-based integration, feature-level integration, and semantic integration. Stage-based integration refers to analyzing different forms of data during different stages of data mining. In this type of integration technique, there is no interaction between the data, making it difficult to achieve intrinsic data integration. Feature-level integration involves extracting features from each heterogeneous data and analyzing and processing them. Semantic integration requires understanding each dataset and the relationships between features across datasets, considering the extracted features of heterogeneous data as interpretable. These heterogeneous data integration techniques consider multiple sources of heterogeneous data on a single node and are not entirely suitable for the integration of heterogeneous scenario data in complex systems involving multiple nodes.

The integration techniques for heterogeneous scenario data can also be classified into spatial grid-based integration, multi-scale temporal integration, and behavior or decision-based integration. Spatial grid-based integration primarily integrates heterogeneous scenario data from a spatial perspective. It applies spatial intelligence techniques and smart grid models to divide the spatial regions of complex system scenarios into grid units. Each grid unit corresponds to a set of values representing the relevant scenario data of the complex system scenario in the grid space. Mathematically, this can be represented as a matrix, and in computer implementation, it can be represented as a two-dimensional array. Multi-scale temporal integration integrates heterogeneous scenario data from a temporal perspective by recording scenario data at different time points according to time scales. It constructs multivariate data sequences and normalizes different types of data and time scales based on the minimum time unit. Integration of heterogeneous scenario data based on the complexity of the system focuses on merging heterogeneous scenario data from a complexity perspective. It classifies scenario data based on dimensions such as economics, society, and nature and further categorizes them into secondary dimensions. It continuously merges similar data within the same dimensions and constructs a multi-level dimensional data space structure.

13.4.2 Techniques for Establishing Cultivation Models

Scenario cultivation models are a type of computer model. A scenario cultivation model should be able to directly and operationally represent and describe the elements, interrelationships, and dynamic evolution of complex system scenarios. This operation should not be seen as a personal skill of the modeler but should become a paradigm. Additionally, establishing a scenario cultivation model is guided by realistic scenarios and should reflect the guiding and regulatory nature of scenario connotations. In other words, the selection of modeling techniques should be oriented toward specific scenario contents based on basic paradigm principles rather than being technology-driven "technicism." Here, we will use the example of water environment governance as a realistic scenario to introduce technical selection thinking in establishing a cultivation model.

1. **Modeling the System Environment**. Whether it is the natural environment or the socio-economic environment, they have a significant influence on the subjects involved in managing complex systems. Therefore, it is necessary to consider the important relationships between the environment and the subjects during scenario modeling. For example, in the natural system of a lake basin, the elements include wind fields, flow fields, temperature, light, nutrients, algae, aquatic animals, sediment, etc. These elements have a significant impact on human decision-making behaviors and decision rules in water environment governance. Therefore, establishing a basin environment model is one of the key aspects of establishing a cultivation model. Specifically, the basin environment model includes the dynamics of natural climatic environments, self-purification equations for nutrient transportation and diffusion in rivers, water dynamics equations for

lakes, nutrient diffusion equations for lakes, algae growth and consumption equations, aquatic animal predation equations, sediment adsorption and desorption equations, aquatic animal mortality equations, and so on. These equations are continuous dynamic equations that vary with time, so they need to be discretized using techniques such as finite element differences for computer solvability.

2. **Modeling the Subjects**. Abstracting real-world scenarios into computer models is a critical foundation for developing scenario cultivation models. Cultivation models are not, and cannot be, complete replicas of the scenarios in complex systems. Instead, they selectively extract the main features of the scenario and construct effective ways to express them. For example, in the natural system of a lake basin, every location in the lake has a corresponding coordinate. The relationship between the lake's coordinates and its morphological scenarios is one to many. The lake has basic static properties, such as the amount of sewage discharge and the growth rate of aquatic organisms. The dynamic changes in the lake require statistical analysis of each point at every time period, e.g., the growth conditions of aquatic organisms, average temperature, and other attributes, with the condition of one point corresponding to state records at multiple time points. Rivers connect to the lake, with multiple rivers flowing into it, including basic properties such as flow speed and water volume, and one river is determined by multiple river section coordinates. There are multiple sewage outlets corresponding to the river, with the location of the sewage outlets determined by their coordinates, including the total sewage discharge amount and the discharge conditions of nitrogen and phosphorus elements.
3. **Design of the Subject Evolution Rules**. The subject is the center of the scenario, and the design of the subject behavior rules is key to establishing the cultivation model. Under the constraints of the subject behavior rules, the subject undergoes continuous iterations in the scenario environment to reveal the evolution trend of the scenario. For example, in the social system of a lake basin, the government, the public, and enterprises have their own environmental behaviors and rules, which mutually influence each other, resulting in different scenario evolutions. It is necessary to establish the correlation patterns between different rules and different scenario evolutions.
4. **Design of the Data Structure of the Cultivation Model**. In the evolution of scenarios, there are generally multiple ways in which the subject evolves, including data interactions between subjects, data interactions between subjects and the environment, etc. These interactions of scenario data can be seen as operations between data structures. By appropriately describing the scenario structure with data structures, the questions of "what to cultivate" and "how to cultivate" in scenario cultivation can be addressed. For example, if the relationships between subjects are represented as a complex network structure, data structure design can be used to express information transmission between subjects and the evolutionary process of subject attributes. For complex scenario models, the model interaction and data interaction can be appropriately separated, resulting in a clearer model construction and higher operational efficiency.

5. **Data Analysis and Expression**. During the scenario evolution process, a large number of intermediate results are generated. Analyzing and processing these intermediate results can yield many meaningful insights and implications, which require the use of statistical analysis and other mathematical methods in the model. In addition to processing intermediate data, it is also necessary to visually represent the specific expressions of the model on the computer during scenario evolution. This requires making the model visualizable to allow people to intuitively see the evolution process of the scenario and reveal the evolutionary patterns of the scenario in a visual manner.

13.4.3 Computer Implementation Techniques for Scenario Cultivation Models

The implementation of scenario cultivation involves various computer techniques, including setting up the environment, variables, boundary conditions, key algorithms, computer languages, and visualization of simulation results. The implementation process of scenario cultivation generally adopts a bottom-up research approach. Beginning with the elemental level of determining intelligent agents, intelligent agents continuously learn and adapt their behaviors and interact with each other based on changing environments, ultimately achieving the evolution and emergence of the cultivation process. This is a dialogue and task assignment between humans and computers, so acceptable and understandable technologies for computers should be chosen in this process [7].

1. **Selection of the Cultivation Environment**. This involves constructing an appropriate scenario cultivation environment and selecting suitable tools, programming languages, and debugging methods [8]. It usually includes specific tasks such as: ① integrating necessary data processing software or utility program modules for scenario cultivation data processing to improve modeling efficiency and increase the effectiveness of the model; ② selecting a reasonable scenario cultivation running mode, such as standalone computing, distributed computing, synchronous collaborative computing, or other computing technologies, based on the scale of scenario cultivation data processing; ③ determining a reasonable scenario cultivation data storage system, such as local disk storage or network storage, according to the data storage requirements of the computing process and experimental results; ④ considering whether to perform offline analysis of intermediate data results in scenario cultivation and whether interfaces with other data analysis software are needed; ⑤ considering whether dedicated devices, such as specialized graphics processing devices, are needed based on special requirements for presenting scenario cultivation results.
2. **Definition and Implementation of Cultivation Boundary Conditions**. Boundary conditions refer to the limiting conditions and control parameters in the scenario cultivation process, such as specifying the number of iterations for evolution or whether the total number of main objects changes. For example, defining

that the total number of main objects remains constant, allowing new main objects to enter the system, and changing the relationships of artificial objects in the system structure. These boundary conditions determine the development and control of the scenario cultivation model. The implementation process of scenario cultivation usually requires providing computable refined models, computational flowcharts, appropriate data structure descriptions, specific steps for scenario cultivation implementation, and major functional modules, as well as the variations or flow processes of data, all of which require definite technical choices.

3. **Design of Model Variables and Initial Data**. Scenario factors variables include several independent variables and dependent variables. In research on scenario cultivation, it is essential to classify variables, define their attributes, and computationally delineate their relationships. This includes quantitative descriptions formulated mathematically and rule descriptions articulated through "if/then" conditions/actions. The effects of changes in independent variables on dependent variables are then observed and evaluated. Initial data represent the starting conditions for scenario cultivation calculations and describe the initial state of the scenario and environment.
4. **Key Algorithms and Mathematical Models**. Key algorithms are designed to correspond to the evolutionary laws of scenario cultivation models at different levels, such as genetic algorithms for the elemental level of intelligent agents and self-organizing algorithms for the design of intelligent agent levels. Mathematical models refer to the mathematical methods used to describe macro, meso, and micro dynamic mechanisms in scenario cultivation models, as well as various mathematical models for evaluating scenario states.

13.5 Key Technology Section Five: Verification Techniques for Scenario Modeling

13.5.1 Basic Definition

The essence of big data-driven complex system scenario modeling is to construct an artificial, abstract scenario system as a parallel mirror image of the realistic scenario system. Scenario cultivation technology aims to construct interactive and immersive virtual digital scenarios within a certain paradigm in a computer environment. The modeling subject believes that once this model is built, functional and computational analysis can be performed on this platform to analyze and optimize complex holistic problems. Alternatively, based on certain purposes and under controllable conditions, the relationships between variables and variables, variables and systems, and systems and systems in the scenario system can be studied through multiple and repeatable methods, discovering the inherent logical relationships between system elements. Moreover, it can simulate many difficult-to-reproduce or potential scenarios in the real world, such as emergency response and scheduling of sudden events or mass incidents, natural disasters, and war simulations, and provide guidelines for rapid response and decision-making in realistic scenarios.

Although the model developers have clear objectives and possess the necessary basic techniques in this process, their deep understanding of scenarios, the appropriateness of the sequential transformation process from realistic scenarios to conceptual scenarios and structured scenarios, the scientificity of the theoretical basis used for research, the accuracy of the design of cultivation technology routes and processes, and the adaptability of key algorithms used all have significant impacts on the quality of the resulting realistic scenario model. The "quality" of the model refers to the possibility, effectiveness, and robustness of using the cultivated parallel mirror image to achieve the above-mentioned objectives. This means that deviations, failures, or even misleading results may occur during the practical application of scenario cultivation. Therefore, like the inspection of product quality, the scenario model cultivated through scenario cultivation needs to undergo certain "quality testing" and, based on the testing, corrections, improvements, and optimizations to enhance the credibility and acceptability of the scenario model. This specifically includes two parts: the evaluation of the scenario cultivation model and the verification of the scenario cultivation model, with verification as the focus.

13.5.2 Modular Verification Techniques

Although the scenario cultivation model is the final result of the overall and full-process cultivation process, this process is composed of structured and ordered modular chains. In other words, the characteristics and attributes of the modeling process have a significant impact on the "quality" of the final model. Therefore, we can improve and enhance the model quality by structurally decomposing the cultivation process and verifying the cultivation model in each step of the process at different levels and stages.

1. Data Verification

The scenario cultivation process involves various types of data from multiple domains and encounters certain challenges. For example, due to inherent uncertainty and the need for privacy and security protection, the authenticity, completeness, and effectiveness of data are affected. In order to ensure the quality of the input data for cultivation during the era of big data, it is necessary to pay attention to data verification techniques for all the large data sets used in scenario cultivation.

2. Step-By-Step Verification

Scenario cultivation verification is a hierarchical and step-by-step verification process, gradually moving from fine-grained to coarse-grained verification. It can be specifically divided into verification of conceptual scenarios, verification of structured scenarios, verification of cultivated scenarios, and so on.

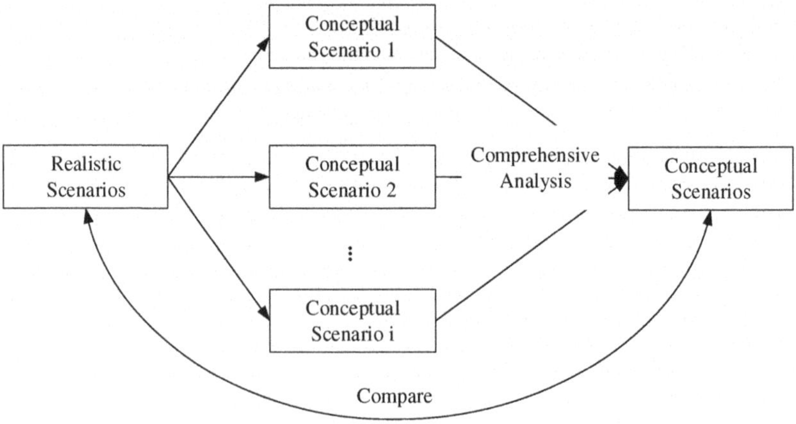

Fig. 13.2 Diagram of conceptual scenarios verification

(1) Verification of conceptual scenarios: in a sense, the most initial and fundamental step in the scenario cultivation process is the mapping of realistic scenarios to conceptual scenarios. Different individuals often have different conceptual scenarios based on their sensory cognition of the same realistic scenario. In other words, a realistic scenario can be transformed into multiple conceptual scenarios. Determining which conceptual scenario better aligns with the realistic scenario and preserves its essential attributes requires the design of multi-agent collaboration and iterative feedback methods to eventually decide, as shown in Fig. 13.2.
(2) Verification of Structured Scenarios: the process of structuring scenarios is the transformation of conceptual scenarios into representations that computers can "understand." Different structured techniques can result in different structured scenarios from the same conceptual scenario. Generally, the verification of this module can be completed through the following steps:

    A. Whether the structured scenario accurately reflects the hierarchical structure and boundaries of the conceptual scenario.
    B. Whether the structured scenario includes the main subjects of the conceptual scenario.
    C. Whether the structured scenario expresses the interaction between the main subjects of the conceptual scenario and the environment.

    In this process, an incremental approach can also be adopted. First, analyze the most important subjects and their interaction relationships, followed by the less important ones.

(3) Verification of Cultivated Scenarios: during the scenario cultivation stage, modeling is done based on the dimensions of time, space, and events, resulting in several smaller models that are gradually integrated into a holistic cultivated model. Considering this, the verification work is generally divided into

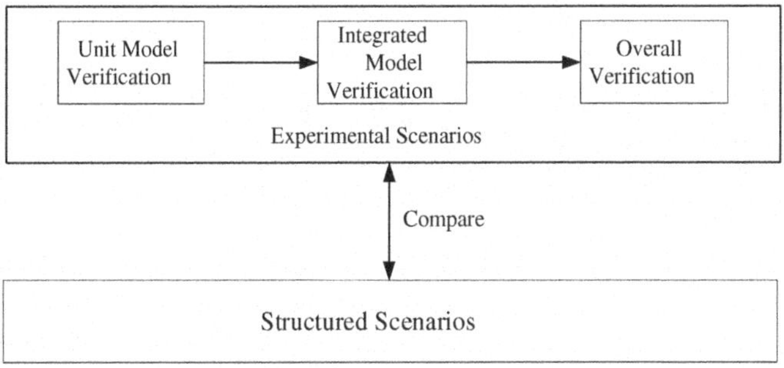

Fig. 13.3 Diagram of experimental scenarios verification

unit model verification, integrated model verification, and overall verification techniques, as shown in Fig. 13.3.

In addition to the iterative process within each stage, the verification of conceptual scenarios, structured scenarios, and cultivated scenarios also requires a larger-scale iterative verification process.

13.5.3 Feedback Learning-Based Verification Techniques

Although we have decomposed the scenario cultivation process into several modules and designed verification stages for each module, the quality of the cultivated results ultimately depends on the holistic functionality and utility of the scenarios. Therefore, we need to evaluate and verify the cultivated results generated by the computer at the overall level above individual modules and continuously improve and optimize them. In other words, the verification discussed so far primarily reflects reductionist thinking, while this part focuses on holistic thinking. For big data-driven scenario modeling or scenario cultivation, this work is more essential and important.

Firstly, let's introduce the concept of feedback learning as a learning method. During cognitive processes or learning, individuals need to continuously improve their cognitive level or correct misconceptions in their acquired knowledge. This requires a corresponding platform that supplements and corrects human cognition or learning. It can be imagined that such a feedback mechanism itself is a way of human learning, or in other words, feedback is a form of learning. The basic paradigm is to construct a feedback mechanism as an evaluation and verification environment. When necessary, the scenario model to be verified can be debugged within this environment. Under certain evaluation criteria and standardized procedures, this environment provides a series of debugging information, such as the quality of the overall scenario, deviations or omissions in comparison to certain indicators, and how to correct deviations and improve the overall quality. This debugging can be repeated under different preset conditions, similar to how people obtain information about their health conditions through multiple health check-ups and learn ways to improve their well-being.

This application of feedback learning to evaluate and improve the quality of holistic scenario cultivation is referred to as feedback learning scenario verification technology. The key now is how to build such an operational technique. The following is a brief introduction to a design concept.

Scenario cultivation is essentially the construction of an artificial virtual scenario system as a parallel mirror image of the realistic scenario system. In this sense, scenario cultivation is also a "simulation" of the realistic scenario system by the computer. The basic connotation of "simulation" here is to use a model system to imitate the real system and achieve a high degree of consistency with the real system, where "real" refers to the form and evolution of the real system.

However, upon careful analysis, there are significant differences between scenario cultivation technology and traditional computer simulation technology. Traditional computer simulation technology constructs a model that is as functionally consistent as possible with the real system through recognition and mechanism analysis of the real system. Therefore, the simulation model takes the real system as the "benchmark" and strives to approach its realism, i.e., to be "authentic." This is the only criterion for evaluating and verifying the quality of simulation systems.

In contrast, due to the inherent uncertainty, adaptivity, self-organization, and emergence of realistic scenarios, the cultivated realistic scenarios become a complex "living" system. In addition to structural mechanisms, there are also semi-structural and non-structural components within the scenarios. The complex form and evolution of scenarios cannot be fully represented by structured dynamic models. Therefore, our understanding of their complex wholeness is often incomplete or even entirely unknown. For example, a scenario may have M-dimensional dynamic mechanisms, but we often only know N dimensions of those mechanisms, where $N < M$. As a result, the remaining M–N dimensions of dynamic mechanisms are unknown to us, and we have no knowledge of their role in the formation of the complex wholeness scenario. This indicates that the future scenarios formed through scenario cultivation can only be one of the dynamic foundational models constructed based on kernel scenario elements rather than the complete representation of the "family" of models. In fact, the future scenarios of complex systems are unknown until they become reality.

Based on the academic ideas mentioned above, we can design the following technology for evaluating and verifying scenario cultivation results based on feedback learning:

Step 1: Use scenario cultivation technology to create a virtual scenario based on the parallel mirror image of the realistic scenario. The analysis, interpretation, and prediction capabilities of this model are the key aspects of evaluating its quality.

Step 2: Select a specific time point as the starting point and evaluate and verify the scenario cultivation within a certain time period from that point until the present. Within this time period, the scenarios have become historical reality, meaning that any scenario is either historical or current, representing the "truth" of history or reality. This step can be further divided into different time lengths, equivalent to obtaining scenario evolution processes with different starting and ending points. Regardless of the process, there is a complete sequence of realistic scenario data, which, in general, creates a scenario "family" with a common kernel scenario and a real evolutionary path.

Step 3: Shift the scenario cultivation models to be evaluated and verified backward and use them as the scenario dynamic models within this data sequence period to carry out various analyses, interpretations, and predictions for verification work. It is not difficult to see that for this data sequence period, this verification work has "degenerated" into a "simulation" verification based on a specific real scenario because, at this time, any scenario in the data sequence period being used for comparison and verification has already become a reality. This step involves comparing the cultivated scenario with the actual historical scenario, assessing the accuracy of predictions and interpretations, and identifying any deviations or omissions.

Step 4: Based on the completion of the above steps, conduct a comprehensive evaluation of the cultivated model and work on improving its quality.

In summary, this section focuses on utilizing the feedback learning approach to investigate whether the scenario cultivation model can reflect the essential attributes of complex systems management and decision-making problems, such as wholeness, adaptability, prevention of overfitting, and improvement of generalization capabilities. It involves constructing a verification technical system based on the consistency of model inputs, model compatibility, and reliability of result outputs. It also explores the credibility of the model and establishes termination criteria for modeling iterations. Figure 13.4 illustrates this technique.

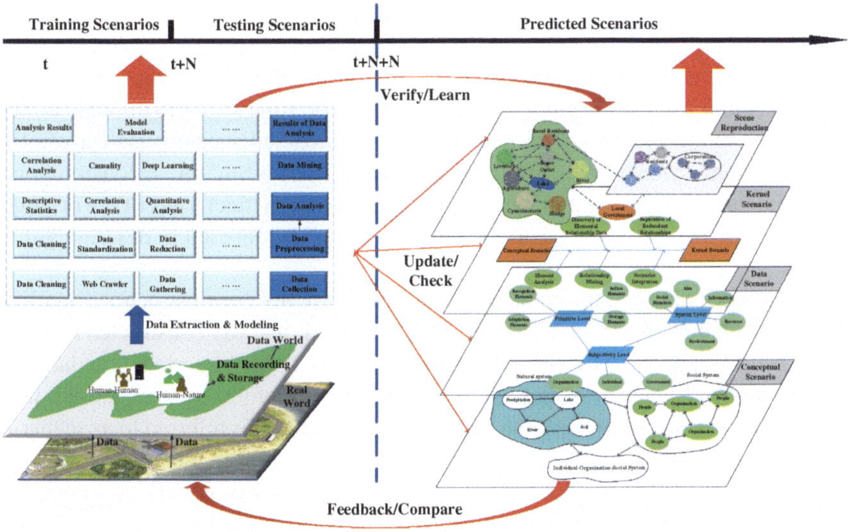

Fig. 13.4 Verification technologies of feedback learning mechanism

13.6 An Academic Concept on Scenario Modeling Cognition

The content presented in this subsection is relatively independent and is better understood within a coherent scientific and linguistic framework. In comparison to the previous sections, it leans more toward operational techniques, with a stronger mathematical component and a greater amount of foundational knowledge. We will only introduce its basic ideas, and not understanding this subsection does not hinder the comprehension of the preceding content.

Big data-driven scenario cultivation technology involves the computer "calculating" a virtual digital scenario that is interactive and immersive relative to the realistic scenario. Essentially, this is a reconstruction of the realistic scenario (system) as a scenario. This viewpoint not only has profound implications in the field of universality learning theory but also provides a new thinking paradigm in the theory of big data-driven scenario modeling. We will now provide an interpretation of this issue.

As mentioned earlier, a realistic scenario is a complex system in the physical world, and the modeling of a realistic scenario is a parallel mirror image of the realistic scenario in the data space or an abstract representation and description of the realistic scenario. In this sense, scenario modeling, as a new form of system, is a reconstructed system of the original, realistic scenario system, and scenario cultivation corresponds to the activities and processes of this system reconstruction.

The understanding of "construction" in the context of scenario modeling is not difficult because the modeled scenario is a system constructed by being driven by big data. Why doesn't it become a "new construction"? Although the modeled scenario is indeed a system with new connotations, in order to emphasize that this new system is an abstract representation and description of the original, realistic scenario system, it retains many essential attributes, holistic structures, interrelationships, spatiotemporal distributions, and evolutionary patterns of the original scenario through kernel scenario techniques. Exactly, in this context, the modeled scenario isn't a wholly unrelated "new system" built from the ground up; instead, it's a "reconstruction" that maintains the fundamental mechanisms and several distinct traits of the original scenario system. Therefore, the term "reconstruction" is more apt in describing this process.

In summary, the current context of scenario (system) "reconstruction" implies two important meanings: firstly, the modeled scenario is a newly constructed realistic scenario, and secondly, the modeled scenario preserves many important foundational mechanisms and features of the realistic scenario. Therefore, to some extent, it is the "reconstruction" of the original scenario system, and thus referred to as scenario (system) "reconstruction."

Furthermore, if we consider a complex system in the physical world as a complex dynamic system, a complex system scenario is the state of that dynamic system, and the space representing all possible states of the system is the phase space. Any scenario, as a possible state of the system, corresponds to a point in the phase space. If the data sequence that represents the scenario is generated by a deterministic dynamic system, the process of recovering and characterizing the original dynamic system based on the data sequence is called phase space reconstruction. Indeed, scenario modeling can be perceived as the reconstruction of phase space in a complex dynamic system. Put differently, the theory of complex dynamic system phase space reconstruction can be employed to discern the methodologies for attaining and analyzing scenario reconstruction. This approach emerges as a pivotal technology within the realm of big data-driven scenario modeling.

The fundamental principle of reconstructing complex system scenarios based on scenario data sequences is that if the reconstructed phase space is approximately equivalent to the actual complex system scenario in certain holistic structural characteristics, the reconstructed phase space can be understood as the abstract mirror image of the actual complex system scenario. The basic idea of phase space reconstruction is that, due to the evolution of a complex system scenario, its trajectory eventually exhibits regular motion and forms a regular trajectory. In this process, the form and evolution of any component of the system are determined by other interacting components, and thus, the information of these related components is implicitly contained in the development process of that component. Consequently, the regularity of the complex system scenario itself can be extracted and recovered from the data sequence of a certain component. To reconstruct an equivalent state space, it is only necessary to examine one or several components and treat their observed quantities at certain fixed time delays as new dimensions, with the delay values considered as new coordinates. This new coordinate can be viewed as a point in a multidimensional state space. Absolutely, by iteratively conducting this process and quantifying the relative delays corresponding to varied instances, a multitude of points can be amassed. It can be substantiated that this methodology can maintain numerous properties of the system's attractor. In essence, this process can reconstitute an abstract depiction of the original dynamic system utilizing a single observed quantity of the system and can furnish an initial estimation of the dimension of the system's authentic phase space.

In this way, phase space reconstruction is the creation of a multidimensional state space based on the scenario data sequence. If it preserves some important geometric structural information of the original system during this process, we refer to it as having geometric invariance. These geometric invariants include eigenvalues of fixed points, the fractal dimension of attractors, and Lyapunov exponents of trajectories. Clearly, if a reconstructed scenario through phase space reconstruction possesses the quality of geometric invariance, it naturally becomes a representation (or description) of the original scenario with relatively good quality in another logical and linguistic system. It aims to accurately and comprehensively represent the essential "genes" of the original, realistic scenario that are inherited and transmitted in the reconstructed scenario [9].

Thus, we need to pay special attention to which geometric invariants have a certain essential "gene" significance for the scenario, such as characterizing the complexity of the scenario or describing the properties of exponential divergence or convergence of adjacent scenario trajectories in phase space. In particular, if we can calculate the geometric invariants of the reconstructed space of the scenario data sequence and

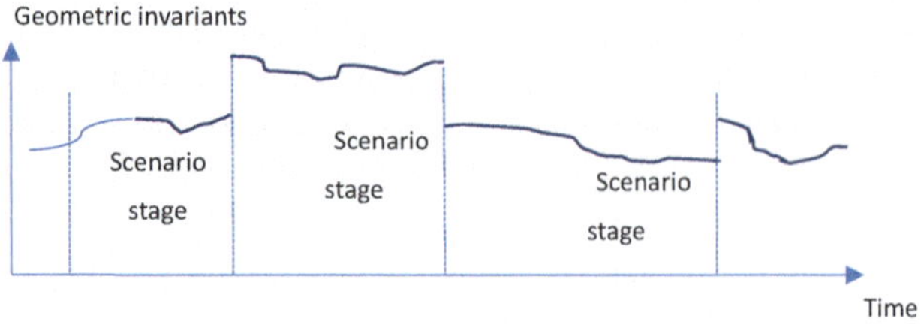

Fig. 13.5 Scenario emergence

consider the scenario evolution stages with almost invariant geometric invariants as scenario stable stages, we can divide the complex system scenario evolution process into different relatively stable stages corresponding to different developmental stages of the realistic scenario, as shown in Fig. 13.5.

It can be seen from this that the time points at which geometric invariants undergo significant changes are also the time points at which complex real-world systems experience scenario mutations. This is because a mutation in the scenario implies that the evolution of the complex system has entered a new turning phase. Correspondingly, essential attributes of the scenario, overall structure, spatiotemporal distribution, and evolutionary laws, which reflect the basic dynamics and behavioral rules of the scenario, will exhibit abrupt changes or emergent properties. This has important implications for recognizing and managing the complexity of the evolution process of complex system scenarios.

13.7 The Process of Big Data-Driven Complex System Scenario Modeling

Under the principles, technical routes, and key technological support of big data-driven complex system scenario modeling, the overall process of big data-driven complex system scenario modeling can be designed and formed without difficulty.

The process of big data-driven scenario modeling is the specific path formed by the integration and logicalization of the basic principles and key technologies of big data-driven scenario modeling. It represents the paradigm of the complete operation of scenario modeling. This process exhibits the following three distinct characteristics:

1. The entire process is highly consistent with the theoretical logic of scenario modeling. Each important stage of the process is supported by corresponding key technologies, which are themselves the technicalization of the principles and can provide holistic functionality.
2. The process starts with data and utilizes a multi-symbol system to explore the scenario information value inherent in the big data set. It follows an ordered transformation and iteration process of data → data scenario information → scenario

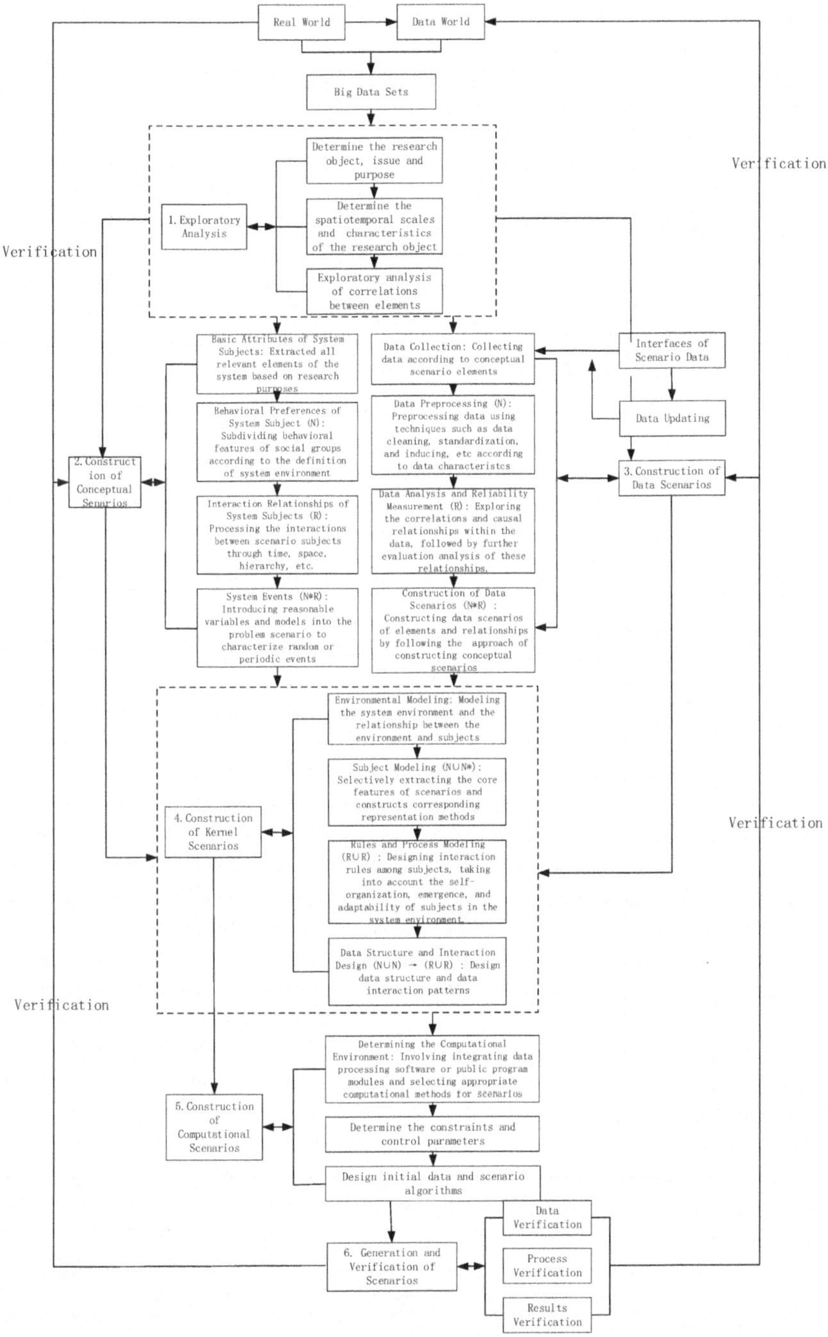

Fig. 13.6 The process of big data-driven scenario modeling

elements → scenario fragments → holistic scenarios, ultimately achieving the externalization of the value of big data scenarios.

3. In terms of reductionism, the process involves reducing the microscopic mechanisms of scenario autonomy and mixed-federation system forms through deductive reasoning. It uses scenario language such as who/what/when/where/why/how to analyze the temporal sequence of scenarios, obtaining local and microscopic complexity cognition. In terms of holism, the process combines deductive reasoning with big data-driven scenario correlation logic. Considering the scarcity of "scenario" samples, it generalizes statistical concepts into "scenario cultivation" methods. By using computer simulation techniques, it allows discrete scenario "slices" to form "a flow of scenario slices" under certain mechanisms and dynamic rules, leading to the emergence and evolution of the uniqueness and macroscopic properties of scenarios (approximation of the overall picture and process restoration).

The kernel scenario extraction and scenario cultivation techniques reflect the uniqueness and unity of the theoretical thinking, basic principles, and operational processes of big data-driven complex system scenario modeling.

The following diagram illustrates the overall process of big data-driven complex system scenario modeling.

The diagram above illustrates the overall process of big data-driven complex system scenario modeling. The process consists of the following steps:

Step 1: Analysis of realistic scenarios. Based on the research requirements, specific, realistic scenarios are defined, and the goals of scenario modeling are clarified to form a basic understanding of the realistic scenarios.

Step 2: Construction of conceptual scenarios. Conceptual scenarios primarily consist of four parts: basic attributes of subjects, behavioral preferences of subjects, interaction relationships among subjects, and events. The basic attributes of subjects are extracted based on research purposes, aiming to include all relevant elements. Behavioral preferences focus on refining the behavioral characteristics of social groups based on environmental definitions to obtain a set of elements. Interaction relationships involve processing the interactions between scenario subjects through time, space, hierarchy, etc., resulting in a set of relationships. Events involve introducing reasonable variables and models into the problem scenario to characterize the basic features of events, thus achieving the transformation from elements to relationship sets.

Step 3: Construction of data scenarios. Data scenarios involve collecting data based on the attributes of conceptual scenario elements. The collected data is then preprocessed using techniques such as data cleaning, standardization, and normalization to obtain a manageable set of elements. Big data technologies are utilized to explore the hidden correlations and causal relationships within the data, followed by further credibility measurement and evaluation analysis of these relationships. This process leads to the extraction and formation of deeper and new sets of relationships. Finally, modeling is performed based on the manageable set of elements and the new set of relationships, completing the construction of data scenarios.

Step 4: Construction of kernel scenarios. Constructing kernel scenarios requires the integration of conceptual scenarios and data, forming a unified whole of "scenarios" and "data." On this basis, environmental modeling, subject modeling, rule and process modeling, and data structure and interaction design are implemented. Environmental modeling involves modeling the system environment and the relationship between the environment and subjects. Subject modeling selectively extracts the core features of scenarios and constructs corresponding representation methods. Rule and process modeling involves integrating various relationships and designing behavioral interaction rules among subjects, taking into account the self-organization, emergence, and adaptability of subjects in the system environment. Data structure and interaction design refer to the evolution of subjects in the scenario, which can be seen as a data interaction process between subjects and between subjects and the environment. Therefore, it requires designing data structures and their interaction patterns.

Step 5: Construction of computational scenarios. The construction of computational scenarios in Step 5 begins by determining the computational environment, which involves integrating data processing software or public program modules. It also includes selecting appropriate computational methods for scenarios, such as standalone computation or distributed computation, and determining the data storage system. Additionally, it is necessary to determine whether intermediate result data requires offline analysis, establish interfaces for interaction with other data software, and decide if specialized devices, such as dedicated graphics processing devices, are needed. After determining the computational environment, the constraints and control parameters during the computational process are defined based on the interaction flowchart and data structure among subjects. Furthermore, the design of independent variables, dependent variables, and control variables in the system is required to describe the relationships between variables and design initial data. Finally, algorithms are designed according to research needs, such as using evolutionary algorithms to optimize subject objectives or employing learning algorithms to simulate human decision-making processes.

Step 6: Verification of scenario computation results. After obtaining the scenario computation results, methods such as feedback learning can be applied to compare and validate the results against realistic scenarios and data scenarios. Iterative verification is performed on the scenario data set, the construction process of the scenario model, and the generated results of the scenario model, aiming to continuously improve the quality of the model.

References

1. Sheng Z. Basic theory of major project management, 1 ed. Nanjing University Press;2020.
2. Tian G, Liu Y, Wang Y. Qualitative research and management discipline construction: based on the reflections of business anthropology. J Manag. 2015;12(01):1–10.
3. Liang YTR, Xu F, Sheng Z. Research on the computable model of multiple scenarios in social systems. J Manage Sci. 2017; 20(01):53–63.

4. Haiyan Wang WL, Chen W. An improved method for measuring statistical dependence between time series. J Nanjing Univ Sci Technol (Natural Science Edition). 2002; 3:325–9.
5. Lin Y, Chen Y, Wang D, et al. A new method to study the correlation between the behaviors of subsystems in complex dynamical giant systems. Appl Math Mech. 2013;34(09):917–28.
6. Suping Li HY. Multivariate time series analysis and its applications. Sci Technol Eng. 2008; 11:2764–9+97.
7. Sheng Z. Experimental theory and application of computing in social sciences. Shanghai Joint Bookstore;2009.
8. Zhang J. A research paradigm for social science computing experiments. Res Explor Lab. 2009;28(07):61–6.
9. Wang C, Zhang F, Chen S. Application of phase space reconstruction theory in foundation pit displacement prediction. Geotech Eng Commun. 2008;01:33–4.

Chapter 14
Examples of Complex Systems Management

This chapter outlines three examples of complex systems management across different fields. The number of examples is limited as, given the complex nature of the systems, we cannot merely introduce related concepts or explain some principles within the examples. We aim to illustrate, as comprehensively as possible, from the background of the problems to how they are analyzed and solved, including the thought process, methods, and conclusions. An incomplete introduction would only serve as an illustrative example and would not provide a full experience of the essence inherent to complex systems management.

14.1 Example 1: Study on Core Decision-Making Paradigm of Major Projects Based on Complex Systems Management

With decisions on two Yangtze bridge engineering projects in China as examples.

In major projects decision-making activities, there exists a category of core decision-making issues that are related to the overall, strategic, and holistic significance of engineering construction. These issues are complex and integral, and any mistake could have serious implications for socio-economic development and the construction project itself. This example, set against the background of two Yangtze Bridge projects in China that set multiple world records, introduces the process of analyzing and designing corresponding decision-making plans according to the thoughts and paradigms of complex systems management.

Z. Sheng, *Outline of Complex Systems Management Theory—Based on Irreversibility of Reductionism Thinking*, International Series in Operations Research & Management Science 360, https://doi.org/10.1007/978-981-97-3361-3_14

14.1.1 Overview of Core Decision-Making Issues in Major Projects

Decision-making is a crucial practice in major projects management, especially those related to the overall situation of major projects construction bear more strategic and holistic significance. For instance, issues like project planning, overall project design, and investment and financing modes are all part of this category because they require clear and accurate answers at the macro level to major questions, such as whether the project should be built, can it be built, how to build it, how much risk is involved, whether there are "bottleneck" technologies, and whether they can be resolved.

Generally, the following three types of practical issues belong to the core decision-making issues in major projects:

1. "Fundamental Determinative" Decision-Making Issues in Engineering Construction: These types of decisions typically have a global impact on the function, quality, and operation of major projects entities, such as decisions related to project location and overall project design schemes.
2. "Demand Innovation" Decision-Making Issues in Engineering Construction: Decisions of this kind often face unpredictable natural and technical challenges and play a pivotal role in engineering construction. They necessitate substantial innovations to resolve issues, like the selection of key technologies for major projects and the design of primary construction technical schemes.
3. "Development Strategic" Decision-Making Issues in Engineering Construction: The engineering functions and construction purposes involved in this category of decisions are directly and closely related to macro strategies, such as national and regional socio-economic development, like the initiation of major environmental renovation projects and the construction of cross-border major projects.

14.1.2 Basic Attributes of Core Decision-Making Issues in Major Projects

Core decision-making issues in major projects imply that such issues are central to the entire project decision-making activity, and resolving such issues is relatively challenging. The term "difficulty" not only embodies certain inherent attributes of the problem itself, but it is also significantly related to the subject's insufficient understanding and capability, thus having a substantial subjective component. However, when conducting scientific research on the core decision-making issues of major projects, it is imperative to eliminate, as much as possible, the various perceptual, subjective, and superficial factors attached to the research subject. The focus should be on the inherent and essential qualities of the research subject itself to reveal the objective laws about the core decision-making issues in major projects. This requires us to deeply refine and abstract the basic attributes of the core decision-making issues in major projects and to conduct scientific research within the scope of these basic attributes.

The term "attribute" refers to the abstraction of a thing's unique, fundamental, and stable qualities. It determines how a thing can be distinguished from other things and serves as a representation of its uniqueness and inherent nature. So, what are the basic attributes of the core decision-making issues in major projects?

(I) Complexity

Regardless of the type of core decision-making issues in major projects, the most crucial aspect of decision-making activities is to propose relevant decision-making schemes. Any decision scheme is a man-made system planned and designed by the decision-making subject and has corresponding functions. However, due to various possible changes in major projects themselves and the composite system of major projects and the environment, the functions planned and designed by the subject cannot always be guaranteed to be realized as scheduled. There might even occur "deviant" functions that the subject did not plan, design, or expect. This reflects that the realization of anticipated functions in core decision-making problems in major projects is much more complicated than in general engineering decisions.

Secondly, as a system form, core decision-making issues in major projects have multiple constituent elements, and the connections between these elements are intricate and complex, displaying a hierarchy. Particularly at the whole level, complex phenomena whose mechanisms are still unclear often emerge. This indicates that the process itself of designing, forming, and implementing core decision-making schemes in major projects both reflects path dependency and is filled with various uncertainties and dynamic evolutions. This leads to the possibility of various complex phenomena occurring during the process of core decision-making activities in major projects.

Lastly, the environment of such decision-making problems is often highly open and dynamic and significantly influences the decision-making process and the release of scheme functions; the diversification and heterogeneity of decision-making subjects may lead to contradictions or conflicts in decision objectives and value preferences. Even if there is no behavioral alienation in the decision-making subjects, they may still be unable to control the complex situations of such problems due to insufficient understanding, lack of ability, or the failure of traditional methods. From the above, it can be seen that the core decision-making problems and processes in major projects are full of system complexities in various forms.

(II) Wholeness

In typical decision-making activities, the classical approach to analysis and resolution is to decompose the decision problem into several parts, understand each part, and thereby understand the whole; if the understanding of the parts is still unclear, one can continue to decompose and research until clarity is achieved. This methodology is known as reductionism. The advantage of the reductionist approach is that it breaks down the whole into finer and finer details for study. However, using reductionism alone is often impractical for core decision-making problems in major projects for the following main reasons:

1. These kinds of decision problems generally have very tight coupling relationships with the major projects environment. Any changes in the environment will profoundly impact the problem, especially since the form and genesis of the problem are often the results of environmental factors. Therefore, if we sever the connection between the problem and its environment, we cannot fully understand and analyze the problem.
2. These kinds of decision problems exist within the context of engineering construction and management activities. The more complex the core decision problem, the more it is related to the context, having "genetic" and "bloodline" connections to it. This requires us to view, think about, and analyze the problem within the overall context in which it resides, identifying decision-making solutions to resolve the problem. It necessitates analyzing and synthesizing the context from top to bottom and bottom to top, and we cannot dissect the context, allowing the problem to be isolated from its context.
3. These kinds of core decision problems usually exhibit various dynamic properties, such as sudden changes, emergence, evolution, etc., and the mechanisms behind these dynamic changes are very complicated. Often, clear causal relationships are not at play; instead, there exist tight, complicated explicit or implicit connections between elements of the problem. These connections can change over time and propagate to other elements in various ways, and the problem is the result of these complex connections.

All of the above tells us that if, in the process of researching and solving major projects core decision-making problems, we solely use reductionism to decompose a complete problem into multiple independent parts and then study each part step by step, this will inevitably dissect the complex relationships and structures between the different parts of the problem and between the problem and its environment, destroying the inherent, indivisible qualities within the problem. This illustrates that major projects core decision-making problems possess a type of quality characteristic that cannot be fully addressed with reductionist methods (or, in short, have irreversibility of reductionism).

Generally, if a thing has irreversibility of reductionism, we say it has "non-additive wholeness" or "complex wholeness." Therefore, major projects core decision-making problems belong to the category of "non-additive wholeness" or "complex wholeness."

In summary, major projects core decision-making problems are filled with various forms of complexity and possess "non-additive wholeness "or "complex wholeness "that cannot be addressed with reductionist methods. Moreover, the complexity within the problems includes elements, connections, and structures directly forming the complexity, the complexity derived, conducted, and induced by wholeness, and the complex wholeness emerging at a higher level formed by the entanglement and coupling of complexity and wholeness. In this regard, we can neither purely dissect wholeness through reductionist methods nor employ any means to "compress" complexity, meaning both wholeness and complexity are inherent, essential attributes of major projects core decision-making problems.

In essence, the fundamental attribute of major projects core decision-making problems is "complex wholeness." By this logic, major projects core decision-making problems are a category of "complex wholeness problems."

14.1.3 Paradigm Shift of Complex System Thinking in Core Decision-Making of Major Projects

Major projects core decision-making activities embody diverse and comprehensive contents and forms due to their complex wholeness, and thus their attributes can be distilled from different perspectives. For instance, core decision-making problems can propose different objectives based on different scales, suggesting that such decisions can be considered multi-scale decisions; the formation of decision-making plans often follows an iterative path, suggesting they can also be seen as a type of iterative decision-making, and so on. However, the most intrinsic attribute is that major projects have long life cycles and complex environmental changes, and the decision-making plans formed by decision-making activities need to sustainably and stably release their due functions over the long lifespan of the project, especially demonstrating robustness against uncertain environmental changes. Therefore, the robustness of decision-making plans concerning situations becomes an important quality and attribute of core decision-making in major projects.

Condensing the attributes of major projects core decisions into situational robustness reflects our essential understanding of the complex wholeness of core decisions. This enables us to find a new starting point to implement the paradigm shift of complex system thinking in major projects core decision-making problems and, based on this, further extend other paradigm shift paths.

An essential concept in complex systems management is the importance of "context," meaning one cannot expect many ready-made basic laws and rules to be the "panacea" for solving various actual problems in complex systems management. It is crucial to deeply analyze and delve into a specific complex systems management problem, forming unique environments and specific dynamic mechanisms. Therefore, based on the paradigm shift of complex system thinking, the unique features of specific major projects core decision-making problems need to be combined with the basic paradigms of complex systems management to form distinctive new paradigmatic points.

14.1.4 Key Points of Paradigm in Major Projects Core Decision-Making Based on Complex Systems Management

Below are the key points of the paradigm in this instance of complex systems management:

(1) Key Point 1: Basic Paradigm

Firstly, the "Paradigm of Major Projects Core Decision-making Based on Complex Systems Management" refers to behavioral guidances and procedural norms of major projects decision-making behaviors under the paradigm of complex system thinking. It can specifically be divided into the following three phases:

1. Generally, in the process of major projects core decision-making, people first perceive the various complexities of the decision-making problems on a physical level as some kind of man-made system, especially the intuitive perceptions of decision-making problems on the physical level. They are the primal substances from which decision-making problems are further abstracted.
2. Since complex systems management is based on the cognition of complex systems, the analysis of major projects core decision-making problems should gradually be abstracted to the level of complex systems. It should employ complex system thinking and discourse systems to express and refine the associative logic and attribute features of core decision-making problems. This is the transition of the physical complexity of decision-making problems to system complexity and also the abstraction and extraction of the core elements and structure of decision-making situations. It is clear that this transition mainly occurs at the level of rational thinking, involving the paradigm shift of decision-making thinking in complex systems, preparing for the further formation of decision-making plans.
3. According to the concepts and principles of management science, the forms and cognition of the system complexity of decision-making problems are once again transformed into corresponding concepts, principles, logic, and discourse systems of management science, forming management science problems that contain the connotations of complex system thinking and fully reflect the essential attributes of complex wholeness decision-making. In this stage, by utilizing means such as theory, empirical research, simulations, and "iterative" paths, solutions to decision-making problems are gradually attained. This is the phase of transformation from system complexity to management complexity.

The transitions of the above three stages constitute the basic paradigm of core decision-making in major projects based on complex systems management (as shown in Fig. 14.1).

(2) Key Point 2: Decomposition and Integration

The integrality of core decision-making problems in major projects should not be understood as just a "monolithic" decision-making problem. In most cases, it

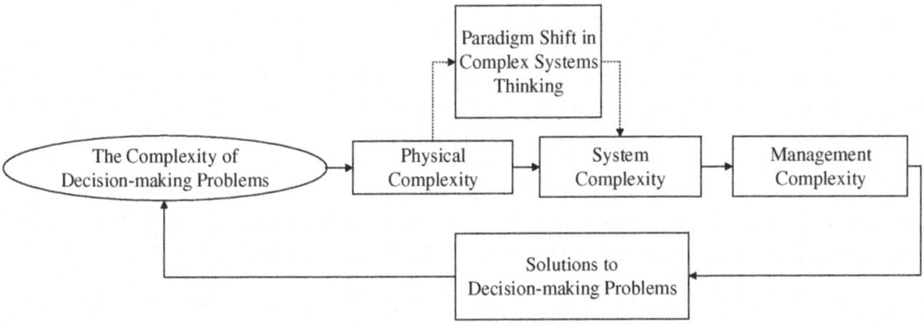

Fig. 14.1 Basic paradigm for decision-making in major projects based on complex systems management

should be understood as a group of decision-making problems that can be appropriately decomposed and have intricate connections with each other. Within this group of problems, there exist hierarchies, sequences, orders, logic, etc., as well as conflicts, synergies, emergencies, and concealments between subproblems, all of which constitute the complex wholeness of core decision-making problems.

Given this situation, we can rely on the principle of engineering virtual variability and decompose the group of decision-making problems into several relatively independent and interrelated subgroups of problems through assumption and decomposition behaviors at a certain scale and granularity to analyze and solve each subproblem. In this process, some subproblems are relatively independent; the decision is essentially concluded once they are individually resolved. However, some problems with close connections to others cannot be considered solvable independently and belong to complex wholeness decision-making problems. Such problems require analysis and resolution based on the integrated methodology of complex systems management.

For instance, during the feasibility demonstration phase of the Hong Kong-Zhuhai-Macau Bridge project, it was decomposed into 29 sub-decision topics, eventually forming 46 decision reports. The subtopics cover a wide range, including the impact of the bridge project on regional socio-economic development and long-term transportation; analysis of the project's short, medium, and long-term roles in the national road network; discussions on the scale, technical standards, alignment, and various bridge-type solutions of the bridge construction; evaluations of the impact on the natural environment during construction and operation phases; analyses of economic benefits of different investment and financing models and corresponding project; and various issues that need coordination and resolution in cross-boundary construction, operation, and management of the project, etc. Each of the 29 decision topics has certain problem boundaries, many of which can be resolved as independent problems, but some important ones, such as the impact of "One Country, Two Systems, Three Laws" on the legal environment of project construction and the conflict between bridge location/type decision and the protection of Chinese White Dolphins, have high levels of complex wholeness and cannot be solved solely by reductionist methods.

(3) Key Point 3: Organizational Adaptability and Flexibility

The execution of major projects core decision-making activities is carried out by decision-making organizations composed of decision-making subject groups. Given that major projects core decision-making problems possess complex wholeness, and the decision-making environment is deeply uncertain, the decision-making organization needs to fully utilize the abilities that emerged from the organization and self-organization to control the complex wholeness problem. Furthermore, it needs to ensure the continuity and executability of the aforementioned control abilities by establishing the operational rules and processes within the decision-making organization. Thus, the major projects decision-making organization can be understood as a platform. Its main function is not to provide specific methods and solutions directly for major projects core decision-making problems but to provide the environment and conditions for forming decision-making plans.

Recognizing the major projects decision-making organization as a type of "platform" reflects the "self-organizing" and "self-adapting" features of the major projects decision-making organization. The emergence of the overall behavioral capabilities of the major projects decision-making organization, especially the emergence of the ability to control the complex wholeness of decision-making problems, is mainly the result of the comprehensive effects of "other-organization" and "self-organization." In this, the mechanisms of "self-adaptation" and "self-organization" play a particularly important role. Therefore, the design and optimization of major projects decision-making organizations are especially important for their ability to emerge or enhance new control mechanisms for complex wholeness of decision-making problems. Generally, the more complex integrative decision-making problems there are, the stronger the adaptability and flexibility of the decision-making organization will be required.

(4) Key Point 4: Iteration and Convergence of Plans

In real major projects core decision-making activities, under the principle of complexity degradation, the overall complexity of the decision-making problem is usually broken down into several relatively independent stages during the plan generation process. This allows the decision-making subject to face only part of the overall complexity at each stage, making it easier to obtain a decision-making plan for a problem with relatively lower complexity at a particular local stage. Then, the plans from each stage are combined to form a sequence of plans, and the iteration of this sequence is used to form plans for the entire stage and the overall problem.

This "iterative" way of generating plans in decision-making practice fully reflects the complexity degradation criteria of complex systems management and also fully embodies the criteria of adaptive selection. It is a universal and effective behavior mode for subjects in actual decision-making activities.

14.1.5 Typical Cases

(1) Case 1: Decision-making on the Overall Plan for the Sutong Bridge Project

The Sutong Bridge is the largest bridge spanning the longest river in China, the Yangtze River, located at its easternmost point. The expected lifetime of the project is 100 years; the process environment is complicated, and the technological difficulty is high. The core decision-making of the bridge project reflects the complex wholeness of multiple levels, domains, and scales.

Based on the analysis of the complex wholeness in the project decision-making, the first step is to analyze the decision-making problem of the overall plan for the bridge project, and the main process is as follows:

Regarding the top-level design of the overall plan, according to the ordered relationship of decision-making problem logic, the overall plan for the bridge project is ultimately refined into a comprehensive selection between two plans: one for the bridge and one for the tunnel.

It should be noted that, in the paradigm of complex systems management, for decision-making problems of the same type or level, equivalent depth comparison is required during plan analysis. "Equivalent depth" means conducting comparable analysis of different plans in terms of inherent content and comprehensive value without being influenced by any prior knowledge or subjective preference of the subject. Moreover, when dealing with the comparison of comprehensive values, it's not about simply "adding up" or "weighted adding up" individual values. Instead, based on the background, function, and complex relationships between different values of different problems, it often adopts the value "emergence" thinking, such as using the "one vote priority" bottom-line thinking principle. For example, during the comprehensive comparison of bridge and tunnel overall plans for the Sutong Bridge, although the relevant value system included many aspects, the bridge plan, which had evident importance in engineering safety value, was ultimately adopted.

In the formation path of the final decision plan, the overall plan of the Sutong Bridge project fully demonstrates the process characteristics of "continuous comparison, gradual approximation, and final determination" (as shown in Fig. 14.2):

1. The first consideration is whether to adopt the bridge plan, the tunnel plan, or a combination of both.
2. After determining the bridge plan, the choice is between a suspension bridge plan and a cable-stayed bridge plan.
4. After deciding on a dual-tower, dual-cable steel box girder cable-stayed bridge plan, the choice is between a five-span continuous steel box girder plan and a seven-span continuous steel box girder plan, as well as the design plan for the bridge axis, etc.

Specifically, initially, a bridge plan was selected through a comprehensive comparison between the feasible tunnel and bridge plans. After deciding on the bridge plan and considering the navigation requirements, the bridge span must be more than

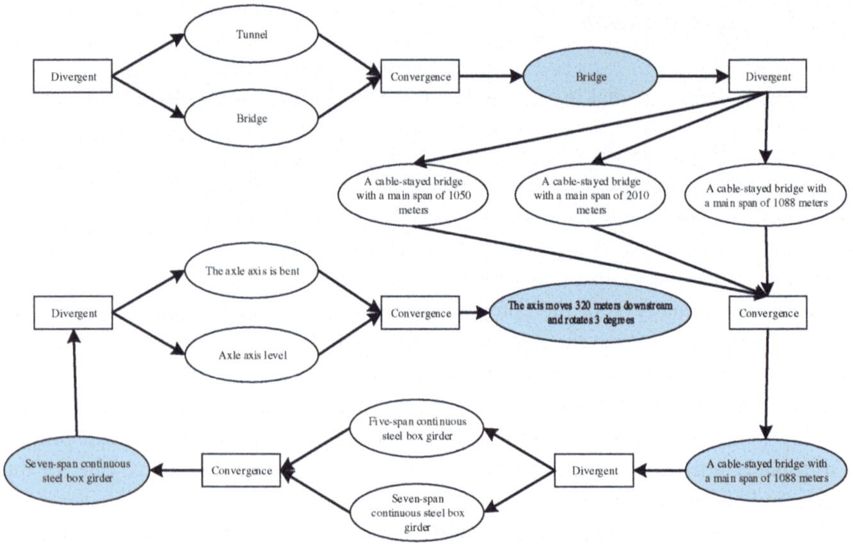

Fig. 14.2 The convergence process of the overall plan decision-making for the sutong bridge project

1081 m. The available bridge plans are mainly large-span suspension bridges and cable-stayed bridges, including the main span 1510 m suspension bridge plan, the main span 2010 m suspension bridge plan, and the main span 1088 m cable-stayed bridge plan. After deciding on the main span 1088 m cable-stayed bridge plan, there is a choice between a 2026 m five-span continuous steel box girder plan and a 2088 m seven-span continuous steel box girder plan. After deciding on the seven-span continuous steel box girder plan, the bridge axis needs to be determined. From the initial plan to the optimized plan and then to the final plan, the axis has been moved 320 m downstream, with a direction rotation of 30°.

It can be seen that the selection of the overall Sutong Bridge plan is a process of progressive plan comparison; that is, after eliminating obviously unsuitable plans, the retained plans undergo equivalent in-depth study and comparison (including detailed design and computational analysis), advancing step by step to form the final plan. The selection of bridge-type plans, as well as main beam, pylon, and cable-stayed structure plans, all employ this method, progressing gradually to confirm through conducting equivalent in-depth studies on plans.

In summary, as a complex wholeness decision-making problem, the complex systems management paradigms such as the decomposition and synthesis of decision problems, the adaptability and flexibility of decision-making organizations, and the comparison, approximation, and convergence of decision-making plans, have been universally and effectively applied in the overall plan decision of the Sutong Bridge. This has greatly improved the quality of decision-making, ensured the scientific nature of the decisions, and played a crucial role in overcoming numerous decision difficulties for the Sutong Bridge.

(2) Case 2: Complex Wholeness Decision-Making on Bridge Location and Chinese White Dolphin Conservation for Hong Kong-Zhuhai-Macao Bridge Project

The Hong Kong-Zhuhai-Macao Bridge is a cross-sea traffic cluster composed of tunnels, islands, bridges, and roads. One of the core decisions of the project revolves around choosing the bridge location—specifically, deciding the positions of the landing points and the alignment schemes that connect Zhuhai, Hong Kong, and Macao in the Lingdingyang waters of the Pearl River estuary. The wholeness nature of this decision arises from several intertwined subproblems that cannot be fully decomposed, and its complexity primarily stems from conflicts of interest between decision-makers, the variability of the natural and ecological environment, and the entangled nature of the decision problems. For instance, the alignment of the bridge has direct implications for the urban planning and transport network layouts of Guangdong, Hong Kong, and Macao. Thus, local governments, each rooted in their perspectives, propose various suggestions for landing points and alignments, inevitably leading to conflicts between individual and overall interests. Additionally, determining the bridge location mandates a comprehensive understanding of meteorological, hydrological, river conditions, seismic, and ecological environments to mitigate decision-making risks, yet the existing capability in these domains is often limited and inadequate. Moreover, the subproblems in the bridge location decision are interconnected and nested; the decision on landing points is contingent on the feasibility of the natural conditions and technical capacities of the bridge alignment, which in turn acts as a premise for the landing point decisions, thus creating mutual constraints. Especially critical is the possibility that the chosen bridge location might traverse the Chinese White Dolphin National Nature Reserve in the Pearl River Estuary. This necessitates proper resolution of the conflicts between bridge construction and the conservation of Chinese White Dolphins.

It's well known that the Chinese White Dolphins inhabiting the Pearl River Estuary in Guangdong are under first-class state protection. Protecting these dolphins is a rigid constraint in the decision-making process and a significant social responsibility associated with the construction of the Hong Kong-Zhuhai-Macao Bridge. Specific considerations under dolphin conservation decisions include whether to allow the bridge to cross the protected area and how to minimize the impact on the dolphins, implying substantial obstacles if the alignment intends to traverse the reserve. The construction and operation of the bridge project must not exert direct or indirect negative impacts or harm to the living environment, survival space, or individuals of the Chinese White Dolphins.

To solve this issue, there are two available approaches: ① Adjust the range/functional zoning of the protected area, meaning the protection of the Chinese White Dolphins would completely conform to the construction of the bridge; ② Adjust the alignment of the bridge to bypass the protected area, meaning the project construction would fully comply with the protection of the Chinese White Dolphins. Based on the recommended engineering alignment selection, if choosing to adjust the alignment of the bridge, both the northern and southern plans will not meet the requirements, significantly limiting the space for choosing engineering solutions. Both mentioned plans are valuable options evaluated through comprehensive assessments by the China Communications Highway Planning and Design Institute; substantial adjustments to these plans would lead to a surge in construction costs and increased technical difficulties, unfavorable for achieving the overall objectives of the project. Therefore, looking from the alignment perspective, the bridge alignment will inevitably traverse the Chinese White Dolphin protected area. Consequently, the alignment decision transforms into how to resolve the direct conflicts arising from the bridge alignment traversing the dolphin-protected area and a series of subsequent issues. Given that the protection of the Chinese White Dolphins and the bridge alignment decisions fall under two different domains—ecological environment protection and national major project construction—each having its rigidity but needing to be considered as a wholeness question, this represents a typical example of complex wholeness decision-making in the Hong Kong-Zhuhai-Macao Bridge project.

Based on the above analysis of this decision-making problem, the following overall strategies are initially implemented from a wholeness level concerning the bridge alignment and the protection of the dolphins: ① Reductionist Strategy: Fully utilizing reductionism, the alignment decision problem is decomposed into finer subproblems, and decisions are made for each, like employing a synchronized asynchronous decision-making method, "landing point first, then alignment, with landing point and alignment complementing each other." ② Integrative Strategy: Given the bridge construction must consider its impact on the Chinese White Dolphins in the Pearl River estuary, after careful analysis, the following flexible strategies were adopted: Initially, in the early stages of the alignment decision, the issue of dolphin protection is "shielded" spatially, i.e., the impact on the dolphins is temporarily disregarded. However, in the later stages of the alignment decision, if the alignment traversing the dolphin-protected area becomes a fixed outcome, the issue of dolphin protection is allowed to "return" spatially, refocusing on how to deal with the dolphin protection issue, achieving wholeness consideration of both decision-making problems.

In specific strategies, "bottom-line thinking" is adopted for the alignment of the bridge in terms of engineering feasibility, which means minimizing the impact on the white dolphins as much as possible while compensating for the loss of the dolphins' living space with other resources. For example, providing financial and ecological compensation to dolphin protection efforts ensures a balance and comprehensive coordination of interests between construction projects and dolphin protection, transferring spatial conflicts to benefit coordination and equilibrium.

The complete process of the above decision-making strategy is shown in Fig. 14.3 and is roughly divided into the following three stages: ① in the "engineering conceptual" level, the protection of white dolphins in the bridge alignment decision is temporarily "shielded" to expand the feasibility of the bridge type and location decision, reflecting the "core scenario" of location in the complex wholeness of bridge type and location decision. In this stage, the decision-makers preliminarily established the landing points in Guangdong, Hong Kong, and Macao and six bridge location corridor plans in three major categories. ② At the "engineering substantive" level, the protection of white dolphins is brought back into the bridge type and location decision problem. The bridge location plans obtained in the first stage serve as input elements for the "return" of white dolphin protection, forming a comprehensive view of complex wholeness in bridge type and location decisions, and decision-making is done on the "panorama" of the problem after return. This stage mainly aims to avoid legal conflicts related to the dolphin protection area and ecological environment during bridge construction and to ensure that the area of the Chinese White Dolphin Protection Area is not reduced due to occupation by bridge construction. The specific process went from initial non-adjustment of the protected area to reconstructing temporary adjustments within the protected area. ③ Given that the bridge has to traverse the dolphin protection area, the adjusted bridge location plan is considered the "bottom line" for the entire bridge type and location decision and takes as many necessary ecological remedial measures as possible for dolphin protection, reducing the impact of the bridge on white dolphins to the minimum and favoring the future survival and reproduction of white dolphins. It can be seen that the entire engineering alignment and the complex wholeness decision-making plan for the protection of white dolphins are gradually iterative and approaching dynamic processes under the complex systems management paradigm.

Viewing the entire process, the complex wholeness decision-making problem of integrating the bridge location of the Hong Kong-Zhuhai-Macao Bridge project and the protection of Chinese white dolphins, which cannot be resolved by simple reductionist methods, undergoes several stages. First, in the intuitive perception stage, it experiences physical complexity, such as the technical complexity of the engineering bridge alignment, severe natural environment, high engineering cost, strict ecological protection requirements, and difficulty in protecting white dolphins. Based on this, various kinds of physical complexity are gradually refined and abstracted to the level of complex systems, forming systemic complexities such as diversified interests of multiple subjects, heterogeneity of engineering value and environmental value, multi-scale, multi-dimensional conflicts in engineering objective systems, and deep uncertainty in decision-making full scenarios. Further integrating these system complexities in the decision-making process, through the analysis of decision problem priorities and resource constraint conditions, the "coincidence of time and space" in two decision problems, in reality, becomes a time-ordered "concomitance of time and space" decision sequence through spatial "shielding" and then "returning." It then integrates different sub-decision problems holistically by exchanging economic resources for spatial resources and ultimately obtains decision-making plans for complex wholeness problems. This reflects the coherent and integrated complex systems management paradigm through the three stages of physical complexity–system complexity–management complexity.

In addition, in terms of decision-making organizational structure, the core decision-makers for different decision problems are different, and the core decision-makers for the same decision problem may also change. The higher the complexity of the decision problem, the higher the hierarchy and authority of the decision-makers. Specifically, in the first phase mentioned above, the core decision-makers for the landing point decision were the Tripartite Coordination Group for the Bridge, the

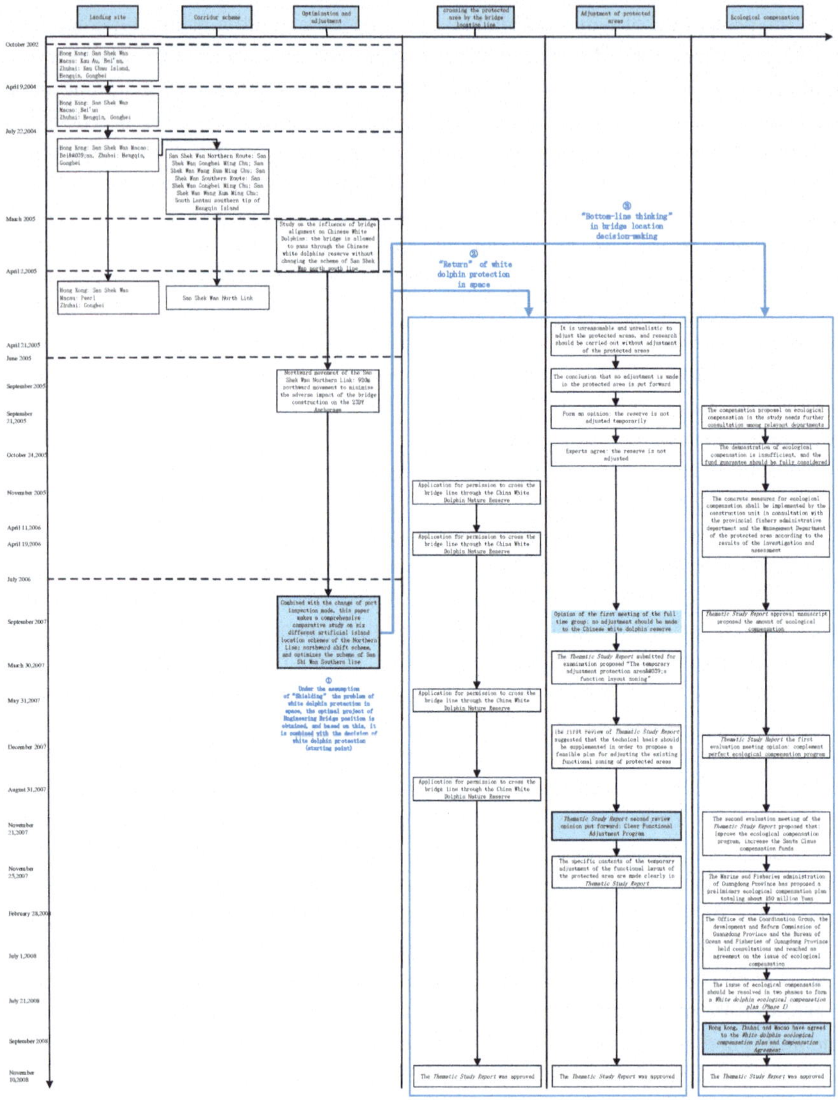

Fig. 14.3 Degradation and integration of complex holistic decision-making regarding the bridge location of the Hong Kong-Zhuhai-Macao Bridge and the protection of Chinese White Dolphins

governments of the three regions, and the China Communications Highway Planning and Design Institute. After determining the landing points on the Hong Kong side as Lantau Island and Pak Mong Tsui, and the convergence of landing points on the Zhuhai and Macau sides, additional decision-makers for further comparison, examination, and deliberation of the landing point combinations and bridge alignment schemes included the National Development and Reform Commission and various relevant departments of the central government (including the Ministry of Transport, Ministry of Water Resources, Hong Kong and Macau Affairs Office, Environmental Protection Bureau, General Equipment Department, Ministry of Agriculture, etc.). In the second phase mentioned above, the primary decision-maker for allowing the bridge alignment to cross the Chinese White Dolphin Conservation Area was the Fisheries Bureau of the Ministry of Agriculture, while the adjustment of the conservation area and ecological compensation was primarily decided by the governments of the three regions. This is partly due to the different responsibilities and authorities of the relevant departments and also because different issues have varying degrees of impact on the interests of the three parties. The placement of the bridge through the Chinese White Dolphin Conservation Area primarily affects the marine ecological environment of the Pearl River Estuary in mainland China and the survival of rare animals. This issue falls under the jurisdiction of the Fisheries Bureau of the Ministry of Agriculture, making it the government department responsible for this issue. Adjusting the conservation area and ecological compensation involves the contributions of the three parties and the allocation of ecological compensation projects and tasks. This relates to the interests of all three parties, so the primary decision-makers for these two decision problems are the governments of the three regions. In the third phase mentioned above, the research on the ecological compensation scheme and the adjustment of conservation area functions was initially mainly led by the South China Sea Fisheries Research Institute. Once the temporary adjustment of the conservation area's internal functional zoning was determined, the scheme for adjusting the conservation area's internal functions was handed over to the Guangdong Provincial Marine and Fisheries Bureau, which also conducted relevant material supplementation. The changes in the composition of the decision-making organization mentioned above primarily occur because, as the specific content of the decision problem changes, decision-makers need to have adaptable and sufficient decision-making authority and expertise. This fully reflects the flexibility and adaptability principles of decision-making organizations in the context of complex systems management paradigms.

14.1.6 Conclusion

Currently, issues like decision-making errors in major project management pose significant challenges to major projects construction in various countries. In response to this, the international academic community of engineering management increasingly recognizes that there is a "tension point" between the practical demands of decision-making in major projects and the innovation of decision-making theories and methods. In this regard, complex systems management is poised to become a new paradigm of thinking, practice, and research in the field of major projects decision management. The theoretical exploration and typical case presented in this instance demonstrate that the logical framework and academic content of complex systems management will greatly assist us in addressing and solving the complex wholeness challenges of core decision-making in major projects in a more scientific manner.

14.2 Example 2: Research on the Reform of Taihu Lake Water Environment Governance Based on Complex Systems Management

The Taihu Lake basin is a region of high socio-economic development in China. The water environment of Taihu Lake plays a crucial role in supporting and safeguarding the development of this area. However, water environmental issues in the Taihu Lake basin have also become increasingly prominent and sensitive. For a long time, the state has carried out large-scale governance on Taihu Lake, and the overall water environment of Taihu Lake has seen considerable mitigation and improvement, with a significant enhancement in the quality of the water environment. However, with the rapid development of China's ecological civilization construction, the requirements for the governance of Taihu Lake continue to rise, gradually revealing the weaknesses and insufficient and unbalanced governance enabling the original governance.

This case study, oriented by a series of profound issues in the current water environment of Taihu Lake, uncovers the complex wholeness attributes of the socio-ecological colossal system of Taihu Lake water environment, explores the paradigm shift in the complex system thinking of Taihu Lake governance; proposes a new mode of Taihu Lake governance reform based on complex systems management. It exemplifies scientific issues such as the basic paradigm of Taihu Lake governance based on complex systems management, complexity analysis and degradation, intrinsic safety objectives, and the digital market allocation of pollution rights trading in the basin, as well as multi-scale evolution models of water environment, GEP-based aquatic ecological compensation mechanisms, and complex wholeness scenario modeling techniques for the water environment. From a new perspective, it discusses how to form a new impetus for governance in the Taihu Lake basin through the transformative reconstruction of governance modes and mechanisms based on the paradigm shift in complex system thinking and promote the realization of the strategic goals of coexistence of human and water environment life.

14.2.1 Introduction

Taihu Lake is a significant water source in the Yangtze River Delta of China. The Taihu Lake basin refers to the entire area traversed by the Taihu Lake water system and its tributaries, encompassed by the watershed of Taihu Lake. Today, the Taihu Lake basin has become one of the most densely populated areas with the largest cities, the most developed economy, and the most vibrant regions in our country.

For a long time, the water environment of Taihu Lake has played a substantial role in promoting and supporting the socio-economic development of the Yangtze River Delta region of China. However, with the rapid socio-economic development and the rise of urban agglomerations, the problem of deterioration in water quality in the Taihu Lake basin has become increasingly serious. Taihu Lake has become a "severely affected area" of eutrophication. For this reason, the state has carried out large-scale water environment governance in the Taihu Lake basin.

"Taihu Lake Water Environment Governance" refers to targeted, organized, and planned interventions and control activities for the water environment of Taihu Lake. This is to enhance the overall coordinated orderliness of the water environment and the socio-economic ecosystem of the basin, mitigate the adverse impact of socio-economic development on the natural environment of the basin, and prevent water environment problems from continuing to expand and intensify.

This case study mainly reveals the complex wholeness attributes of the socio-ecological colossal system of the Taihu Lake water environment and constructs a new understanding of the "human and water environment life community" in the Taihu Lake basin under the guidance of complex systems management thinking. It interprets the necessity, main content, and focus of the transformation in water environment governance in the Taihu Lake basin; refines the basic components and logic of the transformation in water environment governance in the Taihu Lake basin; and analyzes the fundamental points of the transformation in Taihu Lake water environment governance based on complex systems management. From a new perspective of thinking, it provides new momentum for the current reform in water environment governance in the Taihu Lake basin of our country.

14.2.2 Overview and Reflection on Taihu Lake Water Environment Governance

14.2.2.1 Overview of Taihu Lake Water Environment Governance and Real Issues

Since the implementation of the "Overall Plan for Water Environment Governance in the Taihu Lake Basin" in 2008, the water environment issues in the Taihu Lake Basin have generally been significantly alleviated and improved. The water quality of the water network in the Taihu Lake basin has noticeably enhanced, and the average concentrations of the main water quality indicators have all declined, with no occurrences of extensive lake flooding issues for several consecutive years. However, the total amount of water pollution emissions remains high, the ecological environment is heavily burdened, and the risk of outbreaks of extensive blue algae and even lake flooding still exists. Generally, the issues of fragility, repetitiveness, instability, and imbalance in the effects of Taihu Lake governance have always existed, and the water quality in the lake area shows more severe fluctuations (as shown in Fig. 14.4).

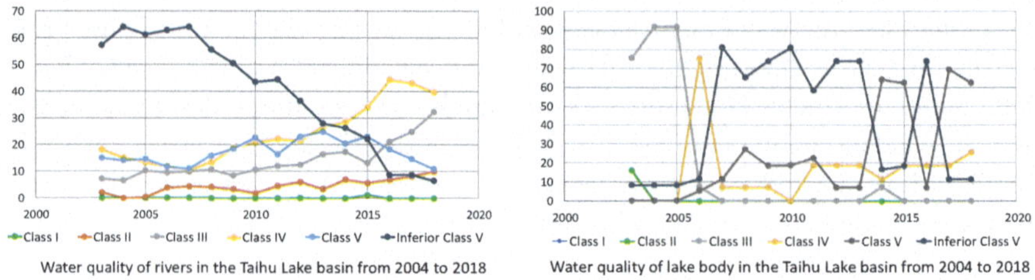

Fig. 14.4 Changes in water quality of rivers and lake body in the Taihu Lake Basin

These harsh realities make us soberly realize that although after more than ten years of intense management, due to the law of diminishing marginal effects, the management of Taihu Lake has entered a challenging stage of overcoming obstacles, the requirements for management are higher, the difficulty is greater, and the weaknesses in mechanism-oriented management as well as the lack and imbalance of mechanistic momentum in management are gradually becoming more prominent and severe.

The situations mentioned above indicate that the water environmental issues in the Taihu Lake basin are complex wholeness problems, intertwined and entangled with the weakening support of water resources, water shortage, damage to water ecology, and severe water pollution, which have arisen in the process of rapid industrialization and urbanization of the basin. They represent the current challenges in front of us regarding how to coordinate and sustain the Taihu Lake water environment—a complex mega-system compounded by regional socio-economic factors. The challenge of how to manage and control our thinking when dealing with the combined complex system of the natural environment and socio-economics is a significant challenge for us.

14.2.2.2 Reflection on Water Environment Governance in the Taihu Lake Basin

Several serious issues still exist in the water environment of Taihu Lake after many years of management, which demonstrates that the water environment of Taihu Lake is a complex system that is highly open and exists across multiple temporal and spatial scales, levels, and mechanisms. It suffers from a lack of systematic and mechanistic momentum for management. This necessitates our serious summarization and reflection, and below are several basic viewpoints from our summarization and reflection:

1. The surface issue of the lake water environment is a "lake disease," but in essence, it is a "societal disease." Humans are both the originators of the "societal disease" affecting the water environment and the healers tasked with treating this "societal disease." However, if people use simplistic thinking or general systematic thinking to view and deal with the complexity of Taihu Lake water environment and its governance, or if they solely use reductionist methods to address the

complex wholeness issues in water environment governance, it will not only be ineffective in solving the problems, but might even exacerbate them.

2. Governance subjects, in the face of the complex wholeness of managing Taihu Lake, often demonstrate inadequacies or even areas of cognitive blindness. At such times, if decisions are made solely based on traditional, limited knowledge and experience, the outcomes may contradict objective laws or fail to reveal deep patterns.
3. Whether it involves clean water, polluted water, or water management within the Taihu Lake water environment, all aspects encompass the values, property rights, allocation of these rights, transfers, and transactions associated with water. Thus, the water management mechanism should incorporate market mechanisms for water resources such as the confirmation of water assets, pricing, transactions, management costs, and compensation mechanisms, rather than solely relying on public administrative power and the rule of law; economic power and market forces are also needed to empower complex water environment governance.
4. Today, modern information technology is one of the significant technologies to enhance the capability of managing the Taihu Lake water environment. For instance, in the current management mechanism, the emergence of digitization of assets and asset digitization must employ modern information technology, such as enabling the internet of value.

The above reflections will guide us from an intuitive understanding of the real issues in Taihu Lake water environment governance to a higher scientific realm, contemplating the essential attributes of the Taihu Lake water environment and its governance and deeply recognizing the necessity of the paradigm shift in complex system thinking for the management of Taihu Lake water environment.

14.2.3 Paradigm Shift in Taihu Lake Governance Thinking Based on Complex Systems

14.2.3.1 Overview of Paradigm Shift

A scientific category or field, with a certain history and accumulation, generally has fundamental ideas, basic theories, methods, and a discourse system acknowledged by the scholarly community within the field, ensuring the uninterrupted academic vitality of this category or field. In this sense, we believe that such a category or field forms a paradigm of scientific research.

The term paradigm, as a scientific concept, can manifest its fundamental principles at three levels: ontology, epistemology, and methodology. It answers questions regarding the reality of existence, the relationship between the knower and the known, and issues related to the research method system, respectively, thereby regulating how research subjects view problems in the field and coordinating their behaviors.

When a scientific field encounters new situations, such as significant changes in object properties or significant shifts in views on the object, causing the original stable paradigm to lose or weaken its problem-solving capability, a new adaptive paradigm may appear; this is known as a paradigm shift. Since our new understanding of the Taihu Lake water environment and its governance is that it's a complex mega-system of the water environment—a regional socio-economic composite, a shift in the thinking paradigm in Taihu Lake water environment governance is inevitable.

14.2.3.2 Key Points of Paradigm Shift in Taihu Lake Governance Thinking

The paradigm of thinking in Taihu Lake water environment governance mainly refers to people's epistemological understanding of the entity, issues, and actions of the Taihu Lake water environment, i.e., how the Taihu Lake water environment is perceived at the level of rational thinking. For instance, what are the essential properties of the Taihu Lake water environment? How can we solve the problems of the Taihu Lake water environment to achieve unity of understanding and reality and ensure effective and sustainable management? Particularly, according to the current environmental status of Taihu Lake, we need to decide based on what principles to make changes in management; this is what is referred to as the paradigm shift in Taihu Lake's governance thinking.

Regardless of which element of Taihu Lake water environment governance is analyzed, the paradigm shift in Taihu Lake water environment governance thinking indicates that there has been a "qualitative" new thinking in people's ontology and epistemology of Taihu Lake water environment governance activities. The core of new thinking has formed the understanding that the Taihu Lake water environment is a complex system, and its governance is a complex system of management; that is, the complex wholeness is the essential property of Taihu Lake water environment governance.

Based on this, the following key points can be summarized regarding the paradigm shift in the complex system thinking of Taihu governance:

1. The management of the Taihu Lake water environment has evolved from simple systems to complex systems and then to socio-ecological complex mega-systems. The essential attribute of this mega-system is the complex wholeness of the socio-ecological environment.
2. In this paradigm shift, the essential value is "the human and water environment life community." Specifically, first, the Taihu Lake water environment itself is a community of life; second, the society of the Taihu Lake basin is another community of life; third, the society and water environment together form a larger community, making the Taihu Lake basin a complex system composed of multiple coupled communities; fourth, each community and the communities among themselves are interdependent, mutually beneficial, symbiotic, and

co-evolving. Disorderly development and reckless exploitation of the basin environment by humans will inevitably harm the sustainable development of human society itself and will ultimately lead to the decay or disintegration of the entire complex system.

3. On the surface, the "pollutants" in the water environment appear to primarily impact the lake water, but in reality, they are products of "negative impact" generated by the socio-economic system of the entire basin. Therefore, water environment governance must consider not only traditional mechanisms such as water quantity, water quality, biology, chemistry, and meteorology but also has a close logical connection with "pollutant" manufacturers, suppliers, logistics networks, water resource transactions, value conversions, and overall water body management activities. Hence, the management of the water environment is a treatment not only for "ecological diseases," but also for "societal diseases."
4. Besides allocating administrative public power and legal authority, water environment governance also needs elements of marketized allocation mechanisms and supply chain management that are in line with market mechanisms, such as mechanisms for cost accounting and compensation in water environment governance, affirming the property rights of water environment data assets, and effective circulation and transaction mechanisms. An open, shared, fair, and trusted circulation platform must be constructed in the water environment governance system. Otherwise, modern technologies like the Internet of Value, big data, and blockchain cannot be socialized in water environment governance. This implies that governance reform needs to incorporate thinking and mechanisms related to market, symbiosis, trust, value, transactions, and alliances.
5. The essence of the reform in Taihu Lake water governance is a comprehensive transformation and restructuring of human behaviors, values, preferences, interest views, and social development models. Therefore, governance strategies should include not only physical-level lake water conservation and improvement projects, but also the transformation of governance subjects' behavior patterns and concepts and the reorganization of a new symbiotic system for the lake water environment at the social level. Thus, the deep-seated ecological, societal, and economic symbiosis behind the "lake disease" should be constructed as the ontology, building new epistemologies and matching methodologies for the water environment scenario governance reform.

It is not difficult to see that, based on the complex systems management thinking, the essential connotation of the paradigm shift in the thinking of Taihu Lake water environment governance is to integrate the epistemology, methodology, and praxiology of Taihu Lake water environment governance into the trajectory of complex system thinking. This has significant guiding importance for improving the effectiveness and sustainability of Taihu Lake governance through the transformation of management thinking.

Until now, we have summed up the current status of Taihu Lake water environment governance and reflected on the limitations of current management thinking, leading to our paradigm shift in the complex system thinking of Taihu Lake water environment governance. The purpose of this paradigm shift is to construct a new and more effective management model, providing a logical starting point for the important topic of "Taihu Lake governance transformation based on complex systems management" discussed below.

14.2.4 *Taihu Lake Governance Transformation Based on Complex Systems Management*

14.2.4.1 Overview of Taihu Lake Water Environment Governance Transformation

The so-called transformation in Taihu Lake water environment governance refers to the practice and actions of governance entities following a paradigm shift in complex system thinking, aiming to solve the challenges existing in the current management system's philosophy, cognition, model, and technology. It is particularly directed toward deciphering the complex wholeness of water environment governance problems, making up for the deficiencies in complexity management, enhancing the ability to control complex entirety, and forming new momentum in management.

Compared to the paradigm shift in the complex system thinking of Taihu Lake governance, the transformation in Taihu Lake water environment governance is more about the "planning" and "execution" of management transformation activities at the practical level, aiming to turn the shift in management thinking into actual governance actions activities. Therefore, the transformation in Taihu Lake water environment governance is mainly about "what to do" and "how to do it" after the shift in thinking paradigm, referred to as Taihu Lake governance transformation. Theoretically speaking, Taihu Lake governance transformation is the practice of complex systems management in Taihu water environment governance or the actual implementation of Taihu Lake governance based on complex systems management.

14.2.4.2 Taihu Lake Governance Based on Complex Systems Management

Taihu Lake governance based on complex systems management has rich content and profound connotations. The basic paradigms of management, knowledge and acquisition paths in management, complexity analysis and solutions in management, generalized management objectives and new connotations in management objectives, and essential safety of lake water environment have all been discussed in previous chapters of this book, so they will not be repeated or detailed here. Below, a few new topics are used to articulate the new connotations of Taihu Lake governance based on complex systems management.

Topic One: Organizational Platform and Modernization of Capabilities in Taihu Lake Governance Transformation

From the perspective of the essence of human practical activities, the transformation in Taihu governance is to construct an artificial system with new water environment governance functions and a certain stable structure. For this, a perfected organizational model is needed to fulfill all related management functions. Typically, this organizational model is integrated from the following three subsystems:

1. Cognitive System: This system mainly aims to understand the complexity of water environment issues in the Taihu Lake basin, assess the degree of complexity and the problems arising from it, and constantly refine the basic paradigm of water environment governance: physical complex wholeness—system complex wholeness—management complex wholeness—formation of complexity management schemes.
2. Coordination System: This system is mainly used for the organizational design and coordination of Taihu Lake governance transformation, the refinement, and integration of management goals, and also includes the interfaces and coordination between the management process and each management subsystem.
3. Execution System: This system is mainly used for the specific operations of the governed system, focusing on the control and execution of specific management schemes.

These three subsystems constitute the organizational platform for Taihu Lake's governance transformation. The functions of the first two parts are not to directly or mainly provide specific methods and schemes for Taihu Lake governance issues but to provide the environment and conditions for forming methods and schemes. Therefore, compared to traditional management functions, their important function is to form modern capabilities to control management complexity. For example, based on the complex wholeness characteristics of management problems and with the aid of modern information technology, they form the excavation and analysis of various weak causality, long association information in the Taihu Lake water environment governance. Through the market allocation of data elements in Taihu Lake basin water environment governance, the orderly flow of data assets and the formation of value streams and value-added networks are realized. Then, under the perspective of complex wholeness, the acquisition and allocation methods of resources in Taihu Lake water environment governance are reconstructed, and on this basis, an environment providing credit passport services for Taihu Lake water environment governance, with environmental credit as the carrier, is constructed. Meanwhile, a modernized governance system driven by proactive government and efficient market duality is designed.

Topic Two: Market Allocation of Digital Assets in Taihu Lake Governance Within the Complex System

According to the paradigm shift of complex system thinking in Taihu Lake governance, the Taihu Lake water environment has natural, social, and economic attributes.

Therefore, the transformation in management needs to follow the laws of nature, social symbiosis, and the rules of value flow and market allocation of elements, requiring the injection of modern management principles such as market, symbiosis, mutual trust, valuation, transactions, and alliances into the current management model. For example, there is a need to research and resolve issues of isolated water environment information (data), difficult traceability, modifiability, and lack of credibility. It is necessary to change the existing centralized credit system based solely on administrative power, realize decentralized, distributed credit systems with multiple autonomous nodes participating, make water environment governance activities a sustainable model compliant with market mechanisms, realize the interconnection and free trading of water asset values, and upgrade the water environment network to a value internet.

Thus, research on the value in water environment governance and the market allocation of digital assets is needed, including the digitization of management elements, the assetization of digital resources, interconnection of digital assets, and the design of smart contracts to better serve the practice of water environment governance transformation. Here, the focus is to explore how to use complex wholeness scenario modeling technology to study the security audit of smart contracts for digital assets in water environment governance and the risk assessment, identification, and early warning of digital assets in water environment governance, providing technical support for effectively avoiding the risks of contract vulnerabilities and improving the robustness of digital asset applications.

Through the above transformative work, targeting the economic "shortcomings" in current water environment governance, a value interconnection and digital asset trading platform are constructed in the transformation of water environment governance.

Topic Three: Complexity Analysis Of long-Timescale Evolution Model of Water Environment

Complex systems management of Taihu Lake governance needs to strengthen the analysis of the complex wholeness of Taihu Lake governance issues, realizing the understanding of multi-scale dynamic evolution laws of Taihu Lake water environment, to further improve risk warning and effective prevention of major extreme water scenarios.

To achieve this, there needs to be a change in the current method of "big data + long-term intensive field measurement" used in Taihu Lake governance and information detection. This method is not only time-consuming and labor-intensive, but the data mainly comes from shallow water areas, lacking information from the underlying and historical traces of the water catchment. Thus, it is difficult to deeply understand the evolution laws of the water environment over long-time scales. The Nanjing Institute of Geography and Limnology of the Chinese Academy of Sciences has adopted the sediment record method. Through the component analysis of lake sediment, it has constructed a dynamic time-varying model of the evolutionary trajectory and rate of Taihu Lake biotic community on a century time scale. Combined with the basin climate—socio-economic information and the coupled model of human behavior and nature, it further accurately analyzes, simulates, and predicts relevant issues of Taihu Lake water environment governance on a century scale. For example, using this method, the analysis of the spatial and temporal differences and the dynamic analysis of driving factors in heavy metal pollution in water quality traced the sources of high pollution elements such as cadmium, nickel, and lead mainly to industrial point sources, atmospheric deposition, and agriculture and fisheries, thus improving the accuracy of determining the direction of water environment governance. Additionally, from the changes in the content of substances such as chlorophyll and carotenoids in the sediments, it can be inferred that the two eutrophication events of algae in 1950 and grasses in 1990 had severe impacts on Taihu Lake in the past hundred years. Such comprehensive scenarios and evolutionary models of the water environment fully embody the comprehensive integration method of complex systems management and play an important role in policy-making such as forecast warning and ecological restoration for Taihu Lake governance.

Topic Four: Ecological Compensation Mechanism for Taihu Lake Water Environment Based on GEP

Over the years, while people have been creating tremendous material wealth, the exploitation of Taihu Lake water resources has also accelerated, disrupting the original balance of the Taihu Lake water environment and leading to the degradation of high-quality water resources. To achieve sustainable water environment governance, it is natural to compensate for the resource value of the water environment as an ecosystem. Therefore, in designing the methods for transformation in the management of the Taihu Lake water environment, we first need to consider that the water environment system has been continually providing essential (resource-based) products and services in the process of human well-being and socio-economic development. These products and services are valuable, and their total value is referred to as the Gross Ecosystem Product (GEP).

The specific method for calculating the GEP involves breaking down the forms of products and services provided by the ecosystem, accounting for the physical or functional quantity of each decomposition item, and then calculating their respective value through certain methods, known as the pricing of ecological products and services. The sum of these is the total value of the ecosystem. This is actually a potential value of the ecosystem and needs to be monetarily assessed to convert it into actual economic value, i.e., the realization of the value of ecological products. Naturally, this requires strengthening the design of water resource value and market-oriented governance within the complex systems of Taihu Lake governance framework, especially the acceleration of the establishment of economic systems and operational mechanisms in the governance of Taihu Lake.

From the above, it can be seen that GEP accounting not only provides a basis for the realization of the value of Taihu Lake aquatic ecological products, but also provides some logical basis for how to implement ecological system value compensation. Constructing the ecological compensation market mechanism is planned to establish a basin-wide pollutant discharge right trading system and an exclusive financial platform for water environment governance in the Taihu Lake basin under the premise of improving the property rights system of aquatic ecological resources. Gradually realize the initiation of pollutant discharge right trading, mutual guarantees, and risk-sharing in water management, governance credit assessment, and water management taxation services.

Topic Five: Modeling Technology for Complex Wholeness Scenarios in a Water Environment

The concept of scenarios is pivotal in the design of Taihu Lake governance techniques within the framework of complex systems. This is because, holistically, the Taihu Lake scenario epitomizes the intricate systemic behavior of the water environment, necessitating the formulation of response strategies for potential scenarios. However, scenarios embody the evolution, multi-scale, emergence, and self-organizing attributes of the water environment, making their reconstruction or prediction challenging. Moreover, a realistic scenario is a representation of the water environment under specific contextual conditions. Thus, any scenario in the Taihu Lake water environment is not only complex, but also unique, rendering it "scarce" in holistic terms. We can only, based on a small number of valuable water environment scenario samples or clues, use computer systems as "soil," plant and cultivate a few realistic scenario concepts and clues as "seeds," letting them grow and eventually obtaining various different scenario "fruits." We then analyze and predict the knowledge and laws of water environment scenarios and their management from the dynamic evolution process formed by these "fruits" and the types and characteristics of these "fruits." We refer to this type of computer simulation method regarding scenario generation in water environment governance research as the scenario cultivation method. Its core essence is modeling technology about complex wholeness.

Scenario cultivation is a computer scenario reconstruction and prediction technology defined under "scenario space." It is based on the dynamics of "one" or "some" scenarios and uses predefined assumptions to embed water environment and management behaviors that share the same essence and dynamic mechanism into "scenario space," enriching our understanding of realistic water environment scenarios by embedding specific unique phenomenon and details into scenario space through context.

From a modeling process perspective, the scenario cultivation method can, in some sense, demonstrate past and present water environment scenarios using computer systems, conduct controllable, repeatable planting on the virtual "counterpart" of realistic water environment scenarios, and tell us about the "yesterday" and "today" of scenarios that have occurred and are occurring in the water environment through the results that grow out. It can also construct the possible "tomorrow" of non-real, virtual water environment scenarios on the computer, showing us the future scenario image of the Taihu Lake water environment—a socio-economic composite system. In this technology, predictions about extreme future scenarios are particularly important, as extreme scenarios have significant implications for the inherent safety and safety risks of complex water environments.

Topic Six: Comprehensive, Integrated Support Platform for Taihu Lake Water Environment Governance

Considering the complex wholeness of the Taihu Lake water governance within a complex system, there is a proposition to develop a water environment governance support platform within the framework of governance reform. The governance support platform is a computer system that includes theory, experimentation, mathematical models, computer simulation, intelligent tools, and integrated human–machine systems; this system assists in conducting related governance issue analysis, governance plan design, governance issue scenario generation, governance process deduction, and governance performance evaluation to enhance the comprehensive capability and level of the governing subjects. The core technology of this platform is a complex wholeness modeling technology and analytical tool driven collaboratively by big data, blockchain, and artificial intelligence, capable of providing a scenario language and modeling symbol system for governance activities.

Based on the basic functions provided by the platform, governing subjects can make the platform's functions selectable, updatable, and extendable according to different governance issue needs. Meanwhile, special research on water environment governance, such as early warning and emergency measures for public incidents like the Taihu Lake blue-green algae outbreak and drinking water safety, can be carried out with the help of the platform.

14.2.5 Conclusion

The governance of the Taihu Lake water environment is a pivotal strategic task in the ongoing construction of ecological civilization in China's Yangtze River Delta. This case study, oriented by the in-depth real scenarios of Taihu Lake water environment governance, confronts new challenges. It integrates new governance thoughts, theoretical frameworks, social rules, and technological innovations, aiming to enhance Taihu Lake's governance capability and improve governance performance, thus offering new thinking paradigms and practical empowerment.

Under the guidance of complex system science principles, this section reveals the complex wholeness properties of the Taihu Lake water environment as a socio-ecological giant complex system. It presents a new understanding of the governance goal concerning the "coexistence of human and water environment life" in the Taihu Lake basin and initiates research on the paradigm shift and governance reform in Taihu Lake based on this understanding.

The new model for Taihu Lake governance reform, based on complex systems management proposed in this section, not only demonstrates the practical role of complex systems management as a new domain in management science in contemporary Taihu Lake water environment governance research, but also signifies the substantial practical importance of complex systems management in enhancing the top-level design and overall strategizing of ecological environment governance.

14.3 Example 3: Big Data-Driven Diffusive Complex System Scenario Modeling Technology

Due to the identity of properties between big data and scenarios, meaning both big data and scenarios share common essential characteristics and logical starting points, it is proposed to use "big data-driven" as a transformation model or a kind of mapping between data space and the real world, driven by the goal of scenario modeling. This mapping can transform a big data set into a "parallel mirror image" of a certain scenario, thereby externalizing the scenario value inherent in big data.

This case study focuses on diffusive complex scenarios in complex systems management activities as the object of modeling, introducing how to realize scenario modeling according to the basic principles and technical route of big data-driven complex system scenario modeling. It further elaborates through the modeling of diffusive scenarios of river pollution in the Taihu Lake basin, conducting a comprehensive scenario analysis of the dynamic evolution process of water environment emergent risks under different emergency measures.

14.3.1 Overview of Diffusive Scenarios

Diffusion is a scientific term used to describe the form of material movement, and based on the differences in diffusion medium, mechanism, and path, it can be divided into several different types, each with different meanings in various academic fields. For example, biological diffusion refers to the process where biological individuals or their disseminating bodies (such as seeds) spread from one place to another. Biological entities expand their habitats through diffusion to acquire more resources and increase their population sizes. In physics, diffusion refers to the phenomenon where material molecules move from high-concentration areas to low-concentration ones until evenly distributed, essentially being the process of material molecules transporting mass during thermal motion. In the field of communication studies, diffusion is a kind of social change, usually manifested as the transmission of information between individuals in society, which can then induce changes in the structure and function of the social system. In essence, diffusion describes the process of material, energy, information, or even individuals migrating, transferring, and spreading between different spatial domains within a system.

A scenario is a coherent "story" unfolded with time, space, and logical order, consisting of entities, events, subjects, objects, relationships, cause-and-effects, changes, backgrounds, and plots. Scenario diffusion is a process of the migration, transfer, and spread of a "story" containing the diffusive states formed by the interaction of various elements in the "story" based on various mechanisms, the ordered change of diffusive states, and possible evolutionary paths. Scenario diffusion is very common in complex systems management activities, for example:

(1) Disease Diffusion Scenario. When a disease is contagious, it can spread among people through interpersonal contact, air transmission, water transmission, food transmission, etc. When the disease spreads to a sufficient number of people, it can lead to disease outbreaks and even trigger large-scale epidemics. The disease diffusion scenario describes the process and situation of disease transmission and diffusion among the population, including the state and evolution process of disease diffusion manifested under the combined effect of various elements such as sources of infection, transmission routes, population characteristics, external environment, and preventive and control measures.
(2) Scenario of Crowd Behavior Diffusion. Due to the social nature of humans and influenced by social norms, social identifications, and social connections, when an individual in a crowd adopts a certain behavior or holds a certain viewpoint, it might be imitated or accepted by others around, gradually forming a kind of collective behavior or way of thinking, such as the affection for new products, the pursuit of public figures, or the agreement with a certain public opinion, causing a diffusion phenomenon within a certain time and space range. The scenario of crowd behavior diffusion describes the overall transmission or popularization of a specific behavior or concept in a crowd over a certain period of time, including the features and changes of elements such as individual numbers, group sizes, relationship strength, and environmental impact.
(3) Population Migration Diffusion Scenario. Influenced by economic, political, environmental, and other factors, like political turmoil or wars, populations may migrate from one region to another on a large scale, causing macro changes in population distribution among different regions. In this migration process, populations, materials, information, etc., will diffuse to new areas along with the migrating crowds. Massive population migration can provide abundant labor to the destination, but it may also lead to changes in social structures and cultural conflicts. This scenario describes the process of such population movement and the induced social changes.
(4) Disaster Impact Diffusion Scenario. Caused by natural disasters such as earthquakes, floods, and pandemics, as well as man-made disasters, public health incidents, etc., the impact of a disaster often extends beyond the immediate disaster site due to geographical location, climate conditions, socio-economic levels, and can have different degrees of diffusion affecting a wider range of people and objects. This scenario describes the process of disaster impact diffusion in time and space after a disaster occurs, including the evolution of the characteristics of disaster diffusion under the influence of disaster types, disaster-bearing entities, affected populations and migration, and emergency measures.
(5) Financial Risk Diffusion Scenario. When a problem or risk occurs in a financial institution, market, or product within the financial market, due to the interconnected and interdependent relationships between various financial instruments and institutions within the financial market, financial risk can diffuse to other areas, even affecting the entire socio-economic system, leading to financial market instability, stock market declines, economic recessions, etc. The financial risk diffusion scenario describes the process of the evolution of the states of

various elements in the financial market when financial risk diffusion triggers a chain reaction in financial institutions, markets, or products, and the financial market operates abnormally and unstably.

(6) Supply Chain Risk Diffusion Scenario. Similar to the financial risk diffusion scenario, the supply chain risk diffusion scenario describes the process where a problem or risk event in one link or part of the supply chain (such as supplier bankruptcy, natural disasters, logistics issues, quality issues, political and legal risks, etc.) diffuses risk throughout the supply chain and even affects the entire market. It represents the evolution of the overall scenario in the supply chain network under the combined action of various factors such as management level, importance, orderliness, ability to withstand risk interference, ability to eliminate risk interference, the probability of risk interference, degree, and time among various corporate entities in the supply chain network.

While the above-mentioned diffusion scenarios exhibit distinct morphological characteristics and dynamic evolutionary paths in complex systems management activities due to their adherence to different diffusion mechanisms, their formation and development share a basic commonality. Specifically, they manifest as follows:

(1) Human agents, as the core elements in diffusion scenarios, are rational and irrational, active and passive, subjective and objective, and are a unity of self-organization and other-organization in scenario diffusion. With high intelligence, autonomy, purposiveness, and adaptive ability to the external environment, humans exhibit different behaviors and reactions under different scenarios. The complexity exhibited by humans will influence the diffusion process through various behaviors, further contributing to the complexity of the evolution of diffusion scenarios. For example, in the diffusion process of public opinion on social networks, people may consider various factors such as the credibility of information, dissemination purposes, value preferences, and the structure of social circles. The interweaving of these factors makes the path, speed, and form of public opinion diffusion complex and unpredictable.
(2) In diffusion scenarios, various scenario elements form certain hierarchies or structures, and this structure also constantly changes with the evolution of the scenario. At the same time, the modes of interconnection and causal relationships between elements have their distinct internal mechanisms. The complexity of these connections also leads to the complexity of the dynamic evolution of diffusion scenarios. For instance, the multilevel, diversified, and cross-cutting relationships between suppliers, manufacturers, customers, and partners in the supply chain create various forms of competition and cooperation, all of which can exert certain driving forces or constraints on the scenario diffusion in some way.
(3) Diffusion scenarios represent a coherent dynamic evolutionary process of diffusion phenomenon in complex systems management activities. Due to the openness of complex systems, the evolution of diffusion phenomena is inevitably influenced by changes in the external environment, and the uncertainty of changes in the external environment further intensifies the uncertainty of the

dynamic evolution path of diffusion scenarios. For example, when earthquakes, floods, hurricanes, and other natural disasters occur, the changes in climate and terrain often have a high degree of unpredictability and instability, leading to further uncertainty in the evolution of disaster diffusion scenarios.

In conclusion, due to the numerous and diversified elements, complex element associations and structures, and uncertain evolutionary paths in diffusion-type scenarios in complex systems management activities, diffusion scenarios exhibit distinct complex wholeness.

14.3.2 Paradigm of Diffusion Scenario Modeling Driven by Big Data

The analysis above indicates that, in reality, people initially develop a certain sense of perception for activities or issues with diffusion characteristics in complex system management. Later, due to certain purposes, there is a demand to use contextual language and modeling symbols to abstractly describe diffusion scenarios as a whole. This is known as diffusion scenario modeling. Different scenarios follow varying dynamic mechanisms; before modeling scenarios, it is crucial to determine the driving tools that can reflect the core dynamic mechanism of the scenario and are suitable for modeling it [1]. The essence of the scenario diffusion process can be seen as the diffusion of matter, energy, information, etc., between different nodes in the system, where the key nodes are the diffusion subjects based on a certain scale and coarse granularity of the research problem. In various diffusion scenarios, the various diffusion, evolution, and emergence phenomena occurring at the overall level are driven by the mutual conduction and interaction between diffusion subjects at the individual level. The diffusion scenario model is a symbolic system describing the comprehensive effects and overall patterns in this diffusion phenomenon.

For some diffusion scenarios, a lot of studies mainly explore the diffusion phenomena at the holistic level through the research paradigm of epidemiology, believing that the mutual conduction mechanism between diffusion subjects is similar to the transmission mechanism of infectious diseases. Such diffusion scenario models are usually constructed using methods like differential equations, difference equations, and Monte Carlo simulations to describe the dynamic evolution and trend prediction in the diffusion process. For example, common disease propagation models, like the SIR model, establish differential equations to describe the transformation relationships among susceptible, infected, and removed populations in a community, predicting the process and trends of disease propagation. Complex contagion theory further proposes that, compared to simple contagion mechanisms like infectious diseases, the contagion mechanisms of information with high risks and high uncertainty among subjects are more complex and require stimulation from multiple different individuals to occur.

Common models for new product market diffusion, like the Bass model, use equations to describe the transformation relationships between innovators and imitators, predicting the diffusion process and trends of new products on the market. The spread of public opinion is influenced by the attributes of the subjects, the social characteristics between subjects, and the characteristics of the content of public opinion. For the deeper complexity exhibited in such scenarios, more parameters and more complex equations are generally introduced to describe the entire diffusion phenomenon.

While the ways of expressing scenarios are diverse, the diffusion phenomenon is driven by the mutual conduction between each diffusion entity at the individual level, so using networks to express scenarios is the most common. Expressing through networks is advantageous for providing a clear and intuitive overview to depict the current state of the diffusion scenario, its future evolution trends, the factors affecting the diffusion development, and the forces driving these factors. Therefore, when constructing diffusion scenario models, it is usually necessary to first establish the mutual conduction mechanism between each entity in the diffusion scenario and the network topology composed of the diffusion entities. The several types of diffusion scenario models listed above fundamentally rely on the academic thought of using well-structured equations to describe the inherent mechanisms behind diffusion scenarios. This method requires pre-setting the quantified relationships of causality between conduction rules and influencing factors during the model setting stage. However, this thought is not very effective for describing diffusion scenarios in complex systems management activities. The main reasons are as follows:

(1) Affected by factors such as the adaptability of subjects, the complexity of associations, and environmental uncertainty, diffusion-type scenarios in complex systems management activities possess deep uncertainty and complex wholeness, causing the overall structure of the scenario to have structural, semi-structural, and non-structural components. Therefore, a singular causal mechanistic model cannot fully reflect the complexity of diffusion-type scenarios.
(2) Many mechanistic models are built based on a specific context and satisfying specific assumptions. Due to the unique contextual characteristics of scenarios, the constructed mechanistic models often exhibit insufficiency in adaptability and universality for other scenarios.
(3) Such models usually use equations to describe the mutual conduction mechanism between different diffusion subjects and do not consider the complex relationships of the subjects at different scales, failing to depict the formation and development of scenarios, especially the complex evolution, such as mutations and emergencies.

Considering the deficiencies in traditional modeling thoughts and the complex wholeness of complex systems management scenarios, data-driven scenario modeling has multiple advantages for diffusion-type scenario modeling due to the theoretical unity of big data and scenarios. For instance, it can deeply mine various implicit relationships between diffusion subjects, continually refine models through iterations, and acquire various estimates for diffusion scenarios.

As an example of big data-driven complex system scenario modeling, this instance mainly studies the basic principles, technical routes, and modeling processes of big data-driven diffusion-type complex wholeness scenario modeling.

(1) Conceptualization and Structuralization Conversion of Diffusion-Type Scenarios

According to the principle of scenario orientation, research on complex systems management scientific problems should delve deeper into considering, contemplating, and analyzing issues within their respective scenarios. This way, the extracted attributes, laws, and solutions of the problem will be closest to the real situation of the actual problem. Therefore, before modeling diffusion-type scenarios, a clear physical form of the real scenario should be oriented, and the complex wholeness of this real scenario should be analyzed, defining the specific connotation and physical boundaries of the real scenario according to the research purpose. Following the principle of scenario complex wholeness, this part specifically includes determining the research object, research perspective, research purpose, research questions, and the spatiotemporal characteristics, environment, and boundaries of the research's real object. For instance, to establish a public opinion diffusion model on social networks, one can determine the diffusion evolution problem of different characteristic public opinions in social networks from the social relation perspective, clarify the spatiotemporal boundaries of the study, determine specific time periods and social groups, and establish a clear understanding of the elements of diffusion media, channels, subject attitudes, and behaviors.

The above is the reflection of diffusion-type scenarios in the real physical world. Scenario modeling needs to further abstract real scenarios at the conceptual level to realize the conceptualization of scenarios, mainly describing real scenarios through direct perception and indirect cognition of real scenarios using abstraction, generalization, and logical deduction. This includes inducing, abstracting, or generalizing the elements, associations, behaviors, structures, and various actions of diffusion-type scenarios, including the overall level of scenarios, subject and organization levels, and basic element levels. For example, the spread of information in social networks can be abstracted as a result of the interaction of network structure and node behavior, and the spread of infectious diseases can be abstracted as a result of the interaction of sources of infection, susceptible populations, and transmission routes.

The conceptualized scenario contains certain semi-structured and non-structured parts; for example, the emotions, attitudes, and preferences of information diffusion subjects toward information cannot be directly used for computer calculations and need to be further expressed and described in symbols, rules, formulas, and logical languages along with those already structured scenarios in the conceptual model, forming structured scenarios, realizing the structuralization of diffusion scenarios. For instance, the structuralization of information diffusion scenarios in social networks includes the symbolization of scenario elements such as the characteristics, properties, attitudes, behaviors, etc., of diffusion subjects, the association relationships between diffusion subjects, the topological structure between diffusion subjects, the rules triggering information dissemination behavior, and the changing characteristics of the external environment.

(2) Big Data Transformation of Diffusion-Type Scenario Datasets

Based on the theoretical unity of big data and scenarios, big data serves as a digital snapshot that best reflects the attributes of scenarios. Therefore, to construct models that are more aligned with real-world diffusion scenarios, it is necessary to collect information (data) sets that describe the dynamic evolution process of diffusion scenarios as comprehensively as possible. For example, when building a model for sentiment diffusion in social networks, various forms of textual data should be collected, including social media, news reports, forum posts, etc. This includes user data with details about attributes, behavior, social relationships, event data with information about the cause, development, and impact of events, and information data about the social context, weather, intervention measures, and more. Data acquisition is mainly achieved through web scraping. The construction of infectious disease spread models also requires the support of a large amount of relevant data, including virus data reflecting features such as virus types, infectiousness, and mutation situations; epidemic report data recording the number of infections, deaths, recoveries, case distribution areas, times, etc.; population data on the number of people, distribution, age structure, mobility, etc.; medical resource data including the number of hospitals, the number of beds, the number of medical staff, etc., and data recording vaccination, among others. The main sources for acquiring the above data include official reports, social media, research institutions, and public data platforms. Integrating all kinds of collected data forms the initial dataset for diffusion scenario modeling.

In the initial data set, various types of real-world data are collected initially, not necessarily for the purpose of scenario modeling. Consequently, the initial data may have various "defects," such as data gaps, inconsistent standards, semantic differences, and significant impurities. Therefore, it is not advisable to use the initial data set directly for diffusion scenario modeling. According to the principle of scenario data tolerance, the initial data set needs to be processed into a scenario big data set that meets the basic requirements of big data-driven scenario modeling. This includes:

1. Data Quality Improvement: Removing duplicate records, addressing missing values through deletion, imputation, or cross-referencing with other data sources, dealing with outliers or extreme values through deletion, replacement, or transformation, and eliminating noise from data through various filtering or correction methods.
2. Data Dimensionality Reduction: Selecting and extracting the most important features and trends from the data through feature selection, extraction, or other methods to reduce data complexity, storage, and computational costs and avoid interference from redundant information and noise in data analysis and modeling.
3. Data Reduction: Sampling or subsampling the data to reduce its size, improving computational efficiency.
4. Data Transformation: Applying various transformations, statistics, and logical operations to the data, such as standardization, logarithmic transformation,

normalization, etc. Converting data from one format to another as needed to meet the requirements of data analysis and modeling.

5. Data Aggregation: Aggregating data according to certain criteria to facilitate summarization and analysis. For example, aggregating data based on time, location, category, etc.

(3) Refinement of Diffusion Core Scenario Data

Based on the principle of complexity degradation, it is necessary to extract a finite number of element data from the "infinite" diffusion scenario data set that best reflects and retains the essential attributes, overall structure, spatiotemporal distribution, and evolutionary patterns of diffusion scenarios, along with their interrelated information, to construct the basic framework and evolutionary rules of the diffusion scenario model. These diffusion core scenario data are the core scenario data in the process of diffusion scenario modeling, embodying the essential characteristics of diffusion scenarios and core knowledge about them. The importance of diffusion core scenario data lies in their role in determining the fundamental dynamics and essential attributes within the diffusion scenario. These data are referred to as the core scenario data for diffusion scenarios.

The dynamic processes of most diffusion scenarios in complex systems management activities can be described as the generation, spread, and interaction of substances, energy, and information among different subjects under specific rules, constraints, and environmental conditions. In other words, the dynamic mechanism of diffusion scenario models is mainly influenced by the essential characteristics of the diffusion source, subject characteristics, topological structure, and external environment. Therefore, the core scenario data for diffusion scenarios mainly consists of data sets that influence the dynamic mechanism of this kind. For example, in the context of information diffusion in social networks, factors such as information attributes (credibility, controversy, sentiment, information type, etc.), subject attributes (personality, attitude, behavior habits, identity, etc.), network topology (subject relationships, network density, centrality, in-degree and out-degree of subjects, etc.), and network environment (inclusive, open, closed, etc.) all influence the diffusion process. Therefore, the refinement of core scenario data for information diffusion in social networks should include big data on individual subject information, historical behavior data, and textual data of diffusion information after data processing.

(4) Cultivation of Diffusion-Based Scenarios

According to the principle of data scenario regularization, the latent value of scenario information carried by core diffusion scenario data should be systematically integrated and aggregated across multiple dimensions, such as time and space. The modeling of diffusion scenarios is the core method and means to achieve this process. It mainly involves abstracting the diffusion scenarios using a comprehensive abstract symbol system, constructing a model framework for fundamental activities and processes within the diffusion scenarios, and modeling the diffusion scenarios. In essence, modeling diffusion scenarios involves creating an artificial virtual scenario model as a parallel mirror image of a specific real-world diffusion scenario while preserving its complex overall attributes. Through the study of the diffusion scenario model, conclusions or predictions applicable to real-world scenarios can be derived.

Under the drive of big data, the modeling of diffusion scenarios mainly involves establishing different data models for the changes in the state of different subjects at a specific scale based on the problem scale. Each data model can simulate the coherent evolution and path of the subject's state. Then, various data models are integrated based on temporal and spatial relationships to achieve the reconstruction, simulation, or prediction of the diffusion phenomenon generated by the interactions between subjects at the overall level. After continuous cultivation in this manner, the constructed diffusion scenario model can give rise to new diffusion scenarios.

In the scenario cultivation phase, the key to constructing a diffusion scenario model lies in designing algorithms capable of simulating the evolution patterns of the subject's state within diffusion scenarios. For example, in the context of information diffusion in social networks, a decision tree model can be established to simulate the decision-making process of whether a subject will forward a particular piece of information. The decision rules of the decision tree are derived from the historical decision-making patterns of subjects found in the core scenario data. Furthermore, based on the social relationships between diffusion subjects, an integrated model with a certain topological structure can be formed to simulate the diffusion process of information across the entire social network. Additionally, if the modeling subjects have a comprehensive and profound understanding of the underlying dynamics of the diffusion scenario, a purely mechanistic model can be established to describe the state change process of each diffusion subject. Some parameters in the model can be determined through the analysis of core scenario data.

(5) Verification of Diffusion Scenario Models

Based on the principle of feedback learning in scenario modeling, it is necessary, during the scenario cultivation process, to functionally test the constructed diffusion scenario model on new datasets and adjust the model according to the consistency of model input, model compatibility, and the credibility of result output, iteratively creating a diffusion scenario model satisfactory in terms of validity and error control. The verification of the diffusion scenario model mainly consists of two parts: verification of the individual unit modules of diffusion entities and verification of the integrated model, each part being evaluated based on different criteria to assess the model's ability to replicate the real scenario. For instance, when using a neural network model to simulate whether supply chain companies will incur risks, the model adjusts the parameters in it by continuously learning from the dataset to minimize the gap between simulation results and the real scenario. The feedback learning process uses the model's predictive accuracy as a proxy indicator of model performance, determining the time to stop feedback learning based on specific accuracy requirements. For the verification of the integrated model, depending on the different research questions, for instance, when focusing on the range of risk diffusion, the accuracy of specific diffusion entities' state prediction (such as entities at the end of diffusion) can be used to assess the performance of the integrated model; when concentrating on the overall diffusion and evolution of risk, the average accuracy of all diffusion entities' state prediction can serve as an indicator of integrated model performance evaluation.

Given all the key points of the above paradigms, the value inherent in the data describing diffusion scenarios can be effectively and steadily transformed into real value constituting the overall form of the diffusion scenario.

14.3.3 Case Study: Prediction and Management of Sudden Diffusion Risks in Rivers

Taihu Lake, the third-largest freshwater lake in China and a significant water source has long faced prominent and sensitive water environmental issues. Addressing water environmental management in the Tai Lake basin presents a complex systems management problem characterized by complex integrality [2]. Among these, rivers, as major entities on a basin scale, play a mediator role in transferring surface runoff into the lake, making the construction of river pollution diffusion scenario models significantly meaningful for early warnings of sudden pollution in water environmental management. The essence of river pollution diffusion scenarios is the entire activities and processes where substances in the waterbody migrate and diffuse along the river network under the combined effects of hydrological systems, biological systems, natural environments, and riverside socio-economic environments. The entire activity and process comply with the basic settings of diffusion scenarios, making it suitable for diffusion scenario modeling processes using big data. Additionally, various IoT monitoring devices nowadays have collected abundant data related to water quality, providing a data foundation for unveiling hidden relationships in conduction mechanisms and river pollution diffusion processes. Thus, in this section, big data-driven diffusion scenario modeling is taken as a typical case to conduct research on the prediction and management of sudden risks in rivers.

The specific driving program and process are shown in Fig. 14.5. Through the modeling of river pollution diffusion scenarios driven by big data, it reconstructs, simulates, and predicts river pollution diffusion scenarios in a comprehensive and orderly direction and further carries out a full scenario analysis of the dynamic evolution process of sudden risks under different emergency measures. The main tasks in different phases are as follows:

14.3.3.1 Realistic Scenario Analysis

At this stage, based on the requirements of the research problem, a specific real scenario is defined, the goals for scenario modeling are clarified, and a basic understanding of the real scenario is formed. This example primarily studies the evolution process of sudden (pollution) risk in rivers and the issues related to various emergency measures to respond to pollution diffusion. The specific real scenario mainly

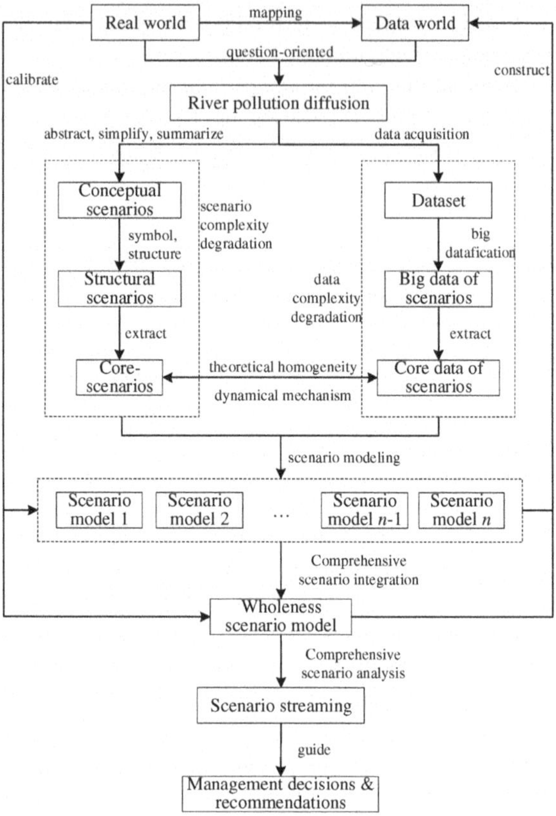

Fig. 14.5 Analytical Process for the Study of Prediction and Control of Sudden Risk of River Pollution

manifests as the pollution diffusion scenario in Tai Lake rivers, aiming to reconstruct historical, mirror actual, and predict future pollution diffusion scenarios in different states in the data space.

The essence of the river pollution diffusion scenario is the evolution result of materials in the water body under the joint action of the hydrodynamic system, biological system, natural environment, and riverside socio-economic environment. Due to the complexity of the inherent dynamics of this evolution, it has the characteristics of complex wholeness. Based on the perception of real scenarios, in physical space, it is confined to sections of the river with monitoring stations in the Tai Lake inflow area.

14.3.3.2 Construction of Conceptual Scenarios

At this stage, the real scenarios of river pollution diffusion are primarily described using abstract summarization and logical deduction, meaning that elements, associations, behaviors, structures, and functions of diffusion scenarios are inductively, abstractly, or summarily expressed. Conceptual scenarios can be manifested as conceptual models that can be described using qualitative and quantitative methods. The conceptual scenario in this example mainly involves the spatial distribution of river networks, the course of rivers, the spatial location of monitoring stations, types of monitoring, types of pollution, spatial distribution characteristics of pollution, and the natural-economic-social environment around the rivers. Scenario elements include enterprises, rivers, populations, and climate, with most pollution diffusion scenario information described in structured tabular data. The spatial distribution of river networks is mainly described in the form of network topology structures. The course of rivers is described by the directions of entry and exit between network nodes. The description of the spatial location of monitoring stations is based on images abstracted by the modeling subjects in cognitive space, further forming a display in the data space.

14.3.3.3 Construction of Data Scenarios

The construction of data scenarios initially involves data collection based on the attributes of the elements in conceptual scenarios. Given the vast amount of data elements associated with the process scenarios of river pollution diffusion, it's crucial to encompass all relevant information from real scenarios during data acquisition, such as attributes, characterizations, and traces of data. In this example, the primary raw datasets acquired mainly involve river section water quality monitoring data, hydrological monitoring data, meteorological monitoring data, elevation data, spatial distribution data of water systems, spatial coordinate data of monitoring sections, and environmental status bulletins of provinces and cities in the Taihu Lake basin.

Additionally, the collected dataset needs further big data processing. In this example, the missing and abnormal values in structured monitoring data can be supplemented and corrected by consulting other data sources or following certain rules; unsuitable and irrelevant monitoring data are removed; datasets with the same frequency are extracted from the monitoring data with different frequencies; semi-structured elevation data needs to be converted into structured tabular data that includes the elevation corresponding to each grid; and text data are converted into codes with unique identifiers, and the dataset is standardized, among other operations. It is important to emphasize that the big data processing at this stage not only includes type conversion of semi-structured and unstructured data, etc., but also involves data merging and data dimensionality reduction, generating new datasets not included in the original datasets. For example, merging spatial coordinate data of monitoring sections, spatial distribution data of water systems, and elevation data, and then reducing their dimensions based on their spatial relationships to form semi-structured network structure data of diffusion subjects at the primary river level, where the diffusion subjects are the river section areas where the monitoring sections are located.

14.3.3.4 Construction of Core Scenarios

Given the complex and holistic features of river pollution diffusion scenarios, directly achieving the modeling of the entire scenario is extremely challenging. This requires a moderated degradation of the entire process of river pollution diffusion and then wholeness restoration using reductionism methods. Based on the principle of complexity degradation in scenarios, the main idea of this stage is to extract a finite number of scenario elements and their interconnections from the "infinite class" scenario element space, which best reflects and preserves the essential attributes, overall structure, spatiotemporal distribution, and evolutionary laws of the scenario, as the most basic framework and behavioral rules for scenario modeling. It specifically consists of the following two steps:

1. Core Scenario Analysis.

The core scenario of river pollution diffusion is a type of scenario that reflects the basic dynamic mechanism and essential attributes of river pollution diffusion. As previous analysis indicated, the change in river water quality is a complex holistic embodiment under the joint action of the hydrodynamic system, biological system, natural environment, and socio-economic environment of the basin. However, essentially it can be viewed as a diffusion scenario model. The core scenario in this example is the process of a substance in the river continuously migrating in the network composed of water systems until it diffuses, including the interaction relationships and processes between various scenario elements. Each subject in the core scenario transfers material according to its own change rules, and due to the uniqueness of the internal and external environments in which each subject is situated, the change rules of different water systems and river sections are heterogeneous. It is also necessary to degrade the core scenario into multiple sub-core scenarios. Each individual sub-core scenario includes a monitoring section and all upstream monitoring sections directly spatially associated with it to form a river section, representing the material conduction evolution process between these river sections.

For instance, in this conceptual diagram represented by Fig. 14.6, triangles represent monitoring sections, and the general direction of water flow is consistent with the direction of the arrows. At this time, the entire river can be divided into four sub-core scenario areas: A, B, C, and D. The number of diffusion subjects included in each sub-core scenario is different. For example, in D, the river section areas where the two monitoring sections are located act as two separate diffusion subjects.

2. Refining the Essential Element Set of Core Scenarios.

The refined set of essential elements of the scenario kernel should be able to include the complex wholeness form of the core scenarios and also be able to represent the complex evolutionary form and path of the core scenarios. After refining, the set of essential elements of the core scenarios can be used directly for subsequent scenario modeling. In this example, various IoT monitoring devices have tracked and measured the process of river pollution diffusion based on human

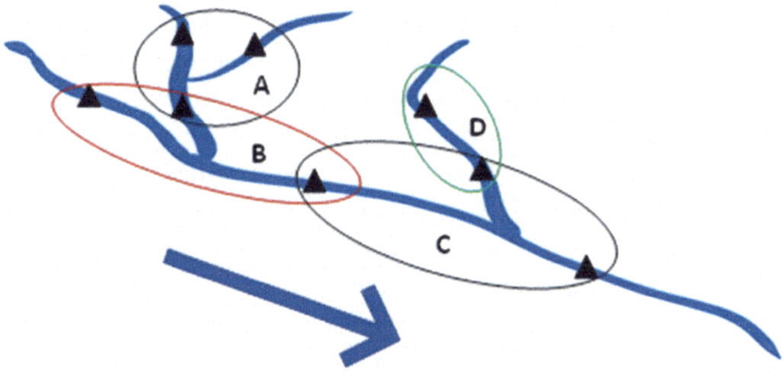

Fig. 14.6 Conceptual Diagram of River Core Scenario Decomposition

activity and behavior, as well as its "trace" in data space. Among them, the simultaneous frequency water quality indicator element dataset of each monitoring section formed by big data reflects its overall characteristics and evolution trend of dynamic mechanics, implying various complex wholeness scenarios in river pollution diffusion. The simultaneous frequency water quality indicator element data of each monitoring section serve as the essential element set of the river pollution diffusion scenario, representing the temporal dimension of this core scenario. In addition, the dynamic mechanics of the river pollution diffusion scenario are also reflected in the spatial hierarchical relationships between different monitoring sections, which cannot be reflected by the water quality indicator element data. Therefore, it is also necessary to use the big data-based monitoring section hierarchical network structure element data as the spatial dimension representation of the core scenario. It needs to be specially explained that the big data set "innately" contains various complex wholeness information in the scenario, but if there is a lack of understanding of the dynamic mechanics of the entire core scenario, blindly selecting the essential element set of the core scenario and mining the implicit relevance between data sets may violate the objective relationships and laws in the scenario in fact. For example, in this example, the water quality of two sections upstream and downstream of the river may have a significant correlation, but generally, the material downstream will not migrate upstream and affect the water quality upstream.

14.3.3.5 Computational Scenario Construction and Verification

In this example, based on the core scenario analysis and refinement in the previous stage, we need to perform the reductive deduction and wholeness integration on the rivers in the core scenario piece by piece, section by section, substance by substance, and spatiotemporal by spatiotemporal. Firstly, based on the essential element set of the core scenarios, we need to build a scenario model for each substance diffusion scenario in each monitoring section. Considering the nonlinear complex relationships in the changes of substances in rivers, constructing neural networks is an effective way to simulate and evaluate nonlinear complex systems. It has features like nonlinear mapping and adaptive learning and has extensive applications in simulating and predicting water environments. Moreover, considering that water quality data usually exist in the form of multivariate time series, the data at a certain moment has a strong correlation with previous data. To consider the temporal information contained in water quality data, we use Long Short-Term Memory (LSTM) neural network, based on deep learning, to construct scenario models for each substance diffusion scenario in the downstream monitoring section of each sub-kernel scenario separately. The scenario modeling stage is mainly divided into model training and model verification, interacting and iterating between training and verification until the model is satisfactory. We use scenario modeling to examine the evolution of Chemical Oxygen Demand (COD) and Total Phosphorus (TP) at Zhihu Port, Binhu District, Taihu Lake, as an example. The specific tasks are as follows:

1. Model Training:

The model framework constructed for scenario modeling comprises a 4-layer network structure (as shown in Fig. 14.7), including 1 input layer, 2 hidden layers, and 1 output layer. The input layer processes the input data into the data dimensions required by the hidden layers, where the hidden layers consist of LSTM units, and the output layer is a fully connected layer providing output results. The output space dimensions for the two hidden layers are set to 36 and 18, respectively. The Adam optimization algorithm is used for weight updates during LSTM backpropagation, and the loss function is Mean Squared Error (MSE). Based on the hierarchical network structure element data of monitoring sections, the Huishan Anqiao monitoring site is determined as the direct upstream point affecting the water quality of Zhihu Port, Binhu District, Taihu Lake. Zhihu Port, Binhu District, Taihu Lake is considered the downstream point. Three years of monitoring section simultaneous high-frequency water quality indicator element data from 2018 to 2020, totaling 6576 data points after big data processing, are used for training. Additionally, 1462 data points from eight months in 2021 are used for model validation. The model takes the water quality indicator element data of the upstream and downstream monitoring sections at the previous time step as input and outputs the specific substance indicator value for the downstream monitoring section at the current time step. The involved indicators include water temperature, pH value, dissolved oxygen concentration, permanganate index, ammonia nitrogen concentration, conductivity, turbidity, total phosphorus, and total nitrogen concentration, totaling 9 items (Fig. 14.7).

2. Model Validation:

The validation process in this example mainly involves two aspects. Firstly, before the model training begins, hyperparameters need to be set in advance. Numerous hyperparameter combinations form a hyperparameter space, and a grid search is employed to generate multiple training models. Each of these models is trained and validated, and the hyperparameter combination that performs best on the validation dataset is selected as the final set of hyperparameters for running the model. Secondly, for each training model with different hyperparameter combinations, the parameters are iteratively updated using the gradient descent method to minimize the loss function.

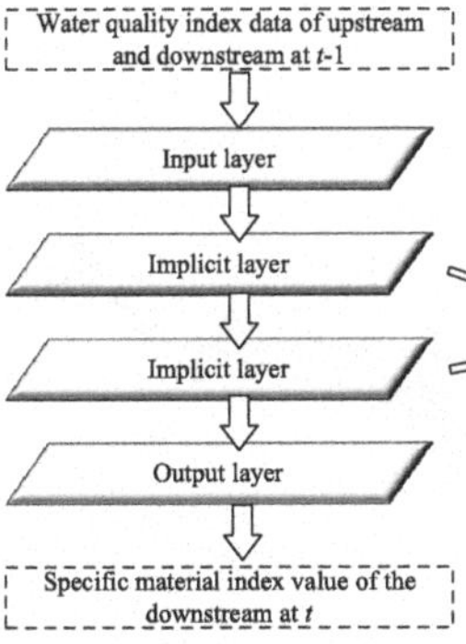

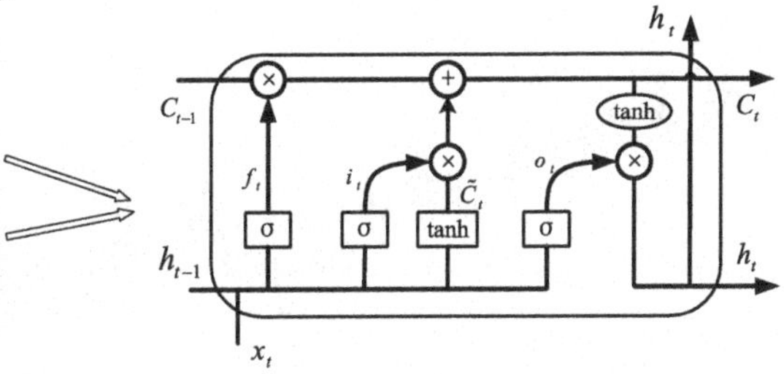

Fig. 14.7 LSTM model framework

Through the iterative process of model training and validation, the optimal hyperparameter combination, the weights, and bias parameters for each gate in the LSTM network are obtained, ensuring that the model output closely approximates the real values. It should be noted that a stopping criterion for model iteration must be set. There are two criteria: first, stop iterating after the hyperparameter space has been fully explored, and second, in a specific hyperparameter combination, to improve the model's generalization ability and prevent overfitting on the old training dataset, stop iterating when the model's loss on the validation dataset no longer decreases, and parameters are no longer updated.

It is particularly important to note that the initialization of LSTM parameters before model training in this example is randomized. Through continuous training and validation on the same dataset, multiple final models with similar performance are generated. The different output results of these models increase the uncertainty in the final predictions, with each result representing a possibility of substance diffusion. All final models can be used for scenario reconstruction of pollution dispersion in the monitoring section Fig. 14.8 shows the prediction results of COD and TP for two of the final models at Zhihu Port, Binhu District, Taihu Lake, on the validation dataset. It can be observed that the trends in the concentration of various substance indicators predicted by the models are roughly consistent with the actual trends. The models exhibit good fitting capabilities for the evolution of substance concentrations in the river, constructing scenario models that closely align with real-world river pollution dispersion. Finally, due to the uniqueness and heterogeneity of each substance's evolution scenario in each monitoring section, it is necessary in this phase to construct corresponding scenario models for each substance in every monitoring section based on the steps outlined above (Fig. 14.8).

14.3.3.6 Scenario Integration and Validation

The evolution scenarios of each substance in every monitoring section are just a tiny part of the complex wholeness scenarios of river pollution dispersion. While

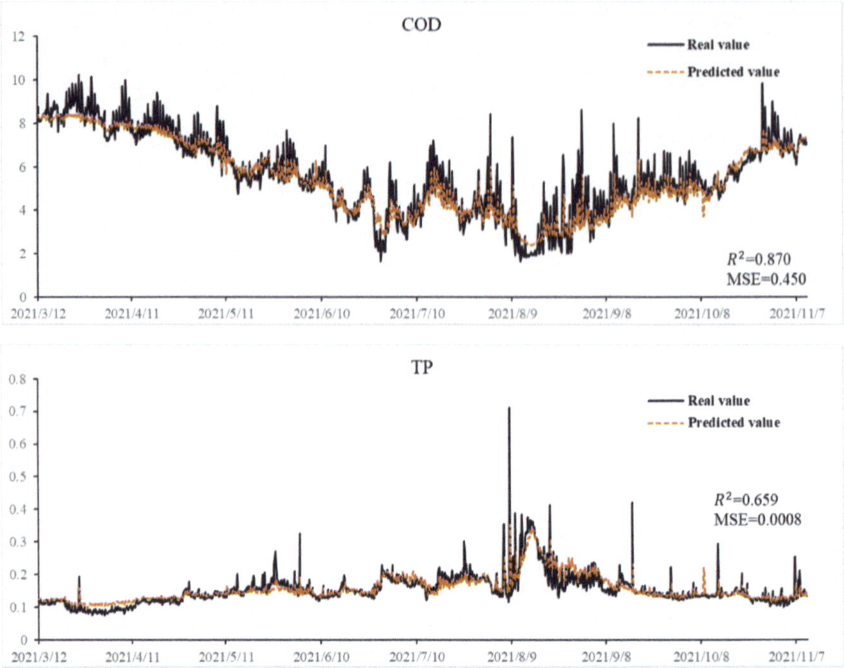

Fig. 14.8 Validation Water Quality Predictions for Zhihu Port, Binhu District, Taihu Lake

the previous stages of comprehensive degradation have helped us understand the pollution dispersion process from a more comprehensible perspective, such understanding is limited and incomplete. It cannot effectively address the interplay and emergence of scenarios at different levels within the problem. Based on the principles of data scenario regularization, this stage mainly involves the process of restoring complex wholeness scenarios of river pollution dispersion. It integrates scenarios comprehensively based on a wholeness understanding and control capability of river pollution dispersion scenarios, aiming to achieve the completeness of river pollution dispersion scenario models. Overall, in the process of calculating scenario integration, technologies and methods such as data integration, spatial integration, temporal fusion, scenario model integration, and algorithm integration are used to establish a logical model of complex wholeness scenarios of river pollution dispersion that reflects the fundamental dynamics of substance evolution in the entire river.

As shown in Fig. 14.9, first, the "pollution dispersion scenario model" submodules that simulate the changes of each substance at monitoring sections are comprehensively integrated into the "sectional pollution dispersion scenario model" unit module for pollution dispersion at that monitoring section. Then, according to the conduction direction in the hierarchical network structure of monitoring sections, the pollution dispersion scenario models of various sectional pollution dispersion are comprehensively integrated into the holistic "river pollution dispersion scenario model."

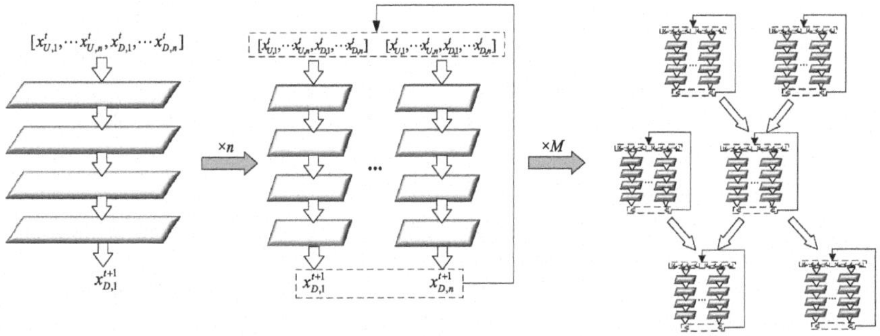

Fig. 14.9 Logical Process of Integrating River Pollution Dispersion Calculation Scenario Models

In the calculation scenario model, the output results of each "pollution dispersion scenario model" submodule not only serve as inputs for the next time step but also as inputs for the "pollution dispersion scenario model" submodule in the associated downstream "sectional pollution dispersion scenario model" unit module. As a result, any numerical changes caused by uncertain disturbances lead to macroscopic scenario emergence and scenario evolution in the continuous data interaction among the scenario models. Similarly, the integrated calculation scenario model also needs validation. At this stage, the loss function output by the model at the lowest downstream monitoring section is used as the evaluation metric for the performance of the integrated calculation scenario model. When the loss value on the validation dataset no longer decreases, the model stops iterating, and parameters are no longer updated.

14.3.3.7 Comprehensive Scenario Analysis

Comprehensive scenario analysis is fundamentally a computer-based scenario reconstruction and prediction technique defined within the "scenario space." It embodies the overall nature of river pollution dispersion phenomena and processes. By reconstructing and predicting scenarios in a holistic and process-oriented manner within the context of the research problem, it represents a research approach to obtain relevant research conclusions [3]. In this stage, the analysis process is based on the fundamental dynamics of "one" or "some" scenarios. Through predefined assumptions, it embeds certain unique information related to a practical uniqueness within a class of pollution control behaviors that share the same essence and dynamics into a specific type of initial scenario space. Scenario cultivation techniques are then applied to cultivate and nurture new scenarios [4]. It's important to emphasize that due to the uncertainty in the output of scenario models, different results correspond to multiple future scenarios in the process of comprehensive scenario analysis. Even with the same initial scenarios, different scenario evolution paths and outcomes can emerge. Each evolved scenario, trajectory, slice, and result represents a "possibility" of river pollution dispersion and could also be the "next" result due to certain human interventions. This reflects the diversity, dynamism, and uncertainty of real-world scenarios of river pollution dispersion. Ultimately, in the comprehensive scenario analysis process, results cultivated through scenario cultivation are summarized and consolidated to obtain conclusions, management recommendations, and problem-solving solutions.

Considering the threat posed to the aquatic environment by increased phosphorus levels leading to eutrophication and cyanobacterial blooms, this example focuses on scenario cultivation of river pollution dispersion models based on variations in phosphorus levels and differences in emergency measures. It is assumed that a sudden phosphorus leakage incident occurred in the upstream region of the Huishan Anqiao area in August 2021. Phosphorus entered the river after surface infiltration, causing a sudden increase in pollutants in the river. It is assumed that the pollutants reached the Huishan Anqiao monitoring section at 8 a.m., with total phosphorus concentration showing an exponential increase, peaking at noon, reaching nearly 50 times the normal value. After gaining control of the leakage incident, the concentration at this point exponentially decreased afternoon and returned to the normal level for monitoring by noon the next day. To prevent the impact of pollutants on downstream water quality, management authorities in the Huishan Anqiao area used chemical phosphorus removal to reduce pollutant concentrations. The experimental scenario parameters are shown in Table 14.1. The phosphorus removal rate is related to phosphorus concentration, input quantity, and method technology. In this experiment, the phosphorus removal rate is fixed, which is a post-event outcome. When dynamically adjusting the phosphorus removal rate, the rate decreases by 1% every half hour. Two phosphorus removal events are set at different locations, with the upstream event occurring earlier. Additionally, when the water quality entering the release point returns to the normal average level (assumed to be 0.2 mg/L in this experiment), the phosphorus removal process ends. Therefore, the downstream release point ends earlier during the two phosphorus removal events. The prediction period of the model is set to 3 days, and the data frequency is reduced to 30 min using mean interpolation. Based on a scenario evolution model without emergency measures, the evolution of total phosphorus concentration at downstream monitoring sections under different scenarios is shown in Fig. 14.10.

Table 14.1 Basic Scenario Parameter Settings

Scenarios	Number of times phosphorus removal	Start time	Phosphorus removal rate
Scenario 1	0	\	\
Scenario 2	1	10:00	50%
Scenario 3	1	12:00	50%
Scenario 4	1	10:00	Gradual reduction from 80%
Scenario 5	1	10:00	80%
Scenario 6	2	10:00; 14:00	50%; 50%
Scenario 7	2	12:00; 16:00	50%; 50%

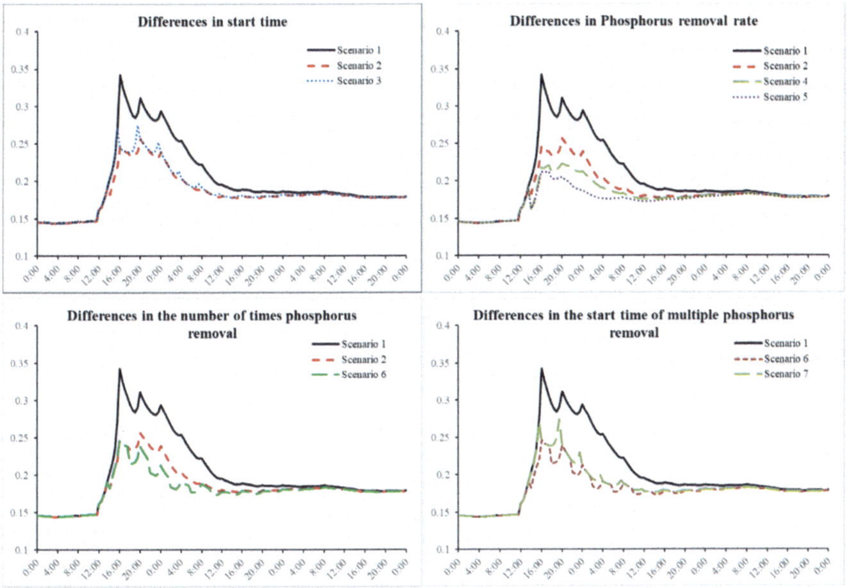

Fig. 14.10 Changes in Pollutant Concentrations under Different Scenarios

The results from Fig. 14.10 indicate the following under different scenarios: In Scenario 1, where no emergency measures are taken, the total phosphorus concentration at the downstream monitoring section reaches its peak at around 16:00 on the same day, approximately 0.342 mg/L. This concentration is 2.34 times higher than the concentration before the pollution incident (approximately 0.146 mg/L). Although the impact on downstream pollutant concentrations in this river section is not substantial after the sudden event, the highest concentration reached is 6.84 times the Water Quality Class III standard for Taihu Lake (0.05 mg/L). Additionally, although the pollution source in the river's pollution dispersion returns to normal levels by 12:00 the next day, it has a longer-term impact on the water quality at the lake's inlet. The pollutant concentration at the lake's inlet decreases in a fluctuating manner but still has not returned to normal levels by the end of the prediction period. When emergency measures are taken to reduce pollutant concentrations, Fig. 14.10a shows that the earlier the measures are initiated, the more effective the reduction of pollutant concentrations at the lake's inlet. Figure 14.10b indicates that when the initiation times are the same, a higher phosphorus removal rate is more favorable for reducing pollutant concentrations at the lake's inlet. In cases where the phosphorus removal rate achievable by the phosphorus removal technology and human resources is limited, single-point discharge has limited control over pollution. Assuming that management personnel establish an additional discharge point downstream after setting up the original discharge point, Fig. 14.10c demonstrates that adding an extra discharge point is more effective in reducing pollutant concentrations at the lake's inlet when the initiation times are the same. Figure 14.10d indicates that when implementing two rounds of phosphorus removal at the same time, an earlier initiation time is more effective. In summary, when a sudden pollution incident occurs, timely response, multi-stage treatment, and improved pollution control technologies can effectively control the spread of pollution in the river and reduce pollutant concentrations at the lake's inlet. To ensure the safety of the drinking water source, although single-point pollution sources can be effectively controlled in the short term, their long-term impact on the lake's water quality should not be ignored.

References

1. Sheng Z. Comprehensive and integrated management of large-scale projects [M]. Beijing: Science Press; 2009.
2. Sheng Z, Yu J. Complex systems management: a new field of management with chinese characteristics [J]. Manag World. 2021; 37(06): 36–50+2.
3. Sheng Z. Fundamental theories of mega infrastructure construction management [M]. Cham: Springer; 2018.
4. Chen J. Multi-role exploration of big data fusion analysis applications [J]. Chin Inf. 2021;(6):100–2.

Epilogue

This book, titled "Outline of Complex Systems Management," is an interpretation of the logical framework of complex systems management theory, a new field in contemporary management studies. It specifically introduces the corresponding knowledge system architecture and core points.

Why is it called an "outline"? "Outline" is a Chinese word, pronounced as "gāng yào." The original meaning of "gang" refers to a large rope used to structure and shape a net, and it later extended to design and plan the crucial parts of something. "Yao" means important and has a similar connotation to "gang." The combination of "gang" and "yao" represents the key points in a concise and comprehensive manner. The outline of complex systems management theory refers to the overall framework, core points, and fundamental aspects of the theory.

Why is this book limited to the "outline" of complex systems management theory? The fundamental reason is that complex systems management, as a globally significant and cutting-edge discipline in the field of management studies, is still in its infancy and early stages. Constructing the theoretical system of complex systems management will be a challenging and enduring process due to the following reasons.

First, globally, there is an increasing number of complex systems management practices emerging in diverse and abundant forms. The seeds of complex systems management theory need to take root in such practical soil, and the maturation and enrichment of practical connotations require a considerable amount of time.

Secondly, the theoretical system of complex systems management, including philosophical thinking, scientific paradigms, theoretical logic, knowledge formation paths, and major issues, will undergo significant transformations. These cannot be accomplished merely by relying on simple repairs and improvements to traditional management thoughts and academic systems. Instead, it requires a paradigm shift in thinking, proposing a series of original "academic propositions" and "knowledge revolutions" and forming an autonomous discourse system, all of which necessitate gradual contemplation and accumulation.

Z. Sheng, *Outline of Complex Systems Management Theory—Based on Irreversibility of Reductionism Thinking*, International Series in Operations Research & Management Science 360, https://doi.org/10.1007/978-981-97-3361-3

Third, new theories must be tested for their correctness, accuracy, effectiveness, and universality through practice and gradually refined through a large number of practical examples.

Fourth, complex systems management involves multiple disciplines and interdisciplinary approaches, necessitating mutual promotion and advancement among different fields. The growth of relevant disciplines has its own timeline, which also affects and constrains the academic development of complex systems management.

This indicates that although the real world has an urgent need for complex systems management theory, techniques, and knowledge systems, we must actively and calmly construct the theory, knowledge, and discourse system, respecting the objective laws of theoretical innovation. Otherwise, if we rush and violate these laws, we cannot guarantee the quality of the theory.

In this sense, even for the author of this book, who has been contemplating the research on complex systems management theory for a long time, in the grand timescale of scientific development, the goal of constructing a theoretical system has only taken a small step on the path of exploration. It has essentially established the thinking principles, theoretical logic, and methodology, designed the basic theoretical framework, and made efforts to explore how to systematize and rationalize it. At most, these can only be considered the "outline" of constructing the theory of complex systems management.

There is an idiom in China called "throwing a brick to attract jade," which means taking the initiative to offer a brick made of clay in exchange for a piece of beautiful jade for appreciation from others. It metaphorically implies that expressing one's immature ideas can elicit brilliant insights from others. This is precisely the attitude that the author of this book holds in the face of such a significant scientific problem as complex systems management theory.

The goal of constructing the theoretical system of complex systems management is grand and complex. From an individual's level of knowledge and ability, even after decades, one can only present an "outline" that is immature and imperfect, like a "brick." However, it is hoped that this outline can pique everyone's interest, leading to further exchanges and collaborations, thus generating numerous "jade-like" new academic perspectives on complex systems management. Ultimately, through collective efforts, the high level and sustained development and progress of complex systems management theory can be achieved. This should be our shared attitude and aspiration.

Upon reflection, since it is an "outline," there will inevitably be various shortcomings and regrets, both in terms of the overall framework and balanced correlations at the macro level, as well as in the profound expressions of the finer details. However, facing such a grand and complex theoretical research topic, an individual's capacity, level, experience, and time are extremely limited. It is impossible to consider comprehensively and delicately. At present, the practice of complex systems management is continuously evolving, and corresponding theoretical research is constantly advancing. As an individual, the author's only feasible approach is to "throw a brick to attract jade" through this book, allowing a community of international management scholars, consisting of more outstanding members, to collectively advance the academic cause of complex systems management from different fields, perspectives, and pathways.

Mrs. Li Li, the author's wife, has long and diligently taken on the heavy responsibilities of family affairs, providing a favorable environment and conditions for the author's academic research work, especially for the writing of this book. The author expresses deep gratitude to her.

Associate Professor Huimin Liu from Nanjing University, Associate Professor Yue Guo from Southern University of Science and Technology, Professor Peng Wu from Curtin University, Professor Jian Zuo from University of Adelaide, Associate Professor Yi-Hsin Lin from Southeast University, Associate Professor Xue Yan from Nanjing Audit University, Associate Professor Ru Liang from Nanjing University, Dr. Enyu Zeng, and Dr. He Chang have provided important assistance. The author extends sincere thanks to them.

Springer Publishing has provided valuable support for the publication of this book. Ms. Emily Zhang and Ms. Aiswarya Sreenivasan have provided meticulous guidance on the editing and publishing of this book. The author expresses heartfelt gratitude to them.

November 30, 2023
Nanjing, China

Bibliography

1. Sheng Z. From systems management to complex systems management—written on the occasion of the 30th anniversary of the founding of the journal of systems management [J]. J Syst Manag. 2022;31(06):1031–4.
2. Yang X, Gao Z, Sheng Z, et al. Complex system management is an important part of the management system with Chinese characteristics [J]. Manage World. 2022;38(10):1–24.
3. Sheng Z, Yu J. Complex systems management: a new field of management with chinese characteristics [J]. Mangag World. 2021;37(06):36–50+2.

Z. Sheng, *Outline of Complex Systems Management Theory—Based on Irreversibility of Reductionism Thinking*, International Series in Operations Research & Management Science 360, https://doi.org/10.1007/978-981-97-3361-3

SPRINGER NATURE

GPSR Compliance

The European Union's (EU) General Product Safety Regulation (GPSR) is a set of rules that requires consumer products to be safe and our obligations to ensure this.

If you have any concerns about our products, you can contact us on ProductSafety@springernature.com

In case Publisher is established outside the EU, the EU authorized representative is:

Springer Nature Customer Service Center GmbH
Europaplatz 3
69115 Heidelberg, Germany

The manufacturer's authorised representative in the EU is Springer Nature Customer Service Centre GmbH, Europaplatz 3, 69115 Heidelberg, Germany. If you have any concerns regarding our products, please contact ProductSafety@springernature.com

Printed and bound by CPI Group (UK) Ltd, Croydon, CR0 4YY

01/10/2026

02225863-0010